THE USE OF POPULATION VIABILITY ANALYSES IN CONSERVATION PLANNING

Ecological Bulletins No. 48

THE USE OF POPULATION VIABILITY ANALYSES IN CONSERVATION PLANNING

Edited by
Per Sjögren-Gulve and Torbjörn Ebenhard

Contents

Ecological Bulletins

ECOLOGICAL BULLETINS are published in cooperation with the ecological journals Ecography and Oikos. Ecological Bulletins consists of monographs, reports and symposia proceedings on topics of international interest, often with an applied aspect, published on a non-profit making basis. Orders for volumes should be placed with the publisher. Discounts are available for standing orders.

Editor-in-Chief and Editorial Office:
Pehr H. Enckell
Joint Editorial Office
Ecology Building
SE-223 62 Lund
Sweden
Fax: +46-46 222 37 90
Email: oikos@ekol.lu.se
http://www.oikos.ekol.lu.se

Technical Editor:
Linus Svensson/Gunilla Andersson

Editorial Board:
Björn E. Berglund, Lund.
Tom Fenchel, Helsingør.
Erkki Leppäkoski, Turku.
Ulrik Lohm, Linköping.
Nils Malmer (Chairman), Lund.
Hans M. Seip, Oslo.

Published and distributed by:
MUNKSGAARD International Publishers Ltd.
P. O. Box 2148
DK-1016 Copenhagen K, Denmark
Fax: +45-77 33 33 33
Email: customerservice@munksgaard.dk
http://www.munksgaard.dk

and

MUNKSGAARD International Publishers Ltd.
Commerce Place
350 Main Street
Malden, MA 02148-5018
USA
Tel: +1- (718) 388 8273, Fax: +1- (781) 388-8274

Suggested citation:
Author's name. 2000. Title of paper. – Ecol. Bull. 48: 000–000.

This volume of ECOLOGICAL BULLETINS can be ordered at
www.oikos.ekol.lu.se

Cover: Aerial photograph of a valley with mosaic habitats in southern Sweden. Photo: Göran Hansson/Naturfotograferna. Bush cricket *Metrioptera bicolor*. Photo: Oskar Kindvall. Wolf *Canis lupus*. Photo: Torbjörn Ebenhard. Field genetian *Gentianella campestris*. Photo: Tommy Lennartsson.

Typeset by ZooBoTech/GrafikGruppen, Torna Hällestad, Sweden, printed by Wallin & Dalholm, Lund, Sweden.

Preface

Population viability analysis (PVA; Shaffer 1990, Boyce 1992) has become a central tool in endangered species management. Risk-based approaches and modeling of scenarios using computer programs are also becoming increasingly common in several fields of environmental science and strategic planning. In conservation biology, risk assessment is now widely employed in PVAs of threatened species to guide management and assess extinction risks (Burgman et al. 1993).

With this focus, the Swedish Environmental Protection Agency (SEPA) and the Swedish Biodiversity Centre organized a two-day symposium, "The Use of Population Viability Analyses in Conservation Planning," in Stockholm in December 1997. The collection of papers in this volume resulted from the symposium. Symposium participants were almost evenly split between academic conservation biologists ("scientists") and managers, teachers and agency personnel ("practitioners"). Symposium round-table discussions highlighted the need for more information on the use of demographic PVAs as well as other promising approaches, such as metapopulation models and nested species subsets models that simply use species presence/absence (occupancy) data. Hence, our objective in these proceedings is to address the application of PVAs in a wider perspective than "traditional" PVAs of endangered species.

The demand for comprehensive, updated and instructive volumes on PVA theory and case studies among conservation practitioners, students, and researchers also became apparent via criticism of PVA by Caughley (1994) and Beissinger and Westphal (1998). As a result, not only this PVA volume but also two other independent volumes (Morris et al. 1999, Beissinger and McCullough 2001) were assembled. Morris et al. (1999) focus on population census data and unstructured demographic models, in which individuals in populations are modeled in the same way irrespective of their age or life stage. Metapopulation models of species presence or absence in habitat patches also are reviewed, but spatially realistic versions of these models largely are not included. Beissinger and McCullough's (2001) book focuses mainly on structured and individual-based demographic models, their parameterization and extensions, and the scope of PVA in endangered species management. In addition, Hanski (1999) provided an in-depth review of metapopulation models and their application to real-world cases. This volume complements the work of Morris et al., Beissinger and McCullough, and Hanski by covering and comparing a broader spectrum of PVA models and their application in conservation planning.

The scope of this volume is intentionally broad. We review applications of spatially realistic occupancy models and age- or stage-structured demographic models as well as individual-based demographic models. These three main PVA approaches initially are compared with other conservation assessment tools in an overview by Akçakaya and Sjögren-Gulve. The three approaches are then reviewed individually by Akçakaya, Lacy and Sjögren-Gulve and Hanski, with a special chapter on plant PVAs by Menges, all using illustrative examples. Fleishman et al. subsequently outline how nested subsets analyses can help discern focal species, taxa whose habitat requirements may encompass those of larger species groups and can be subjected to PVAs. Kindvall then compares the predictive accuracy of simpler occupancy models versus a demographic model for predicting local extinctions and colonizations in a bush-cricket metapopulation. Thereafter, four case studies which use the three model categories are presented (papers by Berglind, Ebenhard, Lennartsson, and Vos et al.). The volume concludes with Gärdenfors' paper on the use of PVAs in the classification of threatened species, and Lacy's presentation of the structure and logic of VORTEX, a widely used demographic and genetic PVA model.

We thank the authors for their cooperation, up-to-date reviews and case studies, and mindfulness of the issues that evolved in group discussions during the symposium. We also acknowledge our deep appreciation for the peer reviews made by various specialists in addition to reviews contributed by authors of this volume. We thank Henrik Andrén (Swedish Univ. of Agricultural Sciences, Grimsö), Lena Berg (SEPA), Barry Brook (Macquarie Univ., Sydney), Susan Harrison (Univ. of California, Davis), Linda Hedlund (SEPA), David Keith (New South Wales National Parks and Wildlife Service), Paul Leberg (Univ. of Southwestern Louisiana, Lafayette), Thomas Nilsson (SEPA), Reed Noss (Conservation Science Inc., Corvallis), Chris Ray (Univ. of Nevada, Reno), Mark Shaffer (Defenders of Wildlife, Washington D.C.), Barbara Taylor (Univ. of California, San Diego), Tim Wade (US EPA) and Thorsten Wiegand (UFZ Centre for Environmental Research Leipzig-Halle) for their constructive comments on drafts of the papers.

Last, but not least, we thank the symposium participants for their discussion input. We are also deeply grateful for funding provided by the Swedish EPA, Swedish Biodiversity Centre and Oikos Editorial Office to publish this volume.

Per Sjögren-Gulve and Torbjörn Ebenhard
Uppsala, September 2000

References

Beissinger, S. R. and Westphal, M. I. 1998. On the use of demographic models of population viability in endangered species management. – J. Wildl. Manage. 62: 821–841.

Beissinger, S. R. and McCullough, D. R. (eds) 2001. Population viability analysis. – Univ. of Chicago Press, in press.

Boyce, M. S. 1992. Population viability analysis. – Annu. Rev. Ecol. Syst. 23: 481–506.

Burgman, M. A., Ferson, S. and Akçakaya, H. R. 1993. Risk assessment in conservation biology. – Chapman and Hall.

Caughley, G. 1994. Directions in conservation biology. – J. Anim. Ecol. 63: 215–244.

Hanski, I. 1999. Metapopulation ecology. – Oxford Univ. Press.

Morris, W. et al. 1999. A practical handbook for population viability analysis. – The Nature Conservancy, New York.

Shaffer, M. L. 1990. Population viability analysis. – Conserv. Biol. 4: 39–40.

Ecological Bulletins 48: 9–21. Copenhagen 2000

Population viability analyses in conservation planning: an overview

H. Reşit Akçakaya and Per Sjögren-Gulve

Akçakaya, H. R. and Sjögren-Gulve, P. 2000. Population viability analyses in conservation planning: an overview. – Ecol. Bull. 48: 9–21.

Population viability analysis (PVA) is a collection of methods for evaluating the threats faced by populations of species, their risks of extinction or decline, and their chances for recovery, based on species-specific data and models. Compared to other alternatives for making conservation decisions, PVA provides a rigorous methodology that can use different types of data, a way to incorporate uncertainties and natural variabilities, and products or predictions that are relevant to conservation goals. The disadvantages of PVA include its single-species focus and requirements for data that may not be available for many species. PVAs are most useful when they address a specific question involving a focal (e.g., threatened, indicator, sensitive, or umbrella) species, when their level of detail is consistent with the available data, and when they focus on relative (i.e., comparative) rather than absolute results, and risks of decline rather than extinction. This overview provides guidelines for choosing a PVA model among three categories, from data-intensive individual-based population models to simple occupancy metapopulation models.

H. R. Akçakaya (resit@ramas.com), Applied Biomathematics, 100 North Country Road, Setauket, NY 11733, USA. – P. Sjögren-Gulve, Dept of Conservation Biology and Genetics, Evolutionary Biology Centre, Uppsala Univ., Norbyvägen 18D, SE-752 36 Uppsala, Sweden (present address: Swedish Environmental Protection Agency, SE-106 48 Stockholm, Sweden).

Practical problems in conservation planning and wildlife management are increasingly phrased in terms of questions about the viability of threatened or indicator species. Because of the nature of these questions, and the natural variation and uncertainty inherent in ecological data, risk-based methods are appropriate for population viability analyses (PVAs). Viability of a species in a given geographic region is often expressed as its risk of extinction or decline, expected time to extinction, or chance of recovery. PVA models attempt to predict such measures of viability based on demographic data (such as censuses, mark-recapture studies, surveys and observations of reproduction and dis-

persal events, presence/absence data) and habitat data.

Although conservation planning is often done at the ecosystem or landscape level (see Alternative methods below), several factors highlight PVA as a central tool for conservation assessments. These factors include the needs of threatened species, recent developments in the use of indicator species (Fleishman et al. 2000), and the potential for rigorous risk assessments using a variety of data types. The aim of this paper is to give a short description of population viability analysis, discuss its advantages and limitations in conservation planning, compare it with other methods of assessment, and discuss the most useful approaches to PVA. We begin by reviewing a number of alternative methods for conservation assessments that provide a context for evaluation of the PVA approaches. We then present a short introduction to PVA and its three main categories of models, and some guidelines for choosing a model. Subsequently we discuss the limitations and advantages of PVA, and end with some recommendations of how to make the method most useful.

Alternative methods for conservation assessments

A number of quantitative methods for assessment are used in conservation planning. These methods form the context in which we evaluate the PVA approach.

Reserve selection algorithms

These methods are designed to select nature reserves, i.e., choose a subset of available habitat patches for protection (e.g., Margules et al. 1988, Pressey et al. 1993, 1996, Pimm and Lawton 1998, Possingham et al. 2000), often using criteria that maximize the number of species included in the reserves. The algorithms are usually based on the presence/absence of species in each habitat patch, and do not explicitly consider the viability of species in habitat patches, or the interaction among populations in different habitat patches (e.g., metapopulation dynamics). The presence of a species in a particular patch does not necessarily indicate that the patch can support a viable population, or that the population will persist even if the neighboring habitat patches are not included in the reserve system. Nevertheless, these methods are useful if the only available data are occurrences.

Habitat suitability models

The aim of habitat suitability (HS) models is to predict a species' response to its environment. The response is usually the occurrence or abundance of the species at a certain locality or the carrying capacity of the habitat. The statistical procedures to obtain the HS model (such as multiple logistic regression) use species occurrence or abundance at each location as the dependent variable and the habitat characteristics as the set of predictive variables (see several chapters of Verner et al. 1986, Straw et al. 1986, Mills et al. 1993, Sjögren-Gulve 1994, Pearce et al. 1994, Fleishman et al. 2000). Most statistical methods require both presence and absence data, while others (such as "climatic envelopes") require only presence data (Elith 2000).

One advantage of habitat suitability models is that they are statistically rigorous and can be validated. They can also be used to explore effects of environmental changes on habitat patch suitability, and to calculate probabilities of species occurrence (see Sjögren-Gulve and Hanski 2000). Another advantage is that they can use all the available habitat data (including point observations, GIS data of various types, satellite images, digital elevation maps, etc.), and incorporate non-linearities of, and interactions among habitat variables. The main disadvantage of habitat suitability models is that suitability is only one component of viability, which also depends on demographic factors. However, habitat suitability models can be integrated with PVA models to identify habitat patches and characterize the spatial structure of metapopulations (e.g., Akçakaya and Atwood 1997).

Gap analysis

A "gap" is the lack of representation or inadequate representation of a plant community or animal species in areas managed primarily for natural values. Identification of a gap indicates potential risk of extinction or extirpation unless changes are made by land stewards in the management status of the element. Gap Analysis Program (GAP) is a process widely used by state agencies in the USA to identify such gaps. The process involves overlaying (intersecting) land cover and species distribution (element occurrence) coverages with the coverage of areas protected or managed primarily for natural values (Scott et al. 1993, Kiester et al. 1996).

The advantages of gap analysis are its widespread use, and its use of all available geographic information. The major disadvantage of gap analysis is that it is not based on population dynamics, and does not utilize available demographic information. Hence, it does not provide a direct measure of viability. Another disadvantage is that it often relies on species-habitat associations and species distribution patterns that are not rigorously determined.

Rule-based and score-based methods for prioritization

These are algorithms for categorizing species in terms of the threat they face (Millsap et al. 1990, Master 1991, IUCN 1994). For example, IUCN rules (IUCN 1994) rules assign species to categories of "Critically endangered", "Endangered", "Vulnerable" and "Lower risk", based on information available on abundance, distribution, population trends, population fragmentation, and extinction risk estimates. They are used widely by international conservation organizations. This method works as a way of classifying threatened species by the risks they face, even if there is little information. For example, IUCN rules are based on many aspects of habitat and demography, but the method is not dependent on a full set of information. Species can often be classified even if information is available only on one aspect (e.g., abundance). IUCN rules can also use PVA results and can explicitly incorporate uncertainties in data (see Akçakaya et al. 2000). One disadvantage of these methods is that the rules and thresholds are necessarily arbitrary. As a result, ranks or classifications of the same set of species with different rule sets may have low correlation (Burgman et al. 1999).

Estimating extinction probability from sighting data

These methods also try to estimate probabilities of extinction, but they work from a record of sightings, rather than the more detailed demographic information that PVA uses (Solow 1993). The quantity estimated with these methods is the probability that the species is already extinct, rather than the probability that it will become extinct by a given future date. Although the meaning of the estimated probability does not exactly coincide with future viability, these methods are useful when the only available data are sightings.

Landscape indices

These include metrics such as patch size distribution, fractal dimension, shape index, and other descriptions of spatial structure, which are calculated from digital raster maps of habitat types in the landscape (for example, the FragStats program; McGarigal and Marks 1995). Although many of these indices may be informative in particular situations, there are three major problems with their general application to conservation issues. First, the objects that form the structure (e.g., patches of forest habitat) are often arbitrarily defined. Second, the spatial scale is often arbitrarily selected. Both the definition of "patch" and the selection of spatial scale require a specific phenomenon or focal species. Third, and most important, the relationship between these metrics and conservation goals may be weak or very restricted (applying to specific populations in specific landscapes).

Other types of landscape indices involve connectivity and dispersal, which are also part of the metapopulation approach. However, these metrics alone may also be ambiguous as conservation goals. For instance, increased dispersal usually increases viability, but not always (see Stacey et al. [1997] and Beier and Noss [1998] for reviews). Even when it does, increasing dispersal may not be the best option (cost may be too high and/or increase in viability may be too low, compared to options related to other aspects such as carrying capacity, fecundity, or survival). The best way to make such metrics relevant to conservation is to use them in metapopulation models and estimate the dispersal parameters of these models.

Ecosystem-based methods

These methods deal with more than target or focal species. Some attempt to consider multiple criteria, dealing with a vast array of issues and factors from fungi species, prescribed fires, tribal rights and tourism, to endangered species and jobs. The assessments are based on various methods, including point scoring sheets, expert opinion, rating systems, etc. Others focus on "emergent" properties such as nutrient cycling or various measures of species diversity.

The clear advantage and appeal of the ecosystem approach is its comprehensiveness. The ultimate goal of most conservation efforts is the preservation of well-functioning, representative, natural ecosystems. Even species-specific methods such as PVA are often used as parts of this overall goal (e.g., by focusing on indicator, sensitive, or umbrella species).

The main disadvantage of the ecosystem approach is the complexity of interactions among species and our lack of understanding of community and ecosystem dynamics. As our understanding increases, conservation practices will hopefully become more ecosystem-based. However, the contingencies and complexities involved may make it impossible to find general laws in ecosystem ecology (Lawton 1999). Currently, ecosystem-based approaches to practical problems in conservation suffer from vagueness and circularity (Goldstein 1999). At their worst, the vagueness of these approaches makes it possible to get almost any answer to practical questions related to management decisions, often to support entrenched views. At their best, they provide a forum for helping stakeholders understand management trade-offs. They are most valuable if the criteria for decision-making can be agreed upon by all interested parties before the assessment is made.

A short introduction to population viability analysis

Population viability analysis is a process of identifying the viability requirements of, and threats faced by, a species and evaluating the likelihood that the population(s) under study will persist for a given time into the future. Population viability analysis is often oriented towards the management of rare and threatened species, with two broad objectives. The short-term objective is to minimize the risk of extinction. The longer-term objective is to promote conditions in which species retain their potential for evolutionary change without intensive management (see also Beissinger and McCullough 2001). Within this context, Box 1 outlines management questions that may be addressed with a PVA.

In addition to the management-oriented objectives in Box 1, PVA is also an excellent tool for organizing the relevant information and assumptions about a species or a

Population viability analysis (PVA) may be used to address the following aspects of management for threatened species and/or focal species, indicative for larger species groups:

1) Planning research and data collection. PVA may reveal that population viability is insensitive to particular parameters. Research may be guided by targeting factors that may have an important effect on probabilities of extinction or recovery.

2) Assessing vulnerability. PVA may be used to estimate the relative vulnerability of populations to extinction. Together with cultural priorities, economic imperatives and taxonomic uniqueness, these results may be used to set policies and priorities for allocating scarce conservation resources.

3) Impact assessment. PVA may be used to assess the impact of human activities (exploitation of natural resources, development, pollution) by comparing results of models with and without the population-level consequences of the human activity.

4) Ranking management options. PVA may be used to predict the likely responses of species to reintroduction, captive breeding, prescribed burning, weed control, habitat rehabilitation, or different designs for nature reserves or corridor networks.

population. By making the assumptions explicit, and highlighting the data deficiencies, it serves as a structured working and learning process. If the PVA focuses on species that are indicative for entire species groups (see Fleishman et al. 2000), its implications for habitat management have wider taxonomic relevance.

The result of a PVA can be expressed in many different forms (see examples in Fig. 1, Akçakaya [2000] and Beissinger and McCullough [2001]). These include extinction risk, time to decline, chance for recovery, persistence time, and local and regional occupancy rate. Which measure is used depends on the question. Most outputs from demographic PVAs are based on three variables: the amount of decline (e.g., 100% or total extinction or partial decline), the probability of decline, and the time frame in which the decline is expected to take place (Akçakaya 1992, 2000). Measures of occupancy model PVAs (see below) include risk of regional extinction, the number or proportion of occupied habitat patches (regional occupancy) projected over time, and extinction and colonization probabilities for individual patches under current environmental conditions (see Sjögren-Gulve and Hanski 2000).

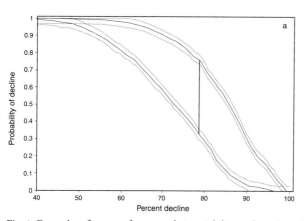
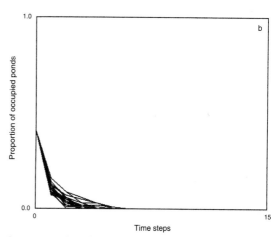

Fig. 1. Examples of outputs from population viability analyses (PVAs) with a structured model and an occupancy model, respectively. a) Risk of decline of a northern spotted owl metapopulation, simulated with a structured PVA model (based on Akçakaya and Raphael 1998). The top curve gives the risk under an assumed timber harvest, and the bottom curve assumes no habitat loss. Each point on the curve shows the probability that the metapopulation abundance will fall by the given percentage from the initial abundance anytime during the next 100 yr. The vertical bar shows the maximum difference between the two curves. In this example, the maximum difference is at a 78% decline. The risk of a 78% decline from the initial abundance is ca 0.33 without habitat loss, and ca 0.77 with habitat loss due to the assumed timber harvest. b) Predicted temporal change in the proportion of 102 ponds occupied by pool frogs *Rana lessonae* when large-scale forestry is omnipresent at the Baltic coast of east-central Sweden. The risk of regional extinction is 0.999 within 15 census intervals (i.e., 53 yr) and medium time to regional extinction is 18 yr (from Sjögren-Gulve and Ray 1996, reprinted with permission © Island Press).

There is no single recipe to follow when doing a PVA, because each case is different in so many respects. Main components of a PVA may include identification of the question (i.e., what issue the PVA is trying to address), data collection, data analysis and parameter estimation, modeling and risk assessment, sensitivity analysis and refinement of the model, monitoring and evaluation (Akçakaya et al. 1999).

Methods of PVA

Various types of models are used in PVAs, each type requires different data, and may answer different questions. The three types of models discussed below range from simple to complex, and demonstrate the trade-off between flexibility (realism) and practicality (data requirements). Simple occupancy models are applicable only to species in metapopulations, either with unoccupied and occupied patches observed at the same time, or with population turnover (i.e., observed local extinctions and recolonizations; see Sjögren-Gulve and Hanski 2000). In the more complex structured (Akçakaya 2000) or individual-based models (Lacy 2000a), single-population models can be considered as a special case of metapopulation models. For a more detailed discussion of single-population models, see Burgman et al. (1988, 1993), Caswell (1989), and Akçakaya et al. (1999).

Occupancy models for metapopulations
The simplest metapopulation approach models the occupancy status of habitat patches in a geographic region (i.e., the presence or absence of the species in these patches). This approach dates back to a model that was originally developed by Levins (1969) and that has been modified and expanded by several authors. The two specific approaches described below are based on this model. For examples of applications of occupancy models, see Sjögren-Gulve and Hanski (2000).

Occupancy models are parameterized using data on the presence or absence of a species in habitat patches from one or more regional inventories. They may be advantageous to demographic models when demographic data are difficult to obtain. However, the management question and the ecology of the species, and not just data availability, should dictate the model used (see Data needs and choosing a model). Occupancy models require that the species has local populations confined to a clearly delimited habitat in a landscape. They ignore local population dynamics, and do not model fluctuations in size or composition of the local populations (sex, age, stage; see Akçakaya 2000). This may be disadvantageous, for instance when population processes not tightly correlated with habitat characteristics are important for local extinctions. Since they model future changes in patch occupancy based on observed instantaneous occupancy or correlates of observed population turnover, their predictions of local extinctions may be considered a less independent assessment than that of demographic models, which are based on survival and fecundity rates among individuals in the populations. An example where occupancy models and a demographic model are compared is provided by Kindvall (2000). Two general types of occupancy models, which are presented in greater detail by Sjögren-Gulve and Hanski (2000), are briefly described in Boxes 2 and 3.

Structured (meta)population models
Structured population models consider factors that may be important for the persistence of local populations by modeling the dynamics of each population occupying a habitat patch. As in the occupancy models discussed above, they also incorporate the spatial structure of the habitat patches (Burgman et al. 1993). In addition, they incorporate internal dynamics of each population (e.g., variation in age structure, immigration, emigration, density dependence, and environmental fluctuations), which often are important determinants of metapopulation persistence (Gilpin 1988, Burgman et al. 1993, LaHaye et al. 1994).

The main advantage of structured population models compared to occupancy models is their flexibility. In modeling the local population dynamics, they can incorporate

Box 2

Occupancy models (I)

Incidence function models (IFM; Hanski 1994, 1999) require data on the areas and geographic locations of suitable habitat patches and the presence/absence of the species in these patches from at least one complete inventory. A habitat-suitability analysis (see above) of the species presence/absence pattern may be required for reliable habitat patch identification and delimitation. Based on these data, colonization and extinction probabilities are estimated for each patch using regression. These estimated probabilities are then used in simulations to predict metapopulation persistence and patch occupancy (e.g. Kindvall 2000, Vos et al. 2000).

Box 3

Occupancy models (II)

State transition models (e.g., Verboom et al. 1991, Sjögren-Gulve and Ray 1996) are conceptually related to the incidence function models discussed in Box 2. They require presence/absence data, but from two or more yearly inventories. Instead of relying on patch occupancy patterns, these models use patterns of patch state transitions. They predict state transitions (vacant to occupied as a result of colonization; and occupied to extinct, as a result of local extinction) from correlated environmental variables. Similar to habitat-suitability models, the patch transitions are modeled using predictive environmental variables discerned by multiple logistic regression (see Sjögren-Gulve and Hanski [2000] and Kindvall [2000]).

several biological factors and can represent spatial structure in various ways; they have been applied to a variety of organisms (see Akçakaya [2000] and Menges [2000] for reviews, and Berglind [2000] and Lennartsson [2000] for examples). Since they model demographic processes, the populations are the focal object rather than the habitat patches. Consequently, the species-habitat association need not be as strong as in occupancy models. Another advantage is that, despite their realism, structured models are based on a number of common techniques or frameworks that allow their implementation as generic programs (such as RAMAS; see Akçakaya 1998). This common framework becomes advantageous when models and viability analyses are needed for a large number of species, and time and resource limitations preclude detailed programming for each species. A third advantage is that structured demographic modeling allows careful risk assessment for species with very few local populations (occupancy models require a larger number), and under circumstances in which no extinctions have occurred and habitat patches are not easily identified.

The main disadvantage of structured models is that they require more data than occupancy models, including stage-specific survival and fecundity rates, and the temporal and spatial variation in these rates. However, for species with weak habitat association, such data may be more easily obtained than observations of population turnover. Another difficulty lies in the estimation of local vital rates for populations that may, in the future, colonize currently vacant patches. In such cases, vital rates are usually estimated as functions of habitat characteristics, based on relationships obtained from occupied patches.

Individual-based (meta)population models
There are various types of individual-based models. In a commonly used approach, the behavior and fate of each individual is modeled in a simulation (DeAngelis and Gross 1992). The behavior and fate (e.g., dispersal, survival, reproduction) of individuals depend on their location, age, size, sex, physiological stage, social status and other characteristics.

The advantage of individual-based models is that they are even more flexible than structured models, and can incorporate such factors as genetics, social structure, and mating systems more easily than other types of models (see Lacy [2000a] and Ebenhard [2000] for examples). One disadvantage of individual-based models is that they are very data-intensive. Only a few species have been studied well enough to use all the power of individual-based modeling. Another disadvantage is that the structure (as well as the parameters) of the models depend on the ecology and behavior of the particular species modeled. Thus, unlike structured models with a common framework, each individual-based model must be designed and implemented separately, making this approach impractical for many species. However, there are generic programs (such as VORTEX; see Lacy 1993, 2000b) that are based on individual-based modeling techniques but with a fixed, age-based structure.

Data needs and choosing a model

The amount of data needed to build a PVA model depends mostly on the question addressed and on the ecology of the species.

The types of data that can be used in a PVA include distributions of suitable habitat, local populations or individuals, patterns of occupancy and extinction in habitat patches, abundances, vital rates (fecundity and survival), as well as temporal variation and spatial covariation in these parameters. Not all of these types of data are required for any one model. For more information about data needs of particular types of PVA models, see Akçakaya (2000), Lacy (2000a) and Sjögren-Gulve and Hanski (2000).

The more data one has, the more detailed models one can build. Including more details makes a model more realistic, and allows addressing more specific questions. However, in most practical cases, available data permit only the simplest models. Attempts to include more details than can be justified by the quality of the available data may result in decreased predictive power and understanding.

The trade-off between realism and functionality depends on the characteristics of the system under study (e.g., the ecology of the species), what you know of the system (the availability of data), and what you want to know or predict about the system (the questions addressed). Even when detailed data are available, models intended to analyze long-term metapopulation persistence may include less detail than those intended to predict next year's distribution of breeding pairs within a local population. In cases where data are available and the ecology of the species implies that more than one type of PVA modeling is appropriate, comparative modeling (e.g., Kindvall 2000, Brook et al. 2000) may shed additional light on management options and strengthen the PVA process and conclusions. It is important to note that there are cases in which exploratory modeling is valuable for its own sake, even in the absence of sufficient data (see Data needs below).

Box 4 lists aspects that should be considered in determining the appropriate model. Different considerations may point to models of different complexity. For instance, the question addressed may require a detailed model whereas the available data can support only a simple model. In such cases, either more data must be collected or the question modified.

Occupancy models may be more advantageous than demographic models in situations where demographic data are not available, and the species occurs in a large number of local populations confined to a distinct type of habitat in the region of concern. In order to do the PVA with an incidence function model, inventory data are needed, including presence/absence of the species and measurements of individual habitat patch characteristics (environmental variables) that may explain its presence/absence pattern. For state transition modeling, a sufficient number of local extinctions (>5) and colonizations (>5) must have occurred between repeated inventories (in different years) that correlate significantly with local patch characteristics (see Sjögren-Gulve and Hanski 2000).

Box 5 presents some further guidelines on conditions under which demographic (structured or individual-based) models are more advantageous than occupancy models. The choice between structured and individual-based models depends on the size of the population(s), the importance of genetics and social interactions, and availability of data (see Akçakaya 2000).

Limitations of population viability analysis

As any other method, PVA has certain limitations, both practical and philosophical.

Single species focus

The focus of a PVA is generally a population or multiple populations of a single species. Its focus on single species is a limitation in cases where the goal is the management and conservation of an ecosystem. In other cases, the single species focus is the strength of PVA: the dynamics of single species are much simpler (and thus better understood) than the dynamics of communities or ecosystems (Lawton 1999). Uncertainties in structure and parameters of single-species models (see below) are magnified when multiple species and their interactions are considered.

One way to deal with the single-species limitation is to select target species that are representative of the community, that are sensitive to potential human impact, and whose conservation is likely to protect other species as well (umbrella species). Such species are sometimes called "indicator" species (see Fleishman et al. 2000). However, it is important not to make the mistake of managing the landscape specifically for the indicator species without ascertaining that the enhancements benefit other species as well (Simberloff 1998). For example, the proverbial "miner's canaries" would be useless as "indicators" if they were given little oxygen masks so that they survive!

Box 4

The following should be considered in determining the appropriate PVA model:
- Model STRUCTURE should be detailed enough to use all the relevant data, but no more detailed.
- Model RESULTS should address the question at hand (e.g., if the question concerns risk of a 50% decline, the model should report such a result).
- The model should have a PARAMETER related to the question (e.g., if the question involves the effect of timber harvest, the model should include parameters that reflect such an effect realistically).
- Model ASSUMPTIONS should be realistic with respect to the ecology of the species and the observed spatial structure (e.g., if there is population subdivision, a metapopulation model should be considered).
- For occupancy modeling, the species must occur as geographically distinct local populations in a landscape or region, and species occurrence or turnover patterns (extinction/colonization) need to correlate significantly with measurable habitat variables (see Boxes 2, 3 and 5).

Box 5

Guidelines for selecting a model

Conditions under which demographic models that incorporate internal dynamics (such as structured models or individual-based models) are more advantageous than occupancy models. Note that these are only general guidelines; there may be exceptions to most of them.

1) Demographic data for building a structured or individual-based model already exist
2) There are reasons to believe that demographic, behavioral or genetic processes are important for local extinction, or the ecology of the species indicates that internal population dynamics are important
3) The species occurs in a small number of populations
4) Suitable but unoccupied habitat patches cannot be easily identified
5) Species occurrence or turnover (extinction/colonization) patterns do not correlate significantly with measurable habitat characteristics (or such data are harder to collect than demographic data)
6) The management question addressed involves a factor related to within-population dynamics (e.g., questions about impacts on different age classes or questions regarding management and conservation actions that affect different life history stages differently)
7) The required answer is in terms of abundance rather than occupancy (e.g., risk of a population decline, or expected time until the population falls below a given threshold abundance)

Data needs

PVAs may need more data than some of the other methods discussed. However, incomplete information does not necessarily preclude meaningful results. First, PVAs can incorporate uncertainties in the data, and in some cases, these uncertainties do not effect the overall conclusion (see below). Second, uncertainties in the data may not affect results when the goal of PVA is comparative, as in ranking management options (Akçakaya and Raphael 1998). Third, there is very significant value in building a model for its own sake. It clarifies assumptions, integrates knowledge from all available sources, and forces us to be explicit and rigorous in our reasoning. It allows us to identify, through sensitivity analyses, which model structures and parameters matter, and which do not (Akçakaya and Burgman 1995). In fact, this modeling process is necessary for determining whether or not there are sufficient data for reaching management decisions. It allows identification of the parameter(s) which deserve highest priority in terms of obtaining more precise estimates. This identification does not refer to the types of data needed for models with different structures, but to the numerical values of the parameters, and to the contribution of each particular parameter to the uncertainty in model results.

Risk criteria

Some uses of PVA involve determining whether the risk faced by a particular species is acceptable. Such questions require a benchmark for "an acceptable level of risk" for the extinction of species. There are some benchmarks used (e.g., IUCN categories; see Gärdenfors 2000), but

none is accepted universally. Obviously, the determination of such benchmarks is a societal issue, outside the scope of PVA.

Identifying causes of decline

Caughley (1994) contrasted two paradigms in conservation biology: "small population" and "declining population". Under the "small population paradigm", factors threatening species with extinction include stochasticity, catastrophes and genetic degradation; under the "declining population paradigm," they include overkill, habitat loss and fragmentation. In this scheme, PVA and modeling are included under the "small population paradigm". This separation is now seen as artificial (Akçakaya and Burgman 1995, Hedrick et al. 1996, Beissinger and McCullough 2001) because PVAs can and do incorporate systemic pressure (i.e., deterministic decline; e.g., LaHaye et al. 1994), effects of habitat loss (e.g., Akçakaya and Raphael 1998), and overkill (or overharvest; e.g., Kokko et al. 1997). It is important to remember that, as Caughley (1994) emphasized, no modeling effort by itself can determine why a population is declining or why it has declined in the past. This is rather obvious, but it is often forgotten and models are expected to provide answers to questions they were not designed to address. For modeling to be used successfully to evaluate options for management of species, it must be part of a larger process and incorporate other methods, including study of natural history, field observations and experiments, analysis of historical and current data and long-term monitoring. The challenge that PVA modelers take is to incorporate all the relevant factors and impacts in their model.

Advantages of population viability analysis

PVA is one of the central tools for conservation planning and evaluation of management options. Compared to other methods reviewed above, PVA has several advantages.

Relevance to conservation of biodiversity

PVA has direct relevance to biodiversity conservation. An increasing number of species are presently threatened or endangered, and PVA results directly relate to the mandates of such laws as the Endangered Species Act. In addition, PVA can be applied to validated focal or umbrella species (Fleishman et al. 2000) to guide conservation efforts for entire nested species groups. Thus, PVAs of selected threatened species and sets of indicative species will be central for efficient conservation planning at local or regional levels, and for measures taken to comply with international treaties such as the UN Convention on Biological Diversity (UNCED 1992). By focusing on species viability, instead of relying only on subjective rules-of-thumb or opinions, or only habitat data, the risk assessment approach directly relates to the maintenance of viable and well-distributed populations of native species.

Rigor

Unlike some of the other methods, PVA is rigorous and quantitative. Its results can be replicated by different researchers. The assumptions of a PVA can be (and should be) explicitly stated and enumerated; they can also be validated given sufficient data. Validation of stochastic results (such as risk of decline or extinction) requires data for several independent populations, as well as observed trajectories or extinctions for comparison. For example, in a collective comparison of the historic trajectories of 21 populations with the results of the PVAs for these populations, Brook et al. (2000) validated PVAs in terms of their predictions of abundance and risks of decline. In this comprehensive and replicated evaluation, they estimated the parameters from the first half of each data set and used the second half to evaluate model performance. They found that PVA predictions were accurate: the risk of population decline closely matched observed outcomes, there was no significant bias, and population size projections did not differ significantly from reality. Further, the predictions of five PVA software packages they tested were highly concordant. They concluded that PVA is a valid and sufficiently accurate tool for categorising and managing endangered species. Although validation of stochastic results may not be possible in every case, components of a PVA can be validated. For example, the density dependence function can be validated by experimental manipulation of densities, or the habitat relationships that form the basis of the spatial structure of a metapopulation PVA can be validated by using half of the available data to predict the other half (e.g., see Akçakaya and Atwood 1997). In addition, some model results can be validated by comparing predicted values with those observed/measured in the field (e.g., Sjögren-Gulve and Ray 1996, Kindvall 2000, McCarthy and Broome 2000, McCarthy et al. 2000, Vos et al. 2000).

Ability to use all available data and multiple data types

A PVA can use various types of data sets, including presence-absence data, habitat relationships, GIS data on landscape characteristics, mark-recapture data, surveys and censuses. Thus, it is possible to incorporate all available data into the assessment. Such an assessment is more reliable than one that ignores part of the available information. Most of the alternative methods discussed above use a limited range of data types. For example, reserve selection, habitat suitability or gap analysis methods do not use available demographic data.

Incorporating uncertainty

Uncertainty is a prevalent feature of ecological data that is ignored by most methods of assessment. If data for a PVA are unavailable or uncertain, ranges (lower and upper bounds, instead of point estimates) of parameters are used. In addition, uncertainties in structure of the model can be incorporated by building multiple models (e.g., with different types of density dependence). There are various methods of propagating such uncertainties in calculations and simulations (Ferson et al. 1998). One of the simplest methods is to build best-case and worst-case models (e.g., Akçakaya and Raphael 1998). A best-case (or optimistic) model includes a combination of the lower bounds of parameters that have a negative effect of viability (such as variation in survival rate), and upper bounds of those that have a positive effect (such as average survival rate). A worst-case or pessimistic model includes the reverse bounds. Combining the results of these two models gives a range of estimates of extinction risk and other assessment end-points. This allows the users of the PVA results (managers, conservationists) to understand the effect of uncertain input, and to make decisions with full knowledge of the uncertainties.

The uncertainties can also be used in a sensitivity analysis. Results of sensitivity analyses are used to identify important parameters and help guide future fieldwork. For example, PVA models can also be analyzed with respect to their sensitivity to uncertain parameters. Such analyses guide fieldwork by quantifying the expected decrease in the uncertainty of the results with narrower ranges for each parameter (see Akçakaya 2000).

Conservation planning with multiple objectives

Conservation and landscape management decisions often involve multiple objectives such as ecological and economic goals. Population viability analyses do not explicitly incorporate economic factors, because it is often counterproductive (and usually impossible) to assign monetary value to the viability or persistence of a species. However, because of the quantitative nature of PVA results, it is possible to jointly consider ecological and economic objectives, for risk-based (and risk-weighted) decision-making. This can be done by keeping ecological and economical values separate, and presenting the results of the analysis in two dimensions, instead of only one (Fig. 2). Thus, the resulting graph has an x-axis in monetary units (e.g., the cost of implementing a certain management or conservation option), and a y-axis in biological units (e.g., reduction in the risk of extinction of the species). As more money is spent, the viability (chance of long-term survival) increases (possibly reaching an asymptote, depending on the problem). However, different management options have different curves, which may cross. This means that depending on the amount of resources available, one or the other option may be preferable. Such a

graph may be used in several ways: selecting the optimal management given the fixed resource; or estimating resources necessary for a certain level of viability (e.g., moving from "endangered" to "vulnerable"). If there are monetary benefits of conservation, these can either be shown as a different curve, or (better yet) subtracted from the cost beforehand.

When are population viability analyses most useful?

The preceding discussion highlights the conditions under which PVAs are most appropriate and most predictive. We conclude this paper with a summarized checklist of how to optimize the PVA as a conservation tool and design the analysis to get reliable qualitative answers.

Address a specific question involving focal/target species
General mandates such as "Manage this landscape so that everyone benefits" or general questions such as "Why are neo-tropical migrants declining?" are not very suitable to a direct PVA. To address such issues with a PVA, they must be reduced to a set of more specific questions, such as "Which management option would result in the highest chance of recovery of threatened species?" (e.g. Berglind 2000) or "Which set of reserves is best for the persistence of one or several focal (umbrella, indicator) species?" or "What are the long-term implications of an observed population decline for the viability of a neo-tropical migrant?"

Focus on a case with sufficient data
When data are scarce, it is risky to make assessment with any method, including PVA. In these cases, PVAs are most appropriate as exploratory tools, used to identify important assumptions and parameters, and to guide fieldwork.

Use all the available and relevant data
Assessments that use all the available and relevant data, including spatial (GIS) data, presence-absence data, habitat relationships, and demographic data from mark-recapture studies, surveys and censuses, are more reliable than those that ignore part of the relevant information.

Use the appropriate model
Model choice should be based on the availability of data, the question addressed and the ecology of the species (see Data needs and choosing a model above).

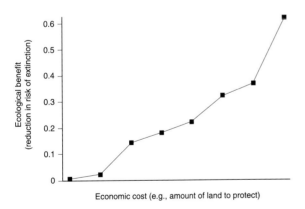

Fig. 2. An example of cost-benefit analysis with the results of a series of PVAs. Each point on the curve gives the result of one PVA, which assumes a certain amount of effort for conservation. This effort is quantified in the x-axis as the cost (e.g., the cost of setting aside a certain amount of land for protection of the species). The result of the PVA is expressed as the reduction in extinction risk from the no-action (i.e., no conservation scenario) and plotted in the y-axis.

State all assumptions explicitly

Modelers usually know the assumptions of their models, but often forget that these assumptions may not be transparent to others. An assessment should explicitly list all the assumptions (even the most obvious ones) related to model structure, parameters and uncertainties.

Validate assumptions and results where possible

Model accuracy (about model precision, see below) can be validated by using data from one half of the study system and making predictions for the other half that are compared to observed values (Kindvall 2000). Alternatively, data from a previous time period can be used for model predictions of the current (observed) situation (e.g. Sjögren-Gulve and Ray 1996, Brook et al. 2000, Vos et al. 2000). Validating assumptions of a model is often difficult; after all, when assumptions can be validated, they become parameters. However, the model should point out to the types of data that may be useful to validate or reject assumptions.

Incorporate data uncertainties, and discuss the implications

All parameters should be specified as ranges that reflect uncertainties (lack of knowledge, measurement errors). See Incorporating uncertainty above.

Analyze the sensitivity of results to assumptions and parameters

Sensitivity analysis identifies important parameters and assumptions. Sensitivity analysis should be geared towards identifying parameters that, if known with a higher precision, would decrease the uncertainty in model results to the largest extent. The importance of a parameter in determining viability depends on both the range of plausible values (its current uncertainty), and practical limitations (such as cost considerations).

Another use of sensitivity analysis is determining the most effective management action (e.g., Crowder et al. 1994, Berglind 2000, Lennartsson 2000). This is often done by evaluating the sensitivity of the model results to each parameter. However, most management actions cause changes in more than one parameter. For example, an effort to increase the survival of newborns affects both the first survival rate and the fecundity in a matrix model based on pre-reproductive census. It most likely affects the survival of other age classes as well. In such cases, it is better to evaluate "whole-model" sensitivity (with respect to management actions) instead of parameter-by-parameter sensitivities (see Akçakaya et al. 1999, Akçakaya 2000).

Report viability results

Results are more reliable and relevant if they are expressed in probabilistic terms (risk of decline) rather than deterministic terms (abundance 10 yr from now) (e.g. Berglind 2000, Ebenhard 2000). When probabilistic results are based on simulations, the number of replications or iterations determines the precision of these results. In most cases, the randomly sampled model parameters are statistically representative if the number of replications is in the 1000 – 10000 range.

Use relative risks (instead of absolute risks)

Risk of extinction, and risk of decline to an unacceptably small population size (quasi-extinction probability results from demographic models; Burgman et al. 1993, Akçakaya 2000), is a frequently reported PVA result. Here, it should be remembered that such results are usually more reliable if they are relative (which option gives higher viability?), rather than absolute (what is the risk of extinction?). As discussed above, relative results may not be as sensitive to uncertainties in the data, even in cases where the uncertainties in the data result in uncertainties in absolute results (e.g., Akçakaya and Raphael 1998).

Focus on risk of decline (instead of risk of extinction)

Because of uncertainties in modeling very small populations, the results are more reliable if risk of decline (rather than total extinction) is used. Thus, results should be expressed as the probability that the population size falls to or below a critical population level for social dysfunction or other severe effects, say, 20 or 50 individuals.

Project population dynamics for short time horizons

Short-term results are more reliable, because uncertainties are compounded with time. If a model is based on 5 yr of data, running simulations for 100 yr makes a lot of assumptions about the average and variation of model parameters. Even if long-term results may be warranted because land-use allocations are irreversible, these assumptions must be kept in mind. Furthermore, if PVA is used for impact assessment, it is important to remember that both very long and very short time horizons may mask the effects of the simulated human impact. This is because in the very near future (say, next year), it is unlikely that the population will fall to very low levels, with or without impact. Over very long time horizons, the risk will be close to one, even without the impact. Thus the difference between the two simulations (with and without impact) will be very small or zero, for very short or very long time horizons. One solution is to select the time horizon that gives the largest difference between impact and no-impact scenarios (i.e., the time horizon for which the model is most sensitive to the simulated impact). PVA models (especially those with long time horizons) should consider the possibility of a trend in average vital rates (e.g., in addition to random fluctuations, fecundity may also have a decreasing trend in its average).

Provide a feedback between fieldwork, modeling and monitoring

It is important that a PVA model, once used to make conservation decisions, is not abandoned. Additional fieldwork should provide data to refine model parameters, and monitoring should check the realism of the model. The revised model should guide further fieldwork (identify important parameters), and monitoring (identify important variables/outcomes).

Allow for adaptive management

Just as a PVA model should evolve as more data become available, the management decisions should also adapt to new PVA results. In some cases, this is not possible. For example in the case of reserve design questions, it may not be possible to change a decision. However, in the case of long-term management actions (for example, translocations, habitat restoration, harvest limits, etc.), the recommendations from a PVA should be revisited whenever new data are used to refine a model.

Acknowledgements – We thank Mark Burgman, Erica Fleishman, Linda Hedlund, Oskar Kindvall, Chris Ray and Mark Shaffer for constructive comments on the manuscript.

References

Akçakaya, H. R. 1992. Population viability analysis and risk assessment. – In: McCullough, D. R. and Barrett, R. H. (eds), Wildlife 2001: populations. Elsevier, pp. 148–157.

Akçakaya, H. R. 1998. RAMAS GIS: linking landscape data with population viability analysis (ver. 3.0). – Appl. Biomath., Setauket, New York.

Akçakaya, H. R. 2000. Population viability analyses with demographically and spatially structured models. – Ecol. Bull. 48: 23–38.

Akçakaya, H. R. and Burgman, M. 1995. PVA in theory and practice (letter). – Conserv. Biol. 9: 705–707.

Akçakaya, H. R. and Atwood, J. L. 1997. A habitat-based metapopulation model of the California gnatcatcher. – Conserv. Biol. 11: 422–434.

Akçakaya, H. R. and Raphael, M. G. 1998. Assessing human impact despite uncertainty: viability of the northern spotted owl metapopulation in the northwestern USA. – Biodiv. Conserv. 7: 875–894.

Akçakaya, H. R., Burgman, M. A. and Ginzburg, L. R. 1999. Applied population ecology: principles and computer exercises using RAMAS EcoLab 2.0. 2nd ed. – Sinauer.

Akçakaya, H. R. et al. 2000. Making consistent IUCN classifications under uncertainty. – Conserv. Biol. 14: 1001–1013.

Beier, P. and Noss, R. F. 1998. Do habitat corridors provide connectivity? – Conserv. Biol. 12: 1241–1252.

Beissinger, S. R. and McCullough, D. R. (eds) 2001. Population viability analysis. – Univ. of Chicago Press, in press.

Berglind, S.-Å. 2000. Demography and management of relict sand lizard *Lacerta agilis* populations on the edge of extinction. – Ecol. Bull. 48: 123–142.

Brook, B. W. et al. 2000. Predictive accuracy of population viability analysis in conservation biology. – Nature 404: 385–387.

Burgman, M., Akçakaya, H. R. and Loew, S. S. 1988. The use of extinction models in species conservation. – Biol. Conserv. 43: 9–25.

Burgman, M. A., Ferson, S. and Akçakaya, H. R. 1993. Risk assessment in conservation biology. – Chapman and Hall.

Burgman, M. A., Keith, D. A. and Walshe, T. V. 1999. Uncertainty in comparative risk analysis of threatened Australian plant species. – Risk Analysis 19: 585–598.

Caswell, H. 1989. Matrix population models: construction, analysis, and interpretation. – Sinauer.

Caughley, G. 1994. Directions in conservation biology. – J. Anim. Ecol. 63: 215–244.

Crowder, L. B. et al. 1994. Predicting the impact of turtle excluder devices on loggerhead sea turtle populations. – Ecol. Appl. 4: 437–445.

DeAngelis, D. L. and Gross, L. J. (eds) 1992. Individual-based models and approaches in ecology: populations, communities and ecosystems. – Chapman and Hall.

Ebenhard, T. 2000. Population viability analyses in endangered species management: the wolf, otter and peregrine falcon in Sweden. – Ecol. Bull. 48: 143–163.

Elith, J. 2000. Quantitative methods for modeling species habitat: comparative performance and an application to Australian plants. – In: Ferson, S. and Burgman, M. (eds), Quantitative methods for conservation biology. Springer, pp. 39–58.

Ferson, S., Root, W. T. and Kuhn, R. 1998. RAMAS Risk Calc: risk assessment with uncertain numbers. – Appl. Biomath., Setauket, New York.

Fleishman, E., Jonsson, B. G. and Sjögren-Gulve, P. 2000. Focal species modeling for biodiversity conservation. – Ecol. Bull. 48: 85–99.

Gärdenfors, U. 2000. Population viability analysis in the classification of threatened species: problems and potentials. – Ecol. Bull. 48: 181–190.

Gilpin, M. E. 1988. A comment on Quinn and Hastings: extinction in subdivided habitats. – Conserv. Biol. 2: 290–292.

Goldstein, P. Z. 1999. Functional ecosystems and biodiversity buzzwords. – Conserv. Biol. 13: 247–255.

Hanski, I. 1994. A practical model of metapopulation dynamics. – J. Anim. Ecol. 63: 151–162.

Hanski, I. 1999. Metapopulation ecology. – Oxford Univ. Press.

Hedrick, P. W. et al. 1996. Directions in conservation biology: comments on Caughley. – Conserv. Biol. 10: 1312–1320.

IUCN 1994. International Union for the Conservation of Nature Red List Categories. – IUCN Species Survival Commission, Gland, Switzerland.

Kiester, A. R. et al. 1996. Conservation prioritization using GAP data. – Conserv. Biol. 10: 1332–1342.

Kindvall, O. 2000. Comparative precision of three spatially realistic simulation models of metapopulation dynamics. – Ecol. Bull. 48: 101–110.

Kokko, H., Lindström, J. and Ranta, E. 1997. Risk analysis of hunting of seal populations in the Baltic. – Conserv. Biol. 11: 917–927.

Lacy, R. C. 1993. VORTEX: a computer simulation model for population viability analysis. – Wildl. Res. 20: 45–65

Lacy, R. C. 2000a. Considering threats to the viability of small populations using individual-based models. – Ecol. Bull. 48: 39–51.

Lacy, R. C. 2000b. Structure of the VORTEX simulation model for population viability analysis. – Ecol. Bull. 48: 191–203.

LaHaye, W. S., Gutierrez, R. J. and Akçakaya, H. R. 1994. Spotted owl metapopulation dynamics in southern California. – J. Anim. Ecol. 63: 775–785.

Lawton, J. H. 1999. Are there general laws in ecology? – Oikos 84: 177–192.

Lennartsson, T. 2000. Management and population viability of the pasture plant *Gentianella campestris*: the role of interactions between habitat factors. – Ecol. Bull. 48: 111–121.

Levins, R. 1969. Some demographic and genetic consequences of environmental heterogeneity for biological control. – Bull. Entomol. Soc. Am. 15: 237–240.

McCarthy, M. A. and Broome, L. S. 2000. A method for validating stochastic models of population viability: a case study of the mountain pygmy-possum (*Burramys parvus*). – J. Anim. Ecol. 69: 599–607.

McCarthy, M. A., Lindenmayer, D. B. and Possingham, H. P. 2000. Testing spatial PVA models of Australian treecreepers (Aves: Climacteridae) in fragmented forest. – Ecol. Appl., in press.

Margules, C. R., Nicholls, A. O. and Pressey, R. L. 1988. Selecting networks of reserves to maximise biological diversity. – Biol. Conserv. 43: 63–76.

Master, L. L. 1991. Assessing threats and setting priorities for conservation. – Conserv. Biol. 5: 559–563.

McGarigal, K. and Marks, B. J. 1995. FRAGSTATS: spatial pattern analysis program for quantifying landscape structure. – Gen. Tech. Rep. PNW-GTR-351. U.S. Dept of Agricult., For. Serv., Pacific Northwest Res. Stn, Portland, Oregon.

Menges, E. S. 2000. Applications of population viability analyses in plant conservation. – Ecol. Bull. 48: 73–84.

Mills, L. S., Fredrickson, R. J. and Moorhead, B. B. 1993. Characteristics of old-growth forests associated with northern spotted owls in Olympic national park. – J. Wildl. Manage. 57: 315–321.

Millsap, B. A. et al. 1990. Setting the priorities for the conservation of fish and wildlife species in Florida. – J. Wildl. Manage. (Suppl.) 54: 5–57.

Pearce, J. L., Burgman, M. A. and Franklin, D. C. 1994. Habitat selection by helmeted honeyeaters. – Wildl. Res. 21: 53–63.

Pimm, S. L. and Lawton, J. H. 1998. Planning for biodiversity. – Science 279: 2068–2069.

Possingham, H., Ball, I. and Andelman, S. 2000. Mathematical methods for identifying representative reserve networks. – In: Ferson, S. and Burgman, M. (eds), Quantitative methods for conservation biology. Springer, pp. 291–306.

Pressey, R. L. et al. 1993. Beyong opportunism: key principles for systematic reserve selection. – Trends Ecol. Evol. 8:124–128.

Pressey, R. L., Possingham, H. P. and Margules, C. R. 1996. Optimality in reserve selection algorithms: when does it matter and how much? – Biol. Conserv. 76: 259–267.

Scott, J. M. et al. 1993. Gap analysis: a geographic approach to protection of biological diversity. – Wildl. Monogr. 123.

Simberloff, D. 1998. Flagships, umbrellas, and keystones: is single-species management passé in the landscape era? – Biol. Conserv. 83: 247–257.

Sjögren-Gulve, P. 1994. Distribution and extinction patterns within a northern metapopulation of the pool frog, *Rana lessonae*. – Ecology 75: 1357–1367.

Sjögren-Gulve, P. and Ray, C. 1996. Using logistic regression to model metapopulation dynamics: large-scale forestry extirpates the pool frog. – In: McCullough, D. R. (ed.), Metapopulations and wildlife conservation. – Island Press, Washington, D.C., pp. 111–137.

Sjögren-Gulve, P. and Hanski, I. 2000. Metapopulation viability analysis using occupancy models. – Ecol. Bull. 48: 53–71.

Solow, A. R. 1993. Inferring extinction from sighting data. – Ecology 74: 962–964.

Stacey, P. B., Johnson, V. A. and Taper, M. L. 1997. Migration within metapopulations: the impact upon local population dynamics. – In: Hanski, I. and Gilpin, M. E. (eds), Metapopulation biology: ecology, genetics, and evolution. Academic Press, pp. 267–291.

Straw, J. A., Wakely, J. S. and Hudgins, J. E. 1986. A model for management of diurnal habitat for American woodcock in Pennsylvania. – J. Wildl. Manage. 50: 378–83.

UNCED 1992. The United Nations Convention on Biological Diversity. – Rio de Janeiro.

Verboom, J. et al. 1991. European nuthatch metapopulations in a fragmented agricultural landscape. – Oikos 61: 149–156.

Verner, J., Morrison, M. L. and Ralph, C. J. (eds) 1986. Wildlife 2000: modeling habitat relationships of terrestrial vertebrates. – Univ. of Wisconsin Press.

Vos, C. C., ter Braak, C. F. J. and Nieuwenhuzen, W. 2000. Incidence function modelling and conservation of the tree frog *Hyla arborea* in the Netherlands. – Ecol. Bull. 48: 165–180.

Ecological Bulletins 48: 23–38. Copenhagen 2000

Population viability analyses with demographically and spatially structured models

H. Reşit Akçakaya

Akçakaya, H. R. 2000. Population viability analyses with demographically and spatially structured models. – Ecol. Bull. 48: 23–38.

This paper presents a review of demographically structured (or, frequency-based) models, in which the individuals in a population are grouped into distinct classes. Structured models are used when vital rates (survival, reproduction, dispersal) of individuals depend on their age or physiological/morphological stage. Variation in these rates (environmental stochasticity) and the effect of abundance (density dependence) are important factors that determine population viability. Metapopulation models are built by adding information about spatial structure (number of populations, and the location, size, shape, and quality of habitat patches they inhabit). At the metapopulation level, the similarity of environmental fluctuations (correlation) and dispersal among patches become important variables determining viability. These modeling methods can be combined with habitat analyses that link landscape data with metapopulation models.

H. R. Akçakaya (resit@ramas.com), Applied Biomathematics, 100 North Country Road, Setauket, NY 11733, USA.

Population viability analyses often use models that simulate the future of the species based on parameters on the ecology and demography of its population(s). In demographically structured (also called frequency-based) models, the individuals in a population are grouped into distinct classes, based on their demographic characteristics, or their location, or both. This paper presents a review of structured models, including factors that such models can incorporate, the types of results they give and their advantages and disadvantages compared to the other types of models reviewed in this volume (Lacy 2000a, Sjögren-Gulve and Hanski 2000).

Structured models that group individuals in a population according to their age (called age-structured models) incorporate age-specific vital rates such as probability of survival form one age class to the next, and fecundity in each age class. Age structure is inadequate to model some species with more complex life histories. These species may

be better modeled with a stage-structured model, in which the individuals are classified according to characteristics such as size, weight, physiological, morphological or developmental state. Examples of stages used in such models include juvenile, subadult, adult; and seed, seedling, sapling, understory, canopy. Vital rates in a stage-structured model are rates of transition from one stage to another, which are used to model survival, growth and reproduction.

Spatial structure refers to the location of individuals, which are grouped into sub- or local populations. Different subpopulations may have different demographic characteristics, such as population size (abundance), carrying capacity, and vital rates.

I begin with a discussion of the type of results from structured models, and the type of questions such results may be used to address (see Outputs from structured models). In the section on How to build structured models, I discuss various factors that are incorporated into structured models, and the methods used for estimation of parameters related to these factors. In later sections, I discuss different uses of structured models, summarize their main assumptions and common data requirements, and review examples of models for different taxa and different types of life histories. In the Discussion section, I summarize the advantages and disadvantages of structured models, and discuss cases and questions for which structured models are most appropriate.

Outputs from structured models

While discussing the details of building structured models in later sections, it is important to keep in mind the types of results such models produce, and how these results can be interpreted. The results of structured models can be expressed in several different ways. Some types of results are based on deterministic measures that attempt to make precise predictions about the fate of a population or species. Examples of deterministic measures include the predicted abundance, and the population growth rate (e.g., the finite rate of increase, λ, estimated as the dominant eigenvalue of the stage matrix; see Caswell [1989] and Burgman et al. [1993], pp. 43–44, 127–132). Other types of results are based on stochastic measures, which incorporate variability and uncertainty by expressing the results in probabilistic terms. These measures include risk of extinction, risk of decline, and time to extinction.

Deterministic measures such as the finite rate of increase only reflect the state of the population at a given time, and may not anticipate the future behavior of the population. For example, a large but fluctuating population may have a higher risk of extinction than a smaller but less variable population. The deterministic growth rate ignores this variability as well as the distribution of individuals to age classes or stages. Because of these two factors, deterministic projections with finite rate of increase will be incorrect in the short-term (up to one generation or so). In the long term, factors such as density dependence and trends in habitat and vital rates may cause deterministic predictions to be biased. The only general statement that one can make about population growth rates is that for most species, the long-term average is ca 1. When short-term growth rate is < 1.0, it may mean that the population is declining because of a decrease in average values of vital rates ("systemic pressure" as defined by Shaffer 1981), or it is undergoing fluctuations (and in the period that it was observed, it happened to be declining), or it is returning to an equilibrium as a result of density dependent mechanisms. When short-term growth rate is > 1.0, this may be because of a temporary increase, and may not mean that the population is safe in any sense meaningful for conservation and management.

Natural variability in population dynamics is compounded by uncertainty in the population parameters due to lack of perfect information (i.e. ignorance). The consequent difficulty of making precise predictions has shaped the language of population viability analysis (e.g., see Shaffer 1990). The conservation-related problems and questions that PVA addresses are usually phrased in terms of probabilities; for example, we may want to assess the probability of extinction or the chance of recovery from a population bottleneck. Stochastic PVA measures are probabilistic expressions of population viability such as these. The concept of minimum viable population (MVP) is also phrased in terms of probabilities (Shaffer 1981), for example as the minimum size of a population that has a 90% chance of surviving for the next 100 yr. Mace and Lande (1991) categorized levels of threat in probabilistic terms; they defined, for example, a "critical" population as one with a 50% chance of extinction in the next 10 yr. Deterministic measures such as population growth rate or population size cannot address conservation questions in these probabilistic terms, but stochastic PVA models do.

Extinction risk assessment

Risk is defined as the probability of an unwanted event (such as an extinction or population decline). Extinction risk assessment attempts to assess the likelihood of a population or species extinction (or decline) by some specified time in the future under various natural conditions and scenarios of management.

The results of risk assessment can be expressed in different terms. One of the commonly used variables is extinction time, i.e., predicted time until a population or species goes extinct. Reporting only the mean extinction time may be misleading, because distribution of extinction times is often skewed, with a large right tail (Fig. 1). Therefore, a

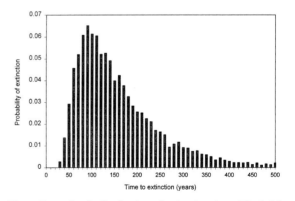

Fig. 1. Example of a distribution of extinction times. The height of each vertical bar gives the probability that the population will go extinct in the 10-yr period ending with the year shown on the x-axis. For example, there is ca 1.4% chance that the population will go extinct between years 30 and 40, and ca 2.9% chance that it will become extinct between years 40 and 50. The median of this distribution is ca 131 yr.

proportion of populations go extinct at times much later than average, having a disproportional impact on the arithmetic mean. This also means there is a substantial probability that the population will go extinct much earlier than the mean extinction time.

This problem can be avoided to some extent by expressing the predicted extinction time in terms of the median extinction time. However, reporting only the median may lead to a wrong perception of the threat the population faces. The level of threat is communicated better by specifying the whole distribution of extinction times, e.g., by plotting is the cumulative probability distribution, which shows the probability of extinction at or before a specific time step. A cumulative distribution of extinction times increases monotonically, and asymptotically reaches a probability of 1.0 for very long time periods (the probability often does not reach 1.0 in models with medium to long time horizons, especially if the population is growing or immigration is assumed).

Extinction time can be made more general and useful by incorporating partial decline, or quasi-extinction (Ginzburg et al. 1982). Thus, the result becomes (the distribution of) the time (e.g., number of years) until the population declines below a predetermined threshold (see Fig. 2). This is useful because most models, and especially structured models, are unreliable at simulating the true dynamics of very small populations consisting of a handful of individuals. This is because there is often insufficient information to incorporate factors such as inbreeding depression and Allee effects (see Adding density dependence below) that affect small populations (see Lacy 2000a).

The threshold used in calculating quasi-extinction times depends on the question addressed and the biology of the species. A particular level may represent an anthro-pogenically significant threshold, as in the case of economically important species. It may also represent a biologically critical level of abundance. For instance, at low abundances, inbreeding depression can grow sharply in severity, or individuals may simply not be able to locate mates, or other Allee effects may come into play that were negligible at higher abundances. It is often very difficult to predict the behavior of a population once it reaches a very low level, because of these complicating demographic and genetic factors. It may be easier and more conservative to calculate quasi-extinction risk, i.e. the probability of a decline to a level where such effects are suspected of becoming dominant factors.

A different type of measure concentrates on the probability of extinction. The simplest measure of this type is simply the extinction risk, which is the probability that a population will go extinct within a specified time period. As with extinction time, it may be better to incorporate a partial decline. This is done by expressing the risk as quasi-extinction risk, which is the probability that the population will fall below a set of threshold population sizes at least once during a fixed time period. If the threshold is zero, then the quasi-extinction risk is the same as the extinction risk.

Alternatively, the risk can be expressed as a function of the amount of decline, i.e., as a function of the threshold abundance. The quasi-extinction risk can also be expressed as the risk of declining by a given amount from the initial population, i.e., as the probability of decline as a function of percentage decline from the initial population size (see Fig. 3).

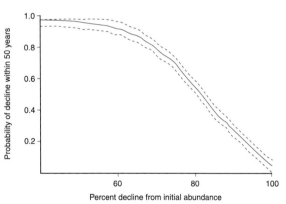

Fig. 3. Example of a risk of decline curve: each point on the curve shows the probability that the population will fall by the given percentage from the initial abundance anytime during the next 50 yr. For example, the curve indicates that the probability of an 80% decline (from initial abundance) in the next 50 yr is ca 0.56. If the total abundance in year 2000 was, for example, 1000 individuals, this means that there is about a 56% risk that the abundance will fall to or < 200 individuals sometime before the year 2050. Dotted curves give 95% confidence intervals based on Kolmogorov-Smirnow statistic, D.

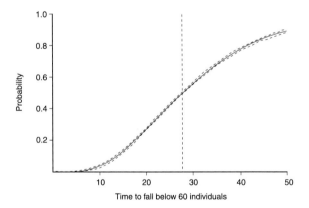

Fig. 2. Example of a cumulative time to quasi-extinction curve: each point in the curve gives the probability that the population will decline to 60 individuals by the end of the time step (given in the x-axis). The vertical line indicates the median time to quasi-extinction (the time step when the probability reaches 0.5), which is 27.6 yr in this example. This means that there is about a 50% risk that the total abundance will fall to or < 60 individuals sometime in the next 28 yr.

These types of outputs can be used to assess the threat faced by a population or species, to compare different types of human impact, to evaluate different conservation or management options, and to plan fieldwork. These will be discussed in more detail below (see How to use structured models).

How to build structured models

Selecting demographic structure and estimating vital rates

In a demographically structured model, the population abundance is represented by a set of numbers (one for each age or stage class), instead of a single number for the whole population. See Box 1.

Using an age-structured model requires that the age of organisms is recorded when collecting the data necessary for estimating vital rates (survival rates and fecundities). Stage structure is more appropriate when: 1) the available information is limited to cruder categories such as juvenile and adult, 2) the demographic characteristics of organisms (their probability of survival or fecundity) is determined by their size or other physiological characteristics, rather than their age, 3) growth is plastic, i.e., some individuals exhibit accelerated or retarded growth relative to others, or even regress (for example, decrease in size).

The first step in determining the stage matrix is to decide on what the stages are. This mostly depends on the life history of species studied. If the stages are defined on the basis of the size of organisms, then the number of stages, and the size limits for each stage must also be decided. This may be a complicated problem. On the one hand, it is nec-

---

Box 1

Formulating an age-structured model of a species that lives for 4 yr. The population abundance N is represented by a vector of 4 numbers:

$$\begin{bmatrix} N_0(t) \\ N_1(t) \\ N_2(t) \\ N_3(t) \end{bmatrix}$$

where $N_i(t)$ represents the number of i-year-old individuals at year t. In models with age or stage structure, vital rates such as survivals and fecundities are often organized in a transition matrix. For an age-structured model, such a matrix is called a Leslie matrix. An example of Leslie matrix (for four age classes) is:

$$\begin{bmatrix} F_0 & F_1 & F_2 & F_3 \\ S_0 & 0 & 0 & 0 \\ 0 & S_1 & 0 & 0 \\ 0 & 0 & S_2 & 0 \end{bmatrix}$$

where F_i and S_i are the fecundity and survival rate, respectively, of the i-th age class. Each row (and column) corresponds to one age class. In a stage-structured model, each row (and column) of the transition matrix corresponds to one stage. The element at the ith row and jth column of the matrix represents the rate of transition from stage j to stage i, including survival, growth and reproduction.

Multiplying the transition matrix with the vector of abundances at year t gives the vector of abundances at year t+1:

$$\begin{bmatrix} N_0(t+1) \\ N_1(t+1) \\ N_2(t+1) \\ N_3(t+1) \end{bmatrix} = \begin{bmatrix} F_0 & F_1 & F_2 & F_3 \\ S_0 & 0 & 0 & 0 \\ 0 & S_1 & 0 & 0 \\ 0 & 0 & S_2 & 0 \end{bmatrix} \times \begin{bmatrix} N_0(t) \\ N_1(t) \\ N_2(t) \\ N_3(t) \end{bmatrix}$$

Repeating this multiplication allows the model to predict the abundance in future years. For more information on matrix models, see Caswell (1989), Burgman et al. (1993) and Akçakaya et al. (1999).

---

essary to define a sufficiently large number of stages so that the demographic characteristics of individuals within a given stage are similar. On the other hand, it is necessary to have a sufficiently large number of individuals in each stage so that the transition probabilities can be calculated with reasonable accuracy (see Vandermeer [1978] and Moloney [1986]).

The estimation of the matrix elements depends on the type of data available. If all individuals in a population can be followed through at least two time steps, and their age class or stage at each time step recorded, these data can be used to estimate each vital rate for each time step (except the last one). If, instead of following each individual, the abundance of individuals in different stages or age classes can be estimated for several time steps, these data can be used in a multiple regression analysis. For methods of data analysis using these approaches, see Caswell (1989) and Akçakaya et al. (1999).

Another commonly used method for estimating abundances and survival rates for an age or stage-structured model is a mark-recapture study (also called a capture-recapture study; see Pollock et al. 1990, Burnham and Anderson 1992, Lebreton et al. 1993). The analysis of mark-recapture data for estimating vital rates is facilitated by specialized software, such as the program MARK (http://www.cnr.colostate.edu/~gwhite/mark/mark.htm), developed by G. White.

When fecundity estimates are based on birth rates or nest surveys, these must be combined with the survival rates to calculate the fecundity element in the stage matrix (see Jenkins 1988, Caswell 1989, pp. 8–15; Burgman et al. 1993, pp. 127–129 and 140–142; and Akçakaya et al 1999, pp. 136–141). This is because, in a matrix model, fecundity is the average number, per individual alive at a given time step (t), of offspring censused at the next time step (t+1). Thus, fecundity depends on: a) the proportion of the age class or stage that are breeders, b) reproductive success (e.g., number of fledglings per nest in a bird species), c) sex ratio (e.g., proportion of females at birth), d) survival of breeders from census to next breeding, and e) survival of newborns from birth to next census.

If, in a stage-structured model, the reproductive stage is "breeders" (instead of, say, "adults"), then (a) is obviously 1.0. If not all adults breed, then the transition rates between stages must reflect this. If only females are modeled (which is common; see Adding sex structure below), then (c) is the proportion of females at birth (i.e., 0.5 if 1:1 sex ratio is assumed). Most matrix models are parameterized either as post-breeding or pre-breeding census, and thus either (d) or (e) is assumed to be 1.0 (but never both). For parameterization of matrix models with post-breeding and pre-breeding assumptions, see Caswell (1989) and Akçakaya et al. (1999).

Fecundity may also be based on regression of the counts of first stage (the stage that individuals start their life in) on the counts of individuals in reproductive stages in the previous time step (see Akçakaya et al. 1999).

Estimating initial abundances

Demographic models predict the abundance of individuals in each age (or stage) class through time, given the initial number of individuals in each age or stage. How much the total initial abundance affects extinction risks depends on other factors, particularly vital rates, density dependence and stochasticity. For some models, it may not have much of an effect. However, the distribution of the initial total abundance to classes (i.e., the initial proportion of individuals in each age or stage) may still have important effects on persistence of a population (Burgman et al. 1994).

There are various methods of estimating total abundance or its distribution to age/stage classes, including censuses, aerial surveys, mark-recapture studies and transect surveys. In variable-distance line-transect surveys, distances to the observed birds are used to fit a function of declining detection probability (Buckland et al. 1993). One program that does such fitting is DISTANCE (Laake et al. 1996; http://www.ruwpa.st-and.ac.uk/distance/). This program is used to analyze distance sampling data, including line transects, point transects (variable circular plots) and cue-counts, to estimate density and abundance of a population.

Adding variability

Natural environments often change in an unpredictable fashion, causing changes in a population's vital rates. Since these rates are partially a reflection of the suitability of the environment for growth, this type of variation is called environmental stochasticity. Extreme environmental events that adversely affect large proportions of a population are called catastrophes. Catastrophes may in some cases be considered to be a source of environmental variation that is independent of the normal year-to-year fluctuations in vital rates, carrying capacities, or other population parameters. When the number of individuals gets to be very small there is another source of variation that becomes important even if the population growth rate were to remain constant. This variation is called demographic stochasticity. For example, if there are 3 individuals in an age class, and if the survival rate for this particular time step is 0.4, the number of survivors can be calculated by assuming that each individual has an independent probability (0.4) of surviving. Thus, the number of survivors may be 3 (with probability 0.4^3) or 0 (with probability $[1-0.4]^3$) or any other number in between (with the corresponding binomial probability).

Environmental stochasticity is usually modeled by sampling the vital rates (fecundity and survival rates) and other model parameters (such as carrying capacity or dispersal rate) in each time step of the simulation from random distributions with given means and variances. The variance of these distributions should be based on temporal variance in the observed rates. This method is called "element selection" because each element of the matrix is selected randomly.

Another method of modeling environmental stochasticity is "matrix selection", i.e., selecting randomly from a set of matrices (see e.g., Cohen 1979, Bierzychudek 1982). For example, if two different matrices are considered to represent good and bad years, one of the two can be randomly selected at each time step of the simulation and used in population projection. The advantage of this method is that it uses only observed vital rates, and does not make assumptions about the type of statistical distribution for these rates. The disadvantage is that, if the number of matrices is small, the range of the observed vital rates may underestimate their long-term ranges. Consequently, the matrix selection method can underestimate long-term variability, and hence the extinction risk. Matrix selection method is used in the simulation program POPPROJ2 (Menges 1992); in addition, the programs RAMAS GIS (Akçakaya 1998) and RAMAS Stage (Ferson 1990) can be used to model both element selection and matrix selection methods.

Demographic stochasticity can be modeled by sampling the number of survivors or dispersers from binomial distributions, and number of offspring (recruits) from Poisson distributions (Akçakaya 1991). Catastrophes are modeled by adding rare, random changes to model parameters or variables. For example, a 100-yr-flood can be modeled by making these changes with a probability of $0.01 \, yr^{-1}$. If a flood kills half the population, its effect can be modeled by multiplying the number of individuals with 0.5 whenever a catastrophe occurs in the model.

There are several other important points to consider when adding variability to a model, see Box 2.

Adding density dependence

As a population grows, the effect of limited resources on population growth results in density-dependent feedback that limits population growth. This type of density-dependence can be incorporated into a matrix model by allowing the matrix elements (vital rates) to decrease as population size increases. Density dependence may have significant effects on the risk of extinction of a population. For example, a density-independent model with growth rate of 1.0 and a density-dependent model may give similar average trajectories, but the variability (hence, the risk of decline) of the density-dependent model will often be less due to the stabilizing effects of density dependence. Thus, in the absence of information, a density-independent model may be used for conservative assessment (Ginzburg et al. 1990).

The simplest type of density dependence involves truncating total abundance at a ceiling. Modeling other types of density dependence requires selecting a function, and

estimating its parameters. Some of the commonly used types of functions can be parameterized in terms of 1) a maximum rate of growth when density is low, and 2) a carrying capacity or an equilibrium abundance (Akçakaya 1998). However, detecting density dependence and estimating its parameters can be a complicated problem (Burgman et al. 1993, pp. 87–91).

Population growth may also be affected negatively as population size reaches very low levels. The factors that cause such a decline (e.g., difficulty in finding a mate or disruption of social functions) are collectively called Allee effects (see Lacy 2000a). If a population declines to a critical level, if only by chance, then Allee effects can pull it down even further. Such phenomena can dramatically influence the risks of quasi-extinction, especially for small or fragmented populations. Modeling Allee effects involves more complicated density-dependence functions. A simple way is to use a quasi-extinction threshold instead of a threshold of zero, i.e., instead of extinction risk, to estimate the risk of decline to a level where Allee effects are suspected to become important.

Adding genetics

Certain simplifying assumptions allow incorporating the effects of genetics to single-population structured models (Mills and Smouse 1994). For metapopulation models, this is complicated by exchange of individuals among populations by dispersal. Thus, realistic models that incorporate genetics are often individual-based (Lacy 2000a). However, for most species, a more fundamental problem is lack of information on the interaction between genetic and demographic factors, including the effects of inbreeding on demographic parameters. In the absence of this crucial information, one precautionary approach might be to model Allee effects, or to use a relatively large threshold for extinction. Another option might be to consider the results as a lower bound estimate of the extinction risk. However, in a study designed to validate PVA models, Brook et al. (2000) found that even models that did not incorporate genetics did not underestimate risks.

Adding sex structure

Many structured models concentrate on a single sex; usually only the females are modeled. In species where one male can mate with several females, the number of males may not affect the total fecundity very much, and only females should be modeled. For other species, it is possible to develop models that include both sexes, but this requires additional information about reproduction. In particular, the contribution of males to fecundity (which is likely to be frequency-dependent) must be known. In models of monogamous species with only one breeding stage, fecun-

Box 2

Important points to consider when adding variability to a structured demographic model

1. The observed variance often includes effects of sampling variance (measurement error) and demographic stochasticity, which must be subtracted from total observed variance. Otherwise the model may overestimate variability, hence the extinction or decline risk.

2. The model should allow putting constraints on parameter values, such as limiting survivals to the [0,1] range, and constraining fecundities to be positive. In stage-structured models, the sum of all survival rates from a given stage (for example, probability of remaining in the same stage and growing to the next stage) must be constrained to be ≤ 1.

3. Even if the sum of such transitions is < 1.0, variation introduced by demographic stochasticity may make the number of individuals surviving from a given stage larger than the number in the stage in the previous time step. This must be corrected.

4. The truncation of sampled values may introduce a bias, so that the realized mean of, say a survival rate, is different from the average value used as the model parameter. This can be corrected by changing the shape of the distribution, by pooling the variance of all survival rates from a stage, or by introducing a different correlation structure. Changing correlation structure applies only in the case of stage-structured models in which transitions from one stage to two or more stages are possible. For example, suppose there are 2 survival rates from a particular stage (say, "small") in a size-structured model: 60% of the individuals stay in the same stage, 35% grow to the next stage (say, "medium"). Because the total of these three transitions (0.95) is large, if each of them is sampled with large variances and with a correlation of 1.0 among them, truncations may occur. However, a large variance, combined with this mean can only mean that in years when a lot of individuals stay in the same stage, there are few that grow to the next stage, and vice versa. This suggests a negative correlation between the proportion that stay in the same stage and the proportion that grow to the next stage (Akçakaya 1998).

5. A high value for the correlation of vital rates among stages (or among populations of a metapopulation; see below) results in higher variability, and higher risk of decline. Thus, when correlations are not known, assuming full correlation rather than independence gives results that are more conservative or precautionary.

6. The number of replications determines the precision (but not the accuracy) of the risk estimates. Each simulation should be run with a minimum of 1000 replications (see Akçakaya and Sjögren-Gulve 2000). The accuracy of the risk estimates is determined by the measurement error in model parameters, including biases in estimation methods. Uncertainty should be incorporated in the form of ranges (lower and upper bounds) of each model parameter.

7. Simulations with catastrophes assume that such a catastrophe did not happen in the period when the demographic data was collected. If data included a catastrophe, then the estimates of demographic rates should exclude those catastrophe years (see Akçakaya and Atwood 1997). Otherwise, the combined effect of modeled catastrophes and environmental stochasticity would overestimate the actual variability in vital rates, and underestimate their means.

dity can be expressed simply as number of offspring per breeding pair, multiplied by the minimum of the number of males and the number of females in the breeding stage. If more than one age class is reproductive, or if breeding is not monogamous, the frequency dependence will be more complicated.

Adding spatial structure

In addition to the factors discussed above that affect the dynamics of populations, there are spatial factors that operate at the metapopulation level. These include, for example, the number and geographic configuration of populations within the metapopulation, distances and the rates of dispersal among populations, and the degree of similarity of environmental fluctuations experienced by different populations.

Dispersal refers to the movement of individuals among populations, which may lead to recolonization of vacant patches (i.e., extinct populations) by immigration from other (extant) populations. Such recolonization would have a positive effect on overall metapopulation persistence. In addition, dispersal may supplement local populations and stabilize local dynamics, thus preventing local extinctions ("rescue effect"; Brown and Kodric-Brown 1977).

Dispersal rates depend on many factors, for example

species-specific characteristics such as the mode of seed dispersal, motility of individuals, ability and propensity of juveniles to disperse, etc. The dispersal rate between any two populations of the same species may also differ drastically, depending on a number of population-specific characteristics such as the distance between the populations, the surrounding habitat, topography, prevailing wind or water currents, the density of the source population, and habitat quality at the target patch.

When vital rates, carrying capacities or catastrophe-related parameters of two populations are different, increased dispersal between them may lead to decreased viability (even as it increases the occupancy rate). One reason this may happen is similar per-capita dispersal rates between a large and a small population in both directions. In this case, the number of dispersers from the large to the small population will overshoot the small population's carrying capacity (and thus not contribute much to its persistence). At the same time, the small number of dispersers from the small population to large population will not compensate for the number that leaves the large population. This effect would not occur if the dispersal rates in the two directions are unequal, which may be the case when there is a large difference in the areas of the two patches (see Akçakaya and Baur 1996, Hill et al. 1996, Kuussaari et al. 1996, Akçakaya and Raphael 1998, Kindvall 1999). Dispersal rates can be estimated by mark-recapture methods (see above).

The effectiveness of dispersal in reducing extinction risks depends largely on the degree of similarity of environmental fluctuations experienced by different populations (i.e., their correlation or interdependence). This is because when all populations decline simultaneously, there is less chance of recolonization of empty patches (Solbreck 1991). However if the fluctuations are at least partially independent (uncorrelated), then when some populations decline or become extinct others may remain extant or even increase, thus providing recolonization opportunities. A range of values for the correlation among populations can be estimated using the correlation of environmental factors causing the variability (e.g., LaHaye et al. 1994).

If the fluctuations are partially or substantially synchronous (correlated), then models based on an assumption of independent population dynamics among patches will underestimate extinction risk. As a result, correlation of environmental fluctuations among populations have important effects on metapopulation persistence and viability (Gilpin 1988, Akçakaya and Ginzburg 1991, Burgman et al. 1993).

Incorporating habitat characteristics

There are various ways of adding spatial structure to a model to account for the spatial factors discussed above. At one extreme are simple occupancy models that are based on the number of occupied and unoccupied habitat patches, ignoring their location (see Sjögren-Gulve and Hanski 2000). At the other extreme are individual-based models that describe the spatial structure with the location of each individual in the population, or the location of territories or home ranges. Between these are spatially explicit metapopulation models that describe the dynamics of each population with structured demographic models, and incorporate spatial dynamics by modeling dispersal and temporal correlation among populations. Both dispersal and correlation between each pair of populations depend on the location of the populations, making these models spatially explicit. One type of spatially explicit metapopulation model is based on a regular grid, each cell of which is modeled as a subpopulation of a metapopulation (e.g., Price and Gilpin 1996). The major limitation of this approach is that the regular pattern of the grid is often arbitrary and does not correspond to the distribution of the species in the landscape.

Another approach expands spatially explicit metapopulation models by incorporating information about habitat relationships of the species and the characteristics of the landscape in which the metapopulation exists (e.g., Akçakaya and Atwood 1997). This method uses a habitat suitability map to determine the spatial structure of the metapopulation (number and location of habitat patches in which subpopulations of the metapopulation live) and population-specific parameters.

The habitat suitability map can be calculated in a number of different ways, including statistical analyses (such as logistic regression) that find the relationship between the occurrence or density of the species and independent variables that describe its habitat requirements. The relationship can be statistically validated by estimating the function from half of the available data, and predicting the habitat suitability of known locations in the other half (see Akçakaya and Atwood 1997).

The habitat suitability map is then used to calculate the spatial structure of the metapopulation. This is done by identifying cluster of cells in a raster map that are suitable (e.g., above a threshold value of habitat suitability), and that are close to other suitable locations. This patch recognition is based on species-specific characteristics such as the home range size, dispersal distance and minimum habitat suitability for reproduction. The demographic parameters (such as carrying capacity and average vital rates) of the population inhabiting each habitat patch can be determined as functions of patch-specific characteristics, such as the total habitat suitability in the patch (Akçakaya 1998). This provides a link between the spatial and demographic components of the model, and makes it easier to parameterize models with large number of populations, based on limited data.

Often, habitat loss or degradation results in deterministic changes (such as temporal trends) in population characteristics that are different from random environmental fluctuations. Such changes can be incorporated by trends in model parameters such as carrying capacities or vital rates (e.g., see Akçakaya and Raphael 1998).

How to use structured models

Structured models can be used to assess the threat of extinction or decline faced by a population or metapopulation. Such assessments can be used to prioritize species, for example in the context of the IUCN (1994) threatened species categories. Other uses of structured models involve sensitivity analyses or comparison of "What if…" scenarios for evaluating management options and assessing human impact. These uses are discussed below.

Sensitivity analysis and planning field research

Because of lack of sufficient data and measurement errors, parameters of a model are often known as ranges instead of single estimates. In such cases, collecting more data makes these ranges narrower, and consequently the results become more certain. Given that there is a cost associated with additional fieldwork, the question is: Which parameter should be estimated better first? In other words: Which of the several uncertain parameters should be given priority for more precise estimation with additional fieldwork?

Such questions can be addressed by sensitivity analyses. There are two common methods of sensitivity analysis: 1) deterministic eigen-analysis is based on the effect of each vital rate on the eigenvalue of the stage matrix (i.e. λ, the finite rate of increase). For example, elasticity analysis quantifies the proportional change in λ caused by a proportional change in each vital rate (de Kroon et al. 1986, Caswell 1989, Burgman et al. 1993). 2) Risk-based sensitivity analysis is based on the effect of changes in vital rates (or other parameters) on population extinction risk or recovery chance.

As discussed above, deterministic methods ignore variability, density dependence and the initial distribution of individuals to stages. In addition, they focus on the deterministic growth rate, rather than the more relevant results such as the risk of extinction.

An important consideration in deciding which parameters are more important to estimate more precisely is the uncertainty in each parameter. Other things being equal, it makes sense to spend more time and money for additional data on a more uncertain parameter. With the risk-based method described above, this consideration is accounted for by changing each parameter to the lower and upper values of its estimated range, instead of changing them plus and minus a fixed percentage. This way, a parameter with a wider range will contribute more to uncertainty about the risk of extinction (other things being equal).

With the deterministic methods (such as elasticities), it is not always possible to take this consideration into account, because those methods are based on linear approximations, which means they assume that growth rate changes linearly with changes in vital rates. This is often a good approximation for small changes, but may not be valid for large ones (e.g., when a survival rate is known as a wide range).

Another disadvantage of the deterministic methods is that they are often applied only to single matrix elements. However, some matrix elements may have to be estimated as products of two vital rates. For example, fecundity may be estimated as the product of maternity (e.g., number of fledglings per adult) and survival of the juveniles until the next census. If we want to decide whether the fieldwork should focus on maternity or juvenile survival (which may require different types of study design), then the sensitivity of the population growth rate to their product (fecundity) is not very useful.

Finally, the risk-based sensitivity approach can be extended to parameters other than those in the stage matrix (average vital rates). Often, the variances of vital rates are known even more poorly than their averages. We may be uncertain about the type of density dependence, or the number of stages to use in the model. In each of these instances, the strategy is the same: change parameter values or model structure to their alternatives and measure the importance of the change by the effect it has on the risks of decline.

Assessing impacts and evaluating management options

Another application of sensitivity analysis involves decisions about which vital rates to focus on in management and conservation efforts. The evaluation of management options requires considerations similar to those for planning field research. The first is the contribution of each vital rate to the expected growth rate, and the chances of decline or recovery of the population. Thus, a formal sensitivity analysis of a model can provide some insight into how best to manage a population.

The second consideration is how much each vital rate (or other model parameter) can be changed with management. Other things being equal, it makes sense to base management actions on those aspects of the life history that are practical to manipulate. Further complicating this issue is the fact that each management or conservation action may affect more than one vital rate. For example, protecting nest locations of a bird species may improve fecundity, and to a lesser extent survival rates, whereas restoring dispersal habitat may improve dispersal rates, juvenile survival, and to a lesser extent adult survival. In these cases, a parameter-by-parameter analysis of sensitivity does not make sense, because the parameters cannot be changed independently (or in isolation from others). It is much better to do a "whole-model sensitivity analysis", and compare management options instead of single parameters. This can be done by developing models for each management or conservation alternative. Each model incorporates changes to all the parameters affected by that particular alternative. The results of these models than can be compared to each other, as well as to a "no-action" scenario (e.g., Berglind 2000).

The same approach can be used when assessing the impact of a disturbance or human activity, by comparing a "no impact" scenario with the results of a model that incorporates effects of the simulated impact on model parameters (see Fig. 4). As with management options, an impact is likely to affect more than one parameter, each to a different degree. Thus, a whole-model sensitivity analysis is more appropriate than one based on single parameters.

Both in planning future fieldwork and in evaluating management actions, a non-biological consideration is the relative cost of each research objective or management action. Population viability analyses do not explicitly incorporate economic factors but, because of quantitative nature of PVA results, it is possible to jointly consider ecological and economical objectives (see Akçakaya and Sjögren-Gulve 2000).

When structured models are used in this comparative mode, the results can be expressed in relative terms, e.g., as the relative increase in the risk of extinction due to an impact. Such relative measures are less sensitive to uncertain-ties in the data compared to absolute measures such as the risk of extinction (Lindenmayer and Possingham 1996, Hanski 1997, Akçakaya and Raphael 1998).

Assumptions of structured models

The type of structured models reviewed in this paper assume discrete time steps, i.e., are based on difference equations. Most of these models address a single species; thus, they cannot explicitly represent competition, predation, mutualism or other interspecific interactions. These interactions can be modeled as constant, fluctuating, or cyclic influences on the demographic parameters.

Structured models group individuals into distinct categories such as age classes, stages, subpopulations, or a combination of these (e.g., stages within subpopulations). Within a group, the individuals are assumed to be identical. For example, the basic assumption of stage-structured models is that the demographic characteristics of individuals are related to their developmental stage. They assume that there is little variation among individuals in the same stage with respect to their demographic characteristics such as chance of surviving, chance of reproducing, and the number of offspring they produce.

Matrix models assume that what an organism will do (its demographic rates) depends only on the stage it is in now, and not on what stage it was in the previous time steps, or how long it had remained in each stage.

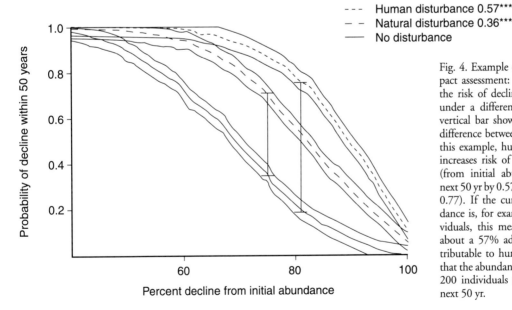

Fig. 4. Example of risk-based impact assessment: each curve gives the risk of decline (as in Fig. 3) under a different scenario. Each vertical bar shows the maximum difference between two curves. In this example, human disturbance increases risk of an 80% decline (from initial abundance) in the next 50 yr by 0.57 (from ca 0.20 to 0.77). If the current total abundance is, for example, 1000 individuals, this means that there is about a 57% additional risk, attributable to human disturbance, that the abundance will fall to or < 200 individuals sometime in the next 50 yr.

Data requirements

The main disadvantage of structured models is that they require more data, such as survival rates and fecundities and their variabilities, compared to unstructured demographic models and occupancy models. However, in many cases (especially for vertebrate populations), the types of data needed are more readily available (from the literature or from data bases maintained by various agencies), than data required for occupancy models (e.g., observations of local extinction and recolonization). If not available, the demographic data required by structured models are more easily collected, and can be obtained by a variety of methods, including censuses, mark-recapture studies, surveys and observations of reproduction and dispersal events. Note that these methods are also used for monitoring, which is an important component of any conservation plan.

Structured models allow various levels of detail. For example, if age-specific data are available, models can be age-structured. If not, or if the demography of the species makes it more appropriate, the model can be stage-structured (including, for instance, only juveniles and adults, instead of each age class). If there are sufficient data, each population can be assigned to a specific set of vital rates (i.e., a different age or stage matrix for each population). If not, the available data can be pooled to estimate a common matrix for all populations. Another option is to calculate vital rates as functions of habitat characteristics. This requires estimation of the vital rates from a few representative populations, and extrapolating the relationship between vital rates and habitat attributes to other populations (see Akçakaya and Raphael 1998 for an example). Similarly, dispersal rates can be specified for each pair of populations, or they can be assigned as a function of the distance between populations, based on studies designed to monitor dispersal among a few populations.

Examples

Structured models have been applied to populations and metapopulations of a variety of organisms, including plants, marine and terrestrial mammals, birds, reptiles, amphibians, fishes and invertebrates. Applications within the last 8 yr included the species listed in Box 3. This list is representative but not comprehensive. For previous applications, see Boyce (1992).

The basic principles of building structured models discussed above apply to most of the cases in Box 3. However, some types of life histories present special problems. For example, in modeling some marine species, delineating populations (stocks) is a difficult task, compounded by the high connectivity of populations in some species through larval dispersal and currents. In marine environments, seasonality (regular cycles with a year) is often less important than long-term regular oscillations of environmental variables.

Some plants and other species have life history stages that are invisible or otherwise difficult to detect. In plants, the seed bank is often modeled as one or more separate stages. Estimating parameters related to these stages often presents difficulties. In addition, time-dependent contingencies and rules of transition (e.g., from seed to seedling following fire) make the application of a simple matrix structure difficult (e.g., Drechsler et al. 1999; also see Menges 2000).

Many of the applications of structured models in Box 3 have aimed at providing management recommendations for rare, endangered or harvested populations. Drechsler et al. (1999) used a stage-structured, stochastic metapopulation model to determine factors affecting the viability of a rare shrub, and to make management recommendations. Maschinsky et al. (1997) used a stage-structured single-population model to compare the dynamics of a endangered plant before and after protection from trampling. Kokko et al. (1997) evaluated harvesting strategies for seal populations in the Baltic with age-structured models. They showed that deterministic models can give a false impression of safety while the same strategies are often judged to be risky when a more realistic, stochastic model is used. Armbruster and Lande (1993) analyzed the effect of culling on the viability of the African elephant with an age-structured, stochastic model. Litvaitis and Villafuerte (1996) used an age-structured, stochastic metapopulation model to evaluate the impact of habitat fragmentation and loss, environmental correlations and management options on the likelihood of persistence of New England cottontails *Sylvilagus transitionalis*. They recommended a management program to maintain a network of suitable habitat patches, and defined the ages and sizes of patches that would be sufficient to achieve acceptably small risks of decline. Akçakaya et al. (1995) used a stage-structured, stochastic metapopulation model to explore the feasibility of translocating helmeted honeyeaters *Lichenostomus melanops cassidix* from their current locations to a new habitat patch. Reed et al. (1998) used the qualitative results of their metapopulation model of the Hawaiian stilt *Himantopus mexicanus knudseni* to argue that the current conservation status of the species should be maintained. Crowder et al. (1994) used a stage-structured model to analyze the effect of turtle excluder devices on the growth rate of loggerhead sea turtle populations. Additional examples are provided by Berglind (2000) and Lennartsson (2000) in this volume.

Box 3

Applications of structured models in population viability analysis. The list is representative but not comprehensive. Some examples are commented further in the text.

Plants
Andropogon spp. (annual grass; Silva et al. 1991, Canales et al. 1994), *Ascophyllum* (seaweed; Aberg 1992), American ginseng and wild leek (Nantel et al. 1996), sentry milk-vetch (Maschinsky et al. 1997), golden heather (Gross et al. 1998), *Banksia goodii* (Drechsler et al. 1999), *Gentianella campestris* (Lennartsson 2000). See Menges (2000) for other examples.

Marine mammals
Mediterranean monk seal (Durant and Harwood 1992), killer whale (Brault and Caswell 1993), northern sea lion (York 1994), Steller sea lion (Pascual and Adkinson 1994), grey and ringed seals (Kokko et al. 1997).

Terrestrial mammals
Roan antelope (Beudels et al. 1992), the African elephant (Armbruster and Lande 1993), rabbit (Palomares et al. 1995), red deer (Benton et al.1995), asiatic wild ass (Saltz and Rubenstein 1995), cottontail rabbit (Litvaitis and Villafuerte 1996).

Birds
Peregrine falcon (Wootton and Bell 1992), California spotted owl (LaHaye et al. 1994), helmeted honeyeater (Akçakaya et al. 1995), red-cockaded woodpeckers (Maguire et al. 1995), golden-cheeked warbler and black-capped vireo (Beardmore et al. 1996a, b), California gnatcatcher (Akçakaya and Atwood 1997), Florida scrub-jay (Root 1998), Hawaiian stilt (Reed et al. 1998), northern spotted owl (Akçakaya and Raphael 1998).

Reptiles and amphibians
Pool frog (Sjögren 1991), salamanders (Gibbs 1993), loggerhead sea turtle (Crowder et al. 1994), desert tortoise (Doak et al. 1994), yellow mud turtle and Kemp's ridley sea turtle (Heppell et al. 1996), spadefoot toad (Hels 1999), sand lizard (Berglind 2000).

Fishes
Striped bass (Cohen et al. 1983, Bulak et al. 1995), cod (Ginzburg et al. 1990), brook trout (McFadden et al. 1967, Marschall and Crowder 1996), *Fundulus* sp. (Munns et al. 1997).

Invertebrates
Marine bryozoan *Membranipora* (Harvell et al. 1990), gorgonian coral *Leptogorgia* (Gotelli 1991), bush cricket (Kindvall 1995, 2000), land snail *Arianta* (Akçakaya and Baur 1996), estuarine polychaetes (Levin et al. 1996), marine snail *Umbonium* (Noda and Nakao1996), pea aphid (Walthall and Start 1997), marine bivalve *Yoldia* (Nakaoka 1997), burrowing mayfly (Madenjian et al. 1998).

Discussion

Advantages and disadvantages of structured models

The main advantage of structured metapopulation models compared to occupancy models is their realism. Structured models are not constrained in terms of the number of populations or turnover rate in metapopulations, making them more flexible. Unrealistic assumptions about the species or (meta)population being modeled can be avoided with structured models, which use all the available demographic data to estimate species viability. As a result, structured models have been applied to a variety of organisms (see Examples, above).

Another advantage, especially as compared to individual-based models, is that, despite their flexibility, structured models are based on a number of common techniques or frameworks that allow their implementation as generic programs. These include the matrix-based approach to model local age or stage-structured dynamics, generating random variates for vital rates to simulate environmental variation and catastrophes, and using replicate trajectories to calculate risk-based viability results. This common framework becomes an advantage when models and viability analyses are needed for a large number of species, and time and resources limitations preclude detailed modeling and programming for each species, as it allows generic programs (e.g., RAMAS; http://www.ramas.com). In contrast, using the full flexibility of individual-based modeling approaches often requires writing a different program for each species modeled. However, some individual-based models (such as VORTEX; see Lacy 2000a, b) solve this problem by having a fixed, age-based structure.

The main disadvantage of structured models compared to occupancy models is that they require more data, such as survival rate and fecundity of each age class or stage and estimates of the temporal and spatial variation in these rates (see Data requirements, above).

When are structured models most appropriate?

The appropriate level of detail for a PVA model depends on the ecology of the species, the availability of data, and the questions addressed (see Akçakaya and Sjögren-Gulve 2000).

Structured models are more appropriate than occupancy models if "internal" (within-population) dynamics are important, there are few subpopulations, or local extinctions and recolonizations are relatively rare (see Akçakaya and Sjögren-Gulve 2000). However, it is important to note that these are only general guidelines, and there may be exceptions. For example, structured models can also be appropriate when there are frequent local extinctions, or over 100 populations, depending on the data, the question, and the species.

Structured models are more appropriate than individual-based models if data at the individual-level (e.g., on dispersal behavior) are not available, or if the population size is not too small. In general, the smaller the population size, the more appropriate are individual-based approaches. What is "too small" depends on: 1) what is known about the population, and about each individual, 2) how complex is the social structure, and 3) question to be addressed.

For example, even a scalar (unstructured) model may be suitable for a population as small as 2 individuals, if the population dynamics and social organization are simple enough (think of a model for a *Paramecium* population). But, for species with complex social structure, and plenty of reliable data, populations that are much larger than this may be better modeled by individual-based models.

The more information there is about each individual (such as age, size, sex, location, reproductive status, social status, genetics, relationship to other individuals, etc.), and about the behavior of the species (social structure, dispersal patterns, reproductive behavior) the more appropriate are individual-based models. If such information is not available, it may be unnecessary, even counter-productive, to attempt to build complex individual-based models.

Individual-based models are especially sensitive to the dispersal behavior of individuals. Small errors in dispersal mortality, mobility and habitat suitability may result in large errors in predicting dispersal success (Ruckelhaus et al. 1997). In structured models, dispersal is modeled at a coarser scale, as the proportion of individuals moving from population to another between two successive time steps, which can be estimated with mark-recapture studies. Dispersal mortality is incorporated into the vital rates (e.g., specified in a stage matrix; see Akçakaya 1998, 2000).

Finally, the questions addressed should also affect the decision about the complexity of the model. If the PVA aims to address questions about future distribution of occupied territories, or the optimal pairing of individuals for captive breeding, then an individual-based approach may be more appropriate. If the goal of the PVA is population or species-level assessment of extinction risk, a structured model may be more appropriate. However, in many cases, both structured and individual-based models can be used. For example, in a study designed to validate the PVA approach, Brook et al. (2000) modeled 21 populations, using both structured programs such as RAMAS and INMAT, and individual-based programs such as GAPPS and VORTEX for each population. They found that the PVA predictions were accurate, and the predictions of different PVA programs were consistent with each other.

References

Aberg, P. 1992. Size-based demography of the seaweed *Ascophyllum nodosum* in stochastic environments. – Ecology 73: 1488–1501.

Akçakaya, H. R. 1991. A method for simulating demographic stochasticity. – Ecol. Model. 54: 133–136.

Akçakaya, H. R. 1998. RAMAS GIS: linking landscape data with population viability analysis (ver. 3.0). – Appl. Biomath., Setauket, New York.

Akçakaya, H. R. 2000. Viability analyses with habitat-based metapopulation models. – Popul. Ecol. 42: 45–53.

Akçakaya, H. R. and Ginzburg, L. R. 1991. Ecological risk analysis for single and multiple populations. – In: Seitz, A. and Loeschke, V. (eds), Species conservation: a population-biological approach. Birkhauser, Basel, pp. 73–87.

Akçakaya, H. R. and Baur, B. 1996. Effects of population subdivision and catastrophes on the persistence of a land snail metapopulation. – Oecologia 105: 475–483.

Akçakaya, H. R. and Atwood, J. L. 1997. A habitat-based metapopulation model of the California gnatcatcher. – Conserv. Biol. 11: 422–434.

Akçakaya, H. R. and Raphael, M. G. 1998. Assessing human impact despite uncertainty: viability of the northern spotted owl metapopulation in the northwestern USA. – Biodiv. Conserv. 7: 875–894.

Akçakaya H. R. and Sjögren-Gulve, P. 2000. Population viability analyses in conservation planning: an overview. – Ecol. Bull. 48: 9–21.

Akçakaya, H. R., McCarthy, M. A. and Pearce, J. 1995. Linking landscape data with population viability analysis: management options for the helmeted honeyeater. – Biol. Conserv. 73: 169–176.

Akçakaya, H. R., Burgman, M. A. and Ginzburg, L. R. 1999. Applied population ecology: principles and computer exercises using RAMAS EcoLab 2.0. 2nd ed. – Sinauer.

Armbruster, P. and Lande, R. 1993. A population viability analysis for African elephants (*Loxodonta africana*): how big should reserves be? – Conserv. Biol. 7: 602–610.

Beardmore, C., Hatfield, J. and Lewis, J. 1996a. Golden-cheeked warbler population and habitat viability assessment report. – Report of an August 1995 workshop arranged by the US Fish and Wildlife Service. Austin, Texas.

Beardmore, C., Hatfield, J. and Lewis, J. 1996b. Black-capped vireo population and habitat viability assessment report. – Report of a September 1995 workshop arranged by the US Fish and Wildlife Service. Austin, Texas.

Benton, T. G., Grant, A. and Clutton-Brock, T. H. 1995. Does environmental stochasticity matter? Analysis of red deer life-histories on Rum. – Evol. Ecol. 9: 559–574.

Berglind, S.-Å. 2000. Demography and management of relict sand lizard *Lacerta agilis* populations on the edge of extinction. – Ecol. Bull. 48: 123–142.

Beudels, R. C., Durant, S. M. and Harwood, J. 1992. Assessing the risk of extinction for local populations of roan antelope *Hippotragus equinus*. – Biol. Conserv. 61: 107-116.

Bierzychudek, P. 1982. The demography of jack-in-the-pulpit, a forest perennial that changes sex. – Ecol. Monogr. 52: 335–351.

Boyce, M. S. 1992. Population viability analysis. – Annu. Rev. Ecol. Syst. 23: 481–506.

Brault, S. and Caswell, H. 1993. Pod-specific demography of killer whales (*Orcinus orca*). – Ecology 74: 1444–1454.

Brook, B. W. et al. 2000. Predictive accuracy of population viability analysis in conservation biology. – Nature 404: 385–387.

Brown, J. H. and Kodric-Brown, A. 1977. Turnover rates in insular biogeography: effect of immigration on extinction. – Ecology 58: 445–449.

Buckland, S. T. et al. 1993. Distance sampling: estimating abundance of biological populations. – Chapman and Hall.

Bulak, J. S., Wethey, D. S. and White III, M. G. 1995. Evaluation of management options for a reproducing striped bass population in the Santee-Cooper system, South Carolina. – North Am. J. Fish. Manage. 15: 84–94.

Burgman, M. A., Ferson, S. and Akçakaya, H. R. 1993. Risk assessment in conservation biology. – Chapman and Hall.

Burgman, M., Ferson, S. and Lindenmayer, D. 1994. The effect of the initial age-class distribution on extinction risks: implications for the reintroduction of Leadbeater's possum. – In: Serena, M. (ed.), Reintroduction biology of the Australasian fauna. Surrey Beatty, Chipping Norton, NSW, Australia, pp. 15–19.

Burnham, K. P. and Anderson, D. R. 1992. Data-based selection of an appropriate biological model: the key to modern data analysis. – In: McCullough, D. R. and Barrett, R. H. (eds), Wildlife 2001: populations. Elsevier, pp. 16–30.

Canales, J. et al. 1994. A demographic study of an annual grass (*Andropogon brevifolius* Schwarz). – Acta Oecol. 15: 261–273.

Caswell, H. 1989. Matrix population models: construction, analysis, and interpretation. – Sinauer.

Cohen, J. E. 1979. Comparative statics and stochastic dynamics of age-structured populations. – Theor. Popul. Biol. 16: 159–171.

Cohen, J. E., Christensen, S. W. and Goodyear, C. P. 1983. A stochastic age-structured model of striped bass (*Morone saxatilis*) in Potomac river. – Can. J. Fish. Aquat. Sci. 40: 2170–2183.

Crowder, L. B. et al. 1994. Predicting the impact of turtle excluder devices on loggerhead sea turtle populations. – Ecol. Appl. 4: 437–445.

de Kroon, H. A. et al. 1986. Elasticity; the relative contribuition of demographic parameters to population growth rate. – Ecology 67: 1427–1431.

Doak, D., Kareiva, P. and Klepetka, B. 1994. Modeling population viability for the desert tortoise in the western Mojave Desert. – Ecol. Appl. 4: 446–460.

Drechsler M. et al. 1999. Modelling the persistence of an apparently immortal *Banksia* species after fire and land clearing. – Biol. Conserv. 88: 246–259.

Durant, S. M. and Harwood, J. 1992. Assessment of monitoring and management strategies for local populations of the Mediterranean monk seal *Monachus monachus*. – Biol. Conserv. 61: 81–91.

Ferson, S. 1990. RAMAS Stage: generalized stage-based modeling for population dynamics. – Appl. Biomath., Setauket, New York.

Gibbs, J. P. 1993. Importance of small wetlands for the persistence of local populations of wetland associated animals. – Wetlands 13: 25–31.

Gilpin, M. E. 1988. A comment on Quinn and Hastings: extinction in subdivided habitats. – Conserv. Biol. 2: 290–292.

Ginzburg, L. R. et al. 1982. Quasiextinction probabilities as a measure of impact on population growth. – Risk Analysis 2: 171–181.

Ginzburg, L. R., Ferson, S. and Akçakaya, H. R. 1990. Reconstructibility of density dependence and the conservative assessment of extinction risks. – Conserv. Biol. 4: 63–70.

Gotelli, N. J. 1991. Demographic models for *Leptogorgia virgulata*, a shallow-water gorgonian. – Ecology 72: 457–467.

Gross, K. et al. 1998. Modeling controlled burning and trampling reduction for conservation of *Hudsonia montana*. – Conserv. Biol. 12: 1291–1301.

Hanski, I. 1997. Habitat destruction and metapopulation dynamics. – In: Pickett, S. T. A. et al. (eds), Enhancing the ecological basis of conservation: heterogeneity, ecosystem function and biodiversity. Chapman and Hall, pp. 217–227.

Harvell, C. D., Caswell, H. and Simpson, P. 1990. Density effects in a colonial monoculture: experimental studies with a marine bryozoan (*Membranipora membranacea* L.). – Oecologia 82: 227–737.

Hels, T. 1999. Effects of roads on amphibian populations. – Ph.D. thesis, Nat. Environ. Res. Inst., Denmark.

Heppell, S. S., Crowder, L. B. and Crouse, D. T. 1996. Models to evaluate headstarting as a management tool for long-lived turtles. – Ecol. Appl. 6: 556–565.

Hill, J. K., Thomas, C. D. and Lewis, O. T. 1996. Effects of habitat patch size and isolation on dispersal by *Hesperia comma* butterflies: implications for metapopulation structure. – J. Anim. Ecol. 65: 725–735.

IUCN 1994. International Union for the Conservation of Nature Red List Categories. – IUCN Species Survival Commission, Gland, Switzerland.

Jenkins, S. H. 1988. Use and abuse of demographic models of population growth. – Bull. Ecol. Soc. Am. 69: 201–202.

Kindvall, O. 1995. Ecology of the bush cricket *Metrioptera bicolor* with implications for metapopulation theory and conservation. – Ph.D. thesis, Swedish Univ. Agricult. Sci., Uppsala.

Kindvall, O. 1999. Dispersal in a metapopulation of the bush cricket, *Metrioptera bicolor* (Orthoptera: Tettigoniidae). – J. Anim. Ecol. 68: 172–185.

Kindvall, O. 2000. Comparative precision of three spatially realistic models of metapopulation dynamics. – Ecol. Bull. 48: 101–110.

Kokko, H., Lindström, J. and Ranta, E. 1997. Risk analysis of hunting of seal populations in the Baltic. – Conserv. Biol. 11: 917–927.

Kuussaari, M., Nieminen, M. and Hanski, I. 1996. An experimental study of migration in the butterfly *Melitaea cinxia*. – J. Anim. Ecol. 65: 791–801.

Laake, J. L. et al. 1996. DISTANCE user's guide V2.2. – Colorado Coop. Fish and Wildl. Res. Unit, Colorado State Univ., Fort Collins.

Lacy, R. C. 2000a. Considering threats to the viability of small populations using individual-based models. – Ecol. Bull. 48: 39–51.

Lacy, R. C. 2000b. Structure of the VORTEX simulation model for population viability analysis. – Ecol. Bull. 48: 191–203.

LaHaye, W. S., Gutierrez, R. J. and Akçakaya, H. R. 1994. Spotted owl meta-population dynamics in southern California. – J. Anim. Ecol. 63: 775–785.

Lebreton, J.-D., Pradel, R. and Clobert, J. 1993. The statistical analysis of survival in animal populations. – Trends Ecol. Evol. 8: 91–95.

Lennartsson, T. 2000. Management and population viability of the pasture plant *Gentianella campestris*: the role of interactions between habitat factors. – Ecol. Bull. 48: 111–121.

Levin, L. et al. 1996. Demographic responses of estuarine polychaetes to pollutants: life table response experiments. – Ecol. Appl. 6: 1295–1313.

Lindenmayer, D. W. and Possingham, H. P. 1996. Ranking conservation and timber management options for Leadbeater's possum in southeastern Australia using population viability analysis. – Conserv. Biol. 10: 1–18

Litvaitis, A. J. and Villafuerte, R. 1996. Factors affecting the persistence of New England cottontail metapopulations: the role of habitat management. – Wildl. Soc. Bull. 24: 686–693.

Mace, G. M. and Lande, R. 1991. Assessing extinction threats: towards a re-evaluation of IUCN threatened species categories. – Conserv. Biol. 5: 148–157.

Madenjian, C. P., Schloesser, D. W. and Krieger, K. A. 1998. Population models of burrowing mayfly recolonization in western Lake Erie. – Ecol Appl. 8: 1206–1212.

Maguire, L. A., Wilhere, G. F. and Dong, Q. 1995. Population viability analysis for red-cockaded woodpeckers in the Georgia Piedmont. – J. Wildl. Manage. 59: 533–542.

Marschall, E. A. and Crowder, L. B. 1996. Assessing population responses to multiple anthropogenic effects: a case study with brook trout. – Ecol. Appl. 6: 152–167.

Maschinsky, J., Fyre, R. and Rutman, S. 1997. Demography and population viability of an endangered plant species before and after protection from trampling. – Conserv. Biol. 11: 990–999.

McFadden, J. T., Alexander, G. R. and Shetter, D. S. 1967. Numerical changes and population regulation in brook trout *Salvelinus fontinalis*. – J. Fish. Res. Board Can. 24: 1425–1459.

Menges, E. S. 1992. Stochastic modeling of extinction in plant populations. – In: Fiedler, P. L. and Jain, S. K. (eds), Conservation biology: the theory and practice of nature conservation, preservation, and management. Chapman and Hall, pp. 253–276.

Menges, E. S. 2000. Applications of population viability analyses in plant conservation. – Ecol. Bull. 48: 73–84.

Mills, L. S. and Smouse, P. E. 1994. Demographic consequences of inbreeding in remnant populations. – Am. Nat. 144: 412–431

Moloney, K. A. 1986. A generalized algorithm for determining category size. – Oecologia 69: 176–180.

Munns, W. R. et al. 1997. Evaluation of the effects of dioxin and PCBs on *Fundulus heteroclitus* populations using a modeling approach. – Environ. Toxicol. Chem. 16: 1074–1081.

Nakaoka, M. 1997. Demography of the marine bivalve *Yoldia notabilis* in fluctuating environments: an analysis using a stochastic matrix model. – Oikos 79: 59–68.

Nantel, P., Gagnon, D. and Nault, A. 1996. Population viability analysis of American ginseng and wild leek harvested in stochastic environments. – Conserv. Biol. 10: 608–621.

Noda, T. and Nakao, S. 1996. Dynamics of an entire population of the subtidal snail *Umbonium costatum*: the importance of annaul recruitment fluctuation. – J. Anim. Ecol. 65: 196–204.

Palomares, F. et al. 1995. Positive effects on game species of top predators by controlling smaller predator populations: an example with lynx, mongooses and rabbits. – Conserv. Biol. 9: 295–305.

Pascual, M. A. and Adkinson, M. D. 1994. The decline of the Steller sea lion in the northeast Pacific: demography, harvest or environment? – Ecol. Appl. 4: 393–403.

Pollock, K. H. et al. 1990. Statistical inference for capture-recapture experiments. – Wildl. Monogr. 107.

Price, M. and Gilpin, M. 1996. Modelers, mammalogists, and metapopulations: designing Steven's kangaroo rat reserves. – In: McCullough, D. R. (ed.), Metapopulations and wildlife conservation. Island Press, Washington, D.C., pp. 217–240.

Reed, J. M., Elphick, C. S. and Oring, L. W. 1998. Life-history and viability analysis of the endangered Hawaiian stilt. – Biol. Conserv. 83: 35–45.

Root, K. V. 1998. Evaluating the effects of habitat quality, connectivity, and catastrophes on a threatened species. – Ecol. Appl. 8: 854–865.

Ruckelhaus, M., Hartway, C. and Kareiva, P. 1997. Assessing the data requirements of spatially explicit dispersal models. – Conserv. Biol. 11: 1298–1306.

Saltz, D. and Rubenstein, D. I. 1995. Population dynamics of a reintroduced asiatic wild ass (*Equus hemionus*) herd. – Ecol. Appl. 5: 327–335.

Shaffer, M. L. 1981. Minimum population sizes for species conservation. – BioScience 31: 131–134.

Shaffer, M. L. 1990. Population viability analysis. – Conserv. Biol. 4: 39–40.

Silva, J. F. et al. 1991. Population responses to fire in a tropical savanna grass *Andropogon semiberbis*: a matrix model approach. – J. Ecol. 79: 345–356.

Sjögren, P. 1991. Extinction and isolation gradients in metapopulations: the case of the pool frog (*Rana lessonae*). – Biol. J. Linn. Soc. 42: 135–147.

Sjögren-Gulve, P. and Hanski, I. 2000. Metapopulation viability analysis using occupancy models. – Ecol. Bull. 48: 53–71.

Solbreck, C. 1991. Unusual weather and insect population dynamics: *Lygaeus equestris* during an extinction and recovery period. – Oikos 60: 343–350.

Vandermeer, J. H. 1978. Choosing category size in a stage projection matrix. – Oecologia 32: 79–84.

Walthall, W. K. and Start, J. D. 1997. Comparison of two population-level ecotoxicological endpoints: the intrinsic (r_m) and instantaneous (r_i) rates of increase. – Environ. Toxicol. Chem. 16: 1068–1073.

Wootton, J. T. and Bell, D. A. 1992. A metapopulation model of the peregrine falcon in California: viability and management strategies. – Ecol. Appl. 2: 307–321.

York, A. E. 1994. The population dynamics of northern sea lions, 1975–1985. – Mar. Mammal Sci. 10: 38–51.

Ecological Bulletins 48: 39–51. Copenhagen 2000

Considering threats to the viability of small populations using individual-based models

Robert C. Lacy

Lacy, R. C. 2000. Considering threats to the viability of small populations using individual-based models. – Ecol. Bull. 48: 39–51.

As wildlife populations become smaller, the number of interacting stochastic processes which can destabilize the populations increases: genetic effects (inbreeding and loss of adaptability) and instability of the breeding structure (sex ratio imbalances, unstable age distribution, and disrupted social systems) can decrease population growth and stability. Recent analyses have shown that some populations can be very sensitive to these stochastic processes, at larger population sizes than had been suggested previously, and often in unexpected ways. Interactions among processes can reduce population viability much more so than would be assumed from consideration of isolated factors. For example, in monogamous species, random fluctuations in sex ratio will depress the mean number of breeding pairs in populations with as many as 500 adults. At low population densities, individuals may not be able to find mates, or may not encounter individuals sufficiently unrelated to be accepted as suitable mates. Inbreeding depression of demographic rates can become a significant contributor to population decline in populations with several hundred individuals, even if genetic problems are not the primary threat. Most models of genetic decay in small and fragmented populations assume demographic stability. However, when the increases in demographic fluctuations of small populations are considered, rates of loss of genetic variation and accumulation of inbreeding can be much faster than has been suggested before. These processes can be examined in detailed, individual-based PVA models. Accurate data to parameterize these models, however, are often not available. Thus, we need to interpret cautiously PVA conclusions for populations that are small, highly fragmented, or projected for many generations.

R. C. Lacy (rlacy@ix.netcom.com), Dept of Conservation Biology, Daniel F. & Ada L. Rice Center, Chicago Zoological Society, Brookfield, IL, 60513 USA.

There are many kinds of threats to the viability of populations of wildlife. The processes which have driven many once-abundant populations down to one or few small populations in scattered remnants of habitat include direct exploitation (over-harvest), habitat destruction and fragmentation, degradation of habitat quality, introduction of exotic species, and chains of extinction (Caughley 1994). Often, after precipitous declines occur, conservation biologists and governmental agencies establish recovery actions to try to prevent local extirpation of populations or the ultimate extinction of the taxon.

As wildlife populations become smaller, additional threats to stability and persistence arise, which can exacerbate the difficulty of stopping or reversing a decline. These problems of small populations generally result from stochastic or random processes. In any sampling process, the predictability of an outcome decreases as the sample size is reduced. Many aspects of population dynamics are inherently sampling processes, rather than completely determined events, including: mate acquisition, breeding success, sex determination, transmission of genetic alleles, survival, and dispersal. The uncertainty in such processes can lead to instability in population dynamics. Moreover, fluctuations in demographic and genetic processes cause depression in long-term rates, because the geometric means and other appropriate compound measures of population performance are less than the arithmetic means. Finally, reductions in growth rates and fluctuations in rates can interact synergistically, causing increasing instability and more rapid decline, until the ultimate stability is reached when the population becomes extinct. These processes were termed "extinction vortices" by Gilpin and Soulé (1986), and their examination constitutes the core of most population viability analyses (PVA) (Soulé 1987, Boyce 1992, Lacy 1993/1994).

Caughley (1994) argued that there is a dichotomy in conservation biology, between those who follow a "declining population paradigm," examining deterministic causes of population decline, and those who follow a "small population paradigm," examining the processes that further imperil populations after they have become small. Caughley called for more theory to guide the declining population approach, more data to support the small population approach, and better use of the strengths of the two approaches to guide conservation, but he also questioned whether too much emphasis has been given to small population processes in wildlife conservation. Hedrick et al. (1996) argued that the problems of small populations have at times been under-appreciated, but also that the processes causing population declines and the processes affecting populations that have become small are inter-linked in complex ways, so that PVA and conservation biology must encompass both of Caughley's paradigms.

The problems of small populations have received extensive theoretical treatment (e.g., Soulé 1987), but further assessment of the factors affecting viability of very small populations is needed. I will argue that we often underestimate the importance of these factors in population viability, as the magnitude and even direction of some of these effects may be different than has been commonly supposed. Although I believe that the problems inherent in small populations are more numerous and more severe than is commonly recognized, the same may be true of the causes of population decline. However, because additional threats to population viability arise as populations become small, the kinds of PVA models that are needed for assessing the status and recovery options for a species will change as a species declines. Therefore, it is important to consider carefully which PVA model is most appropriate for a particular analysis (Lindenmayer et al. 1995, Akçakaya and Sjögren-Gulve 2000). An individual-based simulation program that models the stochastic processes of small populations in detail would probably not be the best model for examining viability of a population which numbers in the tens of thousands. Similarly, a population-based structured model which ignores factors such as fluctuations in sex ratio, mate availability, and inbreeding would probably not be the most accurate model for a population which falls below 100 individuals. Many, perhaps all, presently used PVA models assess only some of the threats facing small populations, and therefore may underestimate probabilities of extinction and difficulties in species recovery.

In this paper, I will describe some of the threats to small populations that are not included in most PVA models. This discussion will provide guidance as to when more detailed, individual-based PVA models may be necessary to represent well the dynamics of small populations. Most of the processes I will discuss are particularly important for species with low intrinsic growth rates and stable social systems, and somewhat less so for those with high fecundity and little structure to the social or breeding system. Therefore, these considerations will be most applicable to mammal and bird species, and the bias toward these groups in examples below reflect my greater experience with PVA of these taxa.

Processes destabilizing small populations

Shaffer (1981) categorized the stochastic threats to small populations: demographic stochasticity, environmental stochasticity, natural catastrophes, and genetic stochasticity. These causes of uncertainty and fluctuation in population size interact, but they are conceptually distinct.

Demographic stochasticity

Demographic stochasticity is the random variation in the numbers of births, number of deaths, and sex ratio in a population that results from the fates of individuals being independent outcomes of probabilistic events of reproduction, mortality, and sex determination (Shaffer 1981). The observed variation across years or across populations with constant probabilities would be distributed as binomial distributions. If fates of individuals are independent, then demographic stochasticity is intrinsic to the population, and is a simple consequence of the sampling that occurs as individuals are subjected to the population rates.

Figure 1 shows the percent of whooping cranes *Grus americana* each year from 1938 to 1994 that failed to return the following year to the wintering grounds in Texas. Although some variation in mortality was due to environmental variation and likely catastrophes (see below), most of the variation in survival across years can be accounted for by demographic stochasticity (Mirande et al. 1991). The reduced variation in survival as the population grew in size from 18 birds in 1938 to ca 150 birds in the 1990s is clearly evident.

Demographic stochasticity has been recognized as a

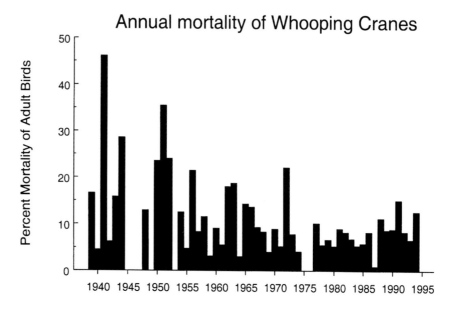

Fig. 1. Percent of whooping cranes observed each year on the wintering grounds which did not survive to return the following year. In years with no bar shown, no birds died. Data from Mirande et al. (1991) and pers. comm.

potential threat to very small populations, but the contribution it can make to population instability has been underestimated. Based on simple calculations of the probability that all individuals will be of the same sex, or die synchronously, it has commonly been stated that demographic stochasticity can cause extinction only when populations fall below ca 10 – 20 individuals (Goodman 1987, Shaffer 1987). However, processes that are not serious problems when acting alone can become significant contributors to population instability and decline when they act synergistically with other threatening processes. The last five dusky seaside sparrows *Ammospiza maritima nigrescens* were all males, an unfortunate but not overly surprising event that effectively eliminated the taxon. The probability that a population of 100 breeders would all be males is essentially zero, so it might be assumed that random fluctuations in sex ratio are unimportant in populations approaching such a size. However, fluctuations in sex ratio can depress population growth significantly, even in such cases. In a population of 100 breeders, we expect ca 50 females and 50 males. But due just to demographic stochasticity, the number of females will deviate by 5 or more (one standard deviation) from this expectation in about one-third of the years. In monogamous species, such as most birds, this means that typically < 50 pairs could be formed. The mean depression in reproduction relative to a population with a constant equal sex ratio can be calculated from the mean absolute deviation of the binomial distribution. Due solely to the random fluctuations in sex ratio, reproduction in a monogamous population with 100 adults would be depressed by ca 8%. This level of reduced population productivity is enough to cause low-fecundity species to switch from positive population growth to long-term population decline and eventual extinction. Brook et

al. (1999) found that interaction of the breeding system with fluctuations in the sex ratio strongly influenced projections for population growth of whooping cranes.

Figure 2 shows the mean depression in breeding caused by random variation in the sex ratio for monogamous populations of various sizes. Even with 500 breeding individuals, the mean number of pairs in the population is 3.6% below what would be available if the sex ratio were fixed at 50:50. This simple example of a threat to the viability of small populations illustrates several important points. First, the random deviation in sex ratio does not in itself cause extinction except in the very smallest of populations, but it can interact with other factors such as the breeding system to depress population growth in a vulnerable population sufficiently to cause extinction. Second, it is unlikely that biologists observing the population would recognize that fluctuation in the sex ratio was a contributing cause of lower reproduction and population decline. Third, it would be possible to incorporate the reduction in breeding as an average effect in a simple life table projection. However, to do so requires that the demographic rates were estimated from a population of the size of the population being currently assessed, and that the population remains constant in size. (This last assumption defeats the purpose of PVA.) The effect of biased sex ratios depressing reproduction in monogamous population of changing size could be modeled as a density-dependent effect on reproduction (Stephens and Sutherland 1999, Courchamp et al. 1999). However, I am not aware of any cases in which an analytical or population-based PVA model incorporated the reduction in breeding, as shown in Fig. 2, for monogamous species that would be expected due to sex ratio fluctuations. Many population models ignore sex ratio and breeding system entirely, projecting numbers of females

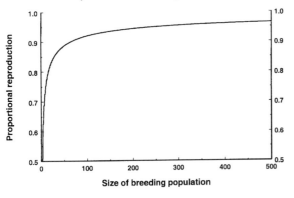

Effect of random fluctuations in sex ratio on reproduction in monogamous species

Y-axis: Proportional reproduction (0.5 to 1.0)
X-axis: Size of breeding population (0 to 500)

Fig. 2. Mean proportion of monogamous pairs that could be formed relative to the case of a constant 50:50 sex ratio, as a consequence of random fluctuations in the sex ratio in breeding populations of varying size.

under the assumption that there are always males available for mating. Individual-based models are well suited for cases in which sex-ratio biases can disrupt breeding, because they automatically generate stochastic variation in the sex ratio. Rules defining the breeding system can then be built into the model.

Thus, if there is not promiscuous breeding, random fluctuations in the sex ratio can depress population growth in even moderate-sized populations. Similarly, random demographic stochasticity in the numbers of births and deaths per year can depress mean population growth, because of variation in the age distribution and other disruptions of optimal breeding. For example, managers of zoo populations are often distressed to find that reproduction is kept well below optimal levels because of temporary imbalances in the sex ratio or the age distribution. Following the advice of conservation biologists, zoo managers have assumed that a population of 50 or more is safe from the threat of demographic stochasticity, but random fluctuations are causing problems for maintaining stable populations of rhinoceroses, spectacled bears, lions, and other species for which there is limited flexibility to accommodate changes in numbers.

Environmental variation

Environmental variation or stochasticity is the variation in demographic rates or probabilities that results from fluctuations in the environment (Shaffer 1981). Thus, local environmental variation causes temporal clustering of births and deaths, which would increase uncertainty and variability in population size, and thereby make a small population more vulnerable to extinction. The kinds of perturbations of the environment which cause variation in birth and death rates include disease, sporadic predation,

irregular food availability, and variable weather. Natural catastrophes are the extreme of environmental variation, in which droughts, floods, fires, disease epidemics, and other local disasters can decimate a population.

Although both demographic stochasticity and environmental variation cause fluctuations in the number of births and deaths in a population, the processes are conceptually distinct and statistically independent. Demographic stochasticity is intrinsic to all populations, regardless of the stability of the environment. As a binomial sampling process, it is highly dependent on the population size. Environmental variation results from variation in habitat quality over time, and is unrelated to population density. The variance in demographic rates caused by environmental variation would be additive with variation due to demographic stochasticity (Goodman 1987).

Environmental variation is not usually affected by the local size of the wildlife population, except in those cases, such as predator-prey interactions, in which the organisms have large effects on their local environment. However, the threat to population viability caused by a given level of environmental variation would be more severe in smaller populations, because smaller populations are closer to extinction. Moreover, the amount of environmental variation would be highly dependent on the total area of habitat occupied by a population. Many environmental stresses are localized, so a population exploiting a large area would benefit from the averaging of any environmental fluctuations that are not synchronous over the entire range. Individuals might use spatial variation in environmental conditions to allow escape from temporal variation in the environment (Kindvall 1996). Even if individuals do not move away from areas with temporarily poor conditions, temporary population declines in some areas would be offset by growth elsewhere.

In PVA models, consideration should be given to whether the amount of environmental variation in the system should change with range contraction and expansion. Unfortunately, data on variation in demographic rates are woefully inadequate for almost all species. Usually, we have no more than crude guesses as to the magnitude of environmental variation for use in population viability analyses. There would be considerable value in a compilation of data across species which would allow generalizations concerning the typical magnitude of fluctuations in demographic rates for species with various life histories, trophic guilds, and habitat types.

Figures 3 and 4 show two examples of fluctuations in natural populations that contrast markedly in the extent to which they are impacted by environmental variation. The population trend for whooping cranes during recovery from a near-extinction shows the reduced relative fluctuations in numbers as the population increased in size. Cranes form long-term monogamous pair bonds, they return to nesting sites for a number of years, they have low fecundity, and they are very long-lived. Hence, they would

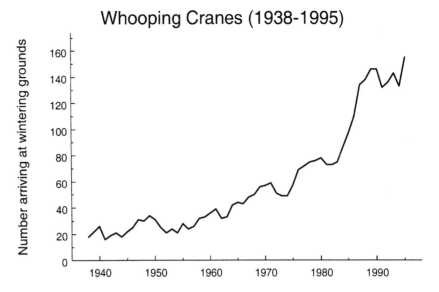

Fig. 3. Number of whooping cranes arriving each winter at the Aransas National Wildlife Refuge, Texas. Data from Mirande et al. (1991).

be expected to show minimal susceptibility to environmental variation. Observed variation in the numbers of hatchlings from 1976 to 1991 was approximately that which would be expected due solely to the demographic variation that would result from a random annual sampling of breeders from the pool of adult birds, each with a constant probability of reproductive success (Mirande et al. 1991). Variation across years in first-year survival was more than three-fold greater than what would be expected due to demographic sampling, indicating that the probability of chick mortality fluctuates across years due to environmental variation. Adult mortality, however, seemed to fall into two classes: in some years, mortality was significantly greater than mortality in other years and deviated significantly further from the mean than can be expected from random demographic stochasticity. The cause of

these years of poor survival is not known, as census counts were made only once a year during those years. However, they would appear to be examples of natural catastrophes. If these catastrophe years are excluded from the data as special cases, the annual variation in adult mortality was only slightly greater than that expected due to random demographic stochasticity (Mirande et al. 1991).

Figure 4 shows the population trends for the palila *Loxioides bailleui*, a finch which is restricted to the mamane forests on the slopes of the major volcanoes of the island of Hawai'i. Although some variation in population size may be caused by imprecise census estimates, the species clearly undergoes striking fluctuations in numbers, even though the mean population size is ca 30 times greater than the current population of whooping cranes. Palila must be sensitive to environmental variation, probably

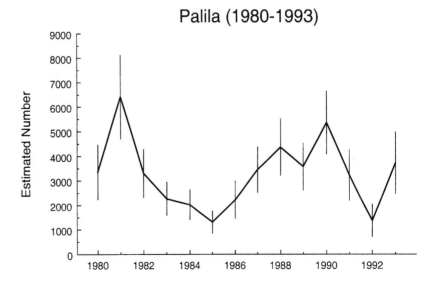

Fig. 4. Estimated numbers of palila, with 90% confidence intervals, from field surveys. Adapted from Ellis et al. (1992).

with respect to both breeding and survival. In part because of the much higher sensitivity to environmental variation, PVA modeling projected a higher probability of extinction for palila (Ellis et al. 1992) than for the much smaller population of whooping cranes (Mirande et al. 1991). The contribution of environmental variation to extinction vortices is well-recognized (Belovsky 1987, Goodman 1987, Foley 1994), but as with demographic stochasticity, we may not always recognize how large the effect can be.

Disrupted breeding systems

There is another cause of demographic instability in small populations, but it does not fit easily within the categories of Shaffer (1981), is rarely considered in PVA models, and may not be fully recognized in conservation plans. At low population densities, the social systems of many species may be disrupted. Such Allee effects are one type of density dependence in reproductive success. Although the importance of understanding and modeling density dependence has been stressed by some authors (Mills et al. 1996, Brook et al. 1997), most of the attention has been given to changes in demographic rates as the population approaches carrying capacity. Yet it is not when a population is near carrying capacity that we need to be concerned about extinction. Allee effects at the low end of density can be incorporated into several of the widely available "generic" PVA models (e.g., RAMAS/Space: Akçakaya and Ferson 1992; VORTEX: Lacy 2000), but this feature seems rarely to be used. Individual-based models can be tailored to provide detailed representations of specific social systems, and this approach was used to look at how the interactions of stochastic processes and pack structure impact viability of wolves (Vucetich et al. 1997) and wild dogs (Vucetich and Creel 1999).

Disruptions of the breeding system can occur for reasons that range from the obvious to the subtle. At very low population densities, animals may be unlikely to encounter any potential mates when they are ready to breed, and non-selfing plants may not be adequately pollinated (Menges 2000). Sumatran rhinoceroses *Dicerorhinus sumatrensis* now exist at densities of only a few per protected area. Field surveys in Malaysia in 1995 found tracks of one juvenile among 35 sets of tracks, and one of 21 adult females captured in the prior decade was pregnant (AsRSG. 1996). If the population were breeding as expected for a rhinoceros species, ca 30% of adult females should be pregnant at any time, and ca 15% of the animals should be under two years of age. Although detailed studies of demography have not been carried out (in part because of the difficulty of studying secretive animals that are at low density in the forest), it is plausible that the scarcity of mates is causing a near cessation of breeding over much of the fragmented range.

Even if some potential mates are available, small populations may provide little opportunity for mate choice. The extent to which a reduced pool of possible mates may be causing breeding delays or failures in small natural populations has not been explored. This problem could be exacerbated if the potential mates are all closely related to the choosing individual or to each other. Ryan (2000) found that, when given a choice of unfamiliar, distantly related females in a Y-apparatus, male *Peromyscus polionotus rhoadsi* mice preferred the less-related female even when differences in kinship averaged only $f = 0.013$, about the level of second cousins. When subsequently paired with one or the other female from the choice test, breeding was delayed and litter sizes were smaller when the only available partner was the one that had been less preferred.

Species with complex social systems may be especially vulnerable to problems resulting from low population density. For example, striped-back wrens *Campylorhynchus nuchalis* have very low breeding success unless they have at least two adult non-breeding helpers sharing in defense against nest-predators, and breeding success is related strongly to the number of such helpers (Rabenold 1990). If such a population were to decline in numbers, recruitment could stop when few birds remained to serve as helpers. One population was rescued from demographic decline when immigrants from nearby populations joined remnant breeding groups (Rabenold et al. 1991). Presumably, extirpation of the local population would have occurred if there had not been nearby sources of immigrants.

Inbreeding depression and loss of genetic diversity

At least two kinds of genetic problems can impact the viability of small populations: reduction of fitness of individuals resulting from inbreeding, and loss of genetic diversity due to random genetic drift (Lacy 1997). There has been much written about (e.g., Frankel and Soulé 1981, Schonewald-Cox et al. 1983, Hedrick and Miller 1992, Frankham 1995a), but also much debate over (e.g., Lande 1988, 1995, Caughley 1994, Caro and Laurenson 1994, Hedrick et al. 1996), the importance of genetics to conservation of wildlife populations.

Inbreeding depression, the reduction in fitness of inbred individuals, frequently occurs when normally outcrossing organisms mate with close relatives. It is commonly believed that it is a problem only for captive populations (which can be substantially buffered from many other risks facing small populations) and for a very few and very small natural populations, or perhaps only as a transient problem that would diminish as selection removes deleterious alleles during repeated generations of inbreeding. Yet, increasing numbers of studies are showing that inbreeding depression can impact population viability to a greater extent, more quickly, and less reversibly than previously supposed (Frankham 1995b, Lacy 1997). Jiménez et al.

(1994) found that inbreeding caused much lower survival of white-footed mice *Peromyscus leucopus* that had been released into a natural habitat than would have been predicted from laboratory measures of inbreeding depression. A population of the greater prairie chicken *Tympanuchus cupido pinnatus*, which had suffered a demographic decline from ca 25 000 birds to < 50 birds over 60 yr, consequently lost substantial genetic variability and suffered reduced fertility and egg viability from inbreeding depression (Westemeier et al. 1998). In Sweden, inbreeding depression of fertility may have been the cause of the rapid decline to extinction of a population of 15 – 20 pairs of the middle spotted woodpecker *Dendrocopos medius* which had been isolated 30 yr before (Pettersson 1985). In Finland, Saccheri et al. (1998) found that local populations of the Glanville fritillary butterfly *Melitaea cinxia* with lower heterozygosity, indicative of greater inbreeding, had lower egg-hatching rate, larval survival, and adult longevity. Apparently as a consequence, these populations had much higher probabilities of extinction.

Following Franklin (1980) and Soulé (1980), only populations with effective sizes below ca 50 have been commonly perceived to be at risk of significant inbreeding depression. To have an effective population size of 50, a typical natural population of a large mammal might need a total population size of ca 200. It is worth reconsidering the likely cumulative effects of inbreeding on the viability of such a population. Inbreeding would accumulate at a rate of 1% per generation. After 10 generations, the 10% cumulative inbreeding may cause a 5% – 20% reduction in survival and in fecundity (Ralls et al. 1988, Lacy et al. 1996, Lynch and Walsh 1998). The consequent reduction in population growth would be sufficient to cause low-fecundity species to decline. Yet, many wildlife managers with responsibility for populations of approximately this size assume that they can ignore effects of inbreeding, and most PVA models for populations of such size omit the impacts of inbreeding on demography.

Many conservation biologists assume that slow inbreeding will not reduce fitness, because selection can remove deleterious alleles during generations of inbreeding (Charlesworth and Charlesworth 1987). Yet, experimental evidence shows that such purging of the genetic load of deleterious alleles from a population often does not work: many populations continue to decline in fitness as they become increasingly inbred (Ballou 1997, Lacy and Ballou 1998) and may go extinct as a consequence (Frankham 1995b). Theoretical work indicates why selection is often ineffective in reducing inbreeding depression. At the small population sizes at which inbreeding occurs, random genetic drift is a much larger force in determining which alleles increase or decrease in frequency than is all but the strongest selection: random loss of adaptive alleles is almost as likely as loss of the deleterious alleles. Except when inbreeding depression is due primarily to a few highly deleterious recessive alleles, inbreeding is more likely to lead to population extinction than to significant reduction of the genetic load (Hedrick 1994). Some populations may be fortunate enough not to carry a genetic load of deleterious alleles which would be expressed under inbreeding, but the evidence suggests that sensitivity to inbreeding may be determined by chance events such as founder effects as much as by any predictable factors (Ralls et al. 1988, Barrett and Kohn 1991, Lacy et al. 1996).

The rate of adaptive evolution of any population is expected to be proportional to its additive genetic variation (and heritability) for the traits under selection (Fisher 1958), and the limited evolutionary potential of domesticated animals with depleted genetic variation has been shown repeatedly. It would be difficult to know whether limited response to selection in rapidly changing habitats is a contributing threat to the persistence of natural populations. Some populations have persisted at small numbers for many generations (e.g., the Javan rhinoceros *Rhinoceros sondaicus*, and, until 1986, the black-footed ferret *Mustela nigripes*), but other populations have rapidly gone extinct when an unusual stress appeared in the environment (e.g., the golden toad *Bufo periglenes* [Pounds and Crump 1994], and the black-footed ferret in 1986 [Clark 1989]). Given the current situation of unprecedented environmental change and an accelerating extinction rate, perhaps more PVA models should consider the maintenance of sufficient genetic variability to ensure ecological and evolutionary flexibility, rather than solely immediate fitness effects and short-term population persistence. One way to accommodate long-term viability into conservation planning would be to use the potential for rapid recovery as the primary measure of population viability, rather than mere population persistence at small numbers. It is also possible to use models that include both the effects of inbreeding on demographic rates and the effects of reduced genetic variability on vulnerability to environmental variation and ability to survive catastrophes. The individual-based model VORTEX (Lacy 2000) can accommodate such dependencies on inbreeding, and similar effects could be built into population-based matrix models as in INMAT by Mills and Smouse (1994). Parameterization of models with genetic-demographic interactions is difficult, but data are increasingly available on the effects of inbreeding on demographic rates (Ralls et al. 1988, Brewer et al. 1990, Saccheri et al. 1998), persistence through environmental stress (Miller 1994, Keller et al. 1994), and population extinction (Frankham 1995b, Saccheri et al. 1998).

Interactions among threatening processes

Although each process described above can individually threaten the viability of small populations, synergistic interactions exacerbate the impacts of many of the processes and are at the center of extinction vortices (Gilpin and Soulé 1986). Interactions among threats are sufficiently complex that they are often omitted from analytical PVA models and from the functions driving demographic projections in simulation models. Most analytical models are constructed by deriving the impact that a factor would have when acting in isolation from other threatening processes. Below are a few examples of interactions among threatening processes that can reduce population viability, often in unexpected and unexpectedly strong ways.

Increased dispersal among patches of habitat is usually assumed to help stabilize a metapopulation. Increased dispersal can restore genetic variation to previously inbred populations, can reduce demographic fluctuations within local populations, can rescue demographically weak populations (Brown and Kodric-Brown 1977), and can lead to recolonization of temporarily extirpated local populations. However, the metapopulation dynamics of small, partly isolated, and frequently extirpated populations can be highly dependent on spatial, temporal, and behavioral aspects of the population structure (Fahrig and Merriam 1994). For example, if a metapopulation declines to a level at which many of its constituent local populations are very small or extinct, then the benefits of dispersal can be replaced by disadvantages. Uncompensated emigration from isolated populations can depress local population growth (Fahrig and Merriam 1985), and suitable habitat that is temporarily empty can act as a population sink where animals fail to find mates (Gyllenberg and Hanski 1992). Consequently, increased dispersal can accelerate decline of sparsely populated metapopulations (Lindenmayer and Lacy 1995).

This collapse into a metapopulation vortex may be more likely if dispersal behaviors evolved in a physical and biotic environment that was different from the current landscape. For example, in largely contiguous habitat with local competition for breeding territories, the optimal dispersal pattern might be for subadults to always disperse, and to disperse in a random direction. When habitat is highly fragmented and often unoccupied, however, it might be adaptive to remain near the natal site unless local densities are very high (Ronce et al. 2000), and to develop dispersal behaviors that more efficiently locate suitable habitat (for example, habitat with an excess of inhabitants of the opposite sex). Many metapopulations may be occupying recently fragmented landscapes for which their evolved dispersal strategies are suboptimal. In PVA models, it might be important to consider that dispersal strategies that are stabilizing at one population density can become destabilizing at different population densities.

Disruption of breeding systems represents another example of interactions between processes. As populations become small, individuals may become closely related to most or all potential mates. In a number of species, individuals have been observed to avoid mating with genetic relatives (Keane 1990). Inbreeding avoidance in a small population could lead to frequent failures to locate any suitable mates. The suppression of breeding might be considered a form of inbreeding depression, but it may not be recognized as such, since the individuals may all have genotypes that otherwise would confer high fitness, and individuals may not be inbred (in fact, it is the avoidance of inbreeding that causes the depression of fitness). Genetic homogeneity leads to an epiphenomenon, with frequency dependent selection causing the "inbreeding depression" at the population level. Some long-isolated populations of beach mice (such as *Peromyscus polionotus leucocephalus*) have low genetic diversity, low frequencies of breeding when mice are paired in captive colonies, and the poor breeding is increasingly exacerbated when experimental populations are further inbred (Brewer et al. 1990, Lacy and Ballou 1998). It is possible that mate choice behavior that evolved to prevent inbreeding is now often preventing breeding, as the mice breed readily when paired with mice from other subspecies (Lacy unpubl.).

Inbreeding interacts with other threats to population viability. For example, Keller et al. (1994) found that inbred song sparrows *Melospiza melodia* were less likely to survive a severe winter, and inbred animals have occasionally been observed to be more vulnerable to other environmental stresses (Miller 1994).

Not only does demographic decline increase inbreeding, which can in turn further depress mean demographic rates, but smaller populations undergo greater relative demographic fluctuations. The increased fluctuations in numbers depress the genetically effective population size (N_e), as inbreeding and genetic drift are more rapid in a population that fluctuates in size than in a stable population of the same mean size. Together with other factors which can reduce breeding success in smaller populations, this can cause the ratio of the effective population size to the total size (N_e/N) to diminish as a population becomes smaller. Hence, while N_e might be 500 when N = 1000, N_e may be 30 when N = 100, and just 10 when N = 50.

In yet another example of the interaction between genetic and demographic threats to population viability, the joint effects of stochasticity in these two processes have been found to lead to consequences which can be opposite those predicted from purely genetic models. A number of authors have reported that a subdivided population will retain more gene diversity over time than will a single panmictic population, because genetic drift will by chance favor different alleles in the various isolated populations (Boecklen 1986, Varvio et al. 1986, Lacy 1987). However, the models on which this conclusion is based assume that populations are constant in size. The apparently beneficial effect of subdivision is partly due to an artifact of the models, in which model constraints create more equal distribution of reproductive success (and therefore higher N_e) when the population is divided into subunits of fixed size (Barton and Whitlock 1997), and partly due to the protection of different alleles as each subpopulation becomes fixed for a random subset of the diversity of the original metapopulation. This second process can be a benefit of population subdivision, but it becomes significant only after populations lose most of their gene diversity and become highly inbred, and it depends on no populations becoming extinct as a result of that inbreeding or other factors.

In individual-based PVA models that model genetic changes, the effects of subdivision on gene diversity are quite different, because the PVA models do not unrealistically constrain the populations to be constant in size. Smaller populations undergo greater fluctuations in number and therefore lose gene diversity much faster than if they were part of a larger breeding population. As a consequence of this greater demographic instability, fragmented metapopulations will usually lose genetic variation (both heterozygosity and number of alleles per locus) more rapidly than does a single, more panmictic, large population. This trend occurs even when the effects of fragmentation are partly offset by dispersal among partly isolated populations. In models of populations of mountain brushtail possums *Trichosurus caninus*, Lacy and Lindenmayer (1995) found that both heterozygosity and number of alleles were reduced more quickly when metapopulations of 100 or 200 possums were fragmented into 2, 5, or 10 subpopulations that exchanged up to 5% of their residents each year. Rates of genetic decay slowed as dispersal rates were increased, but the individual-based stochastic model consistently showed that population subdivision caused faster genetic decay from the metapopulation – a result not predicted from many analytical genetic models.

Characteristics of highly vulnerable species

Summarizing this discussion of the threats to viability of small populations, we can identify some of the characteristics of populations that lead to the most complicated population dynamics as populations become small, and which therefore might require the most detailed and individual-based PVA models. Especially vulnerable species would include those with non-breeding helpers, such as striped-back wrens and naked mole rats, species with cooperative foraging, such as many parrots and social carnivores, and species with group defense behaviors, such as musk ox and many primates. Species with precise mechanisms for mate choice, such as many bird species, could have demographic and genetic problems when that choice becomes limited. Monogamous species will have depressed reproduction when there is demographic stochasticity in the sex ratio.

Species with low fecundity are particularly vulnerable to inbreeding depression (Mills and Smouse 1994), because they can withstand less depression of survival before population growth rates become negative and because they will recover more slowly from population bottle-necks. Moreover, of the few PVA models which consider genetic effects (e.g., INMAT: Mills and Smouse 1994; VORTEX: Lacy 2000; and see Menges 2000 for references on plant PVAs), usually an assumption is made that individuals at the start of the population projection are all unrelated and noninbred. Thus, we may under-appreciate existing or imminent genetic problems in populations which have already lost much genetic variation. For example, early analyses of the remnant population of the Florida panther *Felis concolor coryi* assumed that prior inbreeding would not diminish reproductive rates (Seal and Lacy 1989), even though a majority of the males had only one or no functional testicles (Seal et al. 1992). Golden lion tamarins *Leontopithecus rosalia rosalia* are now restricted to very small remnants of the original Atlantic coastal forest of Brazil, have been reduced to a population of ca 350 animals and scattered smaller populations (Ballou et al. 1998), have low genetic diversity (Forman et al. 1986), and show significant depression of juvenile survival when inbred (Dietz et al. 2000). Thus, any PVA models that assume tamarins presently have adequate genetic diversity may project, perhaps incorrectly, that the population could lose much more variation before genetic problems began to reduce population viability.

PVA modeling of small population processes

As illustrated in the examples presented above, more processes threaten the viability of small populations than are commonly addressed in PVA models. Analytical models and simple population projection models can therefore overestimate population growth, underestimate population fluctuations, and seriously underestimate probabilities of extinction. More accurate PVA of small populations may often require individual-based models that simulate interactions among threatening processes rather than relying on theoretical equations derived under assumptions of simplified population processes acting in isolation. Lindenmayer et al. (2000) found that PVA model predictions matched the observed dynamics of populations of three species of marsupials in a highly fragmented landscape only when the models incorporated distance-dependent and density-dependent dispersal, high mortality during dispersal, and spatial variation in habitat quality. Knowledge of times since isolation of each habitat fragment was also critical, as the metapopulations were projected to lose more component subpopulations before reaching extinction-recolonization equilibria. In an alternative approach, Sjögren-Gulve (1994, Sjögren-Gulve and Ray 1996) used a logistic regression model to incorporate similarly detailed information about the habitat characteristics, spatial arrangement, and surrounding forestry practices on the extinction-recolonization dynamics of pool frog *Rana lessonae* metapopulations. Rather than modeling causal processes, as is done in individual-based models, logistic models can use observed correlates of population transitions to generate predictive models of metapopulation trends. Individual-based models require detailed data on the factors driving population processes, while logistic transition-incidence models require detailed data on the important correlates of population transitions over significant periods of time.

Generalized analytical models can be extremely valuable for discerning many broad trends (e.g., Belovsky 1987), but they do not provide the situation-specific representations that are needed to assess local threats to specific small natural populations nor, usually, the time-specific projections that are needed to understand non-equilibrial systems. Thus, detailed and often individual-based models are more appropriate for comparing management options for endangered species recovery and local conservation planning. We should also retain some skepticism regarding the generality of theoretical results until they have been confirmed to apply to simulated (or, better, real) populations in which the unrealistic assumptions of the theoretical model have been relaxed. There have been recent encouraging confirmations of the ability of PVA models to predict population dynamics (Brook et al. 2000), but more comparisons are needed among analytical results, simulation results from models of varying detail, experimental data, and observations on natural populations.

For a variety of reasons, PVA models for small populations may need to be highly specific with respect to how they model breeding systems, dispersal behavior, and genetic processes. Simple generalizations of population genetics theory may be misleading, because most of that theory was based on large sample approximations. (For example, generalizations about effects of dispersal among small populations often have assumed that an infinite number of such populations exist.) Many metapopulation models assume that the system has reached extinction-recolonization equilibrium. Variation in demographic rates, numbers of individuals, and other population statistics rarely follow a normal distribution. Effects of threatening processes on populations are rarely linear or log-linear, and threshold effects, in which there are sharp discontinuities in effects, are possible. The effects of multiple threats are often synergistic, rather than additive.

Many of the parameters required to build specific, detailed models of small population dynamics can only be estimated well with long-term and extensive data. It is clearly difficult to obtain large samples from small populations, and conservation action may have to precede long-term field studies in order to ensure that the populations persist long enough to permit extended study. Obtaining a complete enumeration of a small population will remove all sampling error in estimation of current parameter values, but it will not necessarily provide sufficiently precise values for predicting future trends. The entire existing population is still only a sample of the universe of all possible populations that could have resulted from the same processes. Given that highly detailed models employing accurate estimates for a large number of parameters might be needed to project the dynamics of small populations well, should PVA models be used to help guide conservation actions? The alternative to using incomplete models with poorly estimated parameters that may overestimate population viability is to use even more general models that will omit many threatening processes and often more seriously underestimate risks, or to rely on intuitive assessments of complex, probabilistic phenomena, something that people are innately poor at doing (Piattelli-Palmarini 1994, Margolis 1996). When planning conservation actions for species that have already declined to near extinction, we should use the best tools available but also recog-

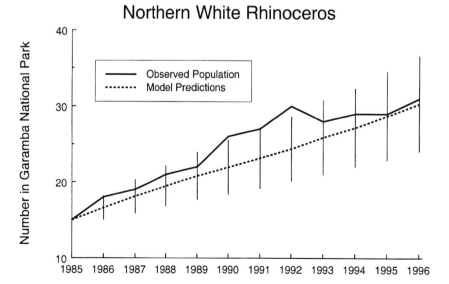

Fig. 5. Observed number (solid line) of northern white rhinoceros in the Garamba National Park, Congo (formerly Zaire), the entire range of the only remaining population of the taxon, and the number projected (dashed line) from the 1985 population. Error bars show the standard deviation across simulations of the projected population size each year. Adapted from Lacy (1996).

nize that our assessments may be crude. Because all PVA models include only a subset of potentially threatening processes, it is possible that many PVAs overestimate viability (Lacy 1993/1994) and that our error will tend to be greatest in the smallest populations – i.e., those which are most critically in need of effective conservation action. Accordingly, margins for error and ongoing monitoring of results should always be part of implementation (see also Akçakaya 2000).

Although the dynamics of small populations can be complex and subjected to many stochastic processes, existing PVA models can provide good representations of dynamics of many such populations. PVA models have predicted well the dynamics of some critically rare species, such as the whooping crane (Brook et al. 1999) and the northern white rhinoceros, *Ceratotherium simum cottoni* (see Fig. 5). Brook et al. (2000) found that PVA models that are sufficiently detailed and for which there are adequate data for estimating parameters can provide unbiased and reasonably accurate predictions for a number of species. The species for which historic population trajectories were modeled in the above studies had been reduced to single populations, rather than existing within metapopulations, and most had relatively simple breeding systems, rather than complex social structures. Also, PVA models may be more reliable when habitat is not limiting population growth than when dynamics near carrying capacity are modeled (Mills et al. 1996, Brook et al. 1997).

How small is small?

The processes and cases described in this paper suggest that, when it comes to assessing whether wildlife populations have declined to dangerously small sizes, small may be bigger than we usually think it is. Isolated populations with fewer than ca 50 breeding adults may suffer from inbreeding depression within a few generations. Larger populations that are fragmented into partially isolated subunits of fewer than ca 50 breeding animals in each may lose variability much faster than would be estimated from the total metapopulation size (Lacy and Lindenmayer 1995). Genetic decay in populations with fewer than ca 500 breeding adults, or even 5000 adults (Lande 1995), may eventually reduce adaptability and potential for recovery (Lacy 1997). Monogamous species and species with complex social systems may have reduced breeding if numbers fall below several hundred adults. Each of these factors tends to be a greater threat in those species, such as many mammals and birds, that have low fecundity and only slow population growth under optimal conditions.

Finally, it may be difficult to know when a population is so small that additional stochastic factors must be included in a PVA to obtain an accurate projection of its dynamics. Therefore, it is often useful to test several models to determine if added complexity substantially alters PVA predictions, and provides a better fit to observed population trends. A good example of this type of exploration is provided by the ongoing work of Lindenmayer and his colleagues on the fauna of fragmented forests in Australia (Lindenmayer et al. 1999, 2000). PVA models should be no more complex than necessary, but to be useful for conservation they must also be detailed enough to model the real population dynamics accurately (Starfield and Bleloch 1986, Starfield 1997, Akçakaya and Sjögren-Gulve 2000, Lacy and Miller 2001).

Acknowledgements – I thank the editors for their many constructive comments.

References

Akçakaya, H. R. 2000. Population viability analyses with demographically and spatially structured models. – Ecol. Bull. 48: 23–38.

Akçakaya, H. R. and Ferson, S. 1992. RAMAS/Space: Spatially structured population models for conservation biology. – Appl. Biomath., New York.

Akçakaya, H. R. and Sjögren-Gulve, P. 2000. Population viability analyses in conservation planning: an overview. – Ecol. Bull. 48: 9–21.

AsRSG. 1996. Report of a meeting of the IUCN SSC Asian Rhino Specialist Group (AsRSG). – Sandakan, Malaysia, 29 Nov. – 1 Dec. 1995.

Ballou, J. D. 1997. Ancestral inbreeding only minimally affects inbreeding depression in mammalian populations. – J. Heredity 88: 169–178.

Ballou, J. D. et al. 1998. Leontopithecus II. The second population and habitat viability assessment for lion tamarins (*Leontopithecus*). – Conservation Breeding Specialist Group (SSC/IUCN), Apple Valley, Minnesota.

Barrett, S. C. H. and Kohn, J. R. 1991. Genetic and evolutionary consequences of small population size in plants: implications for conservation. – In: Falk, D. A. and Holsinger, K. E. (eds), Genetics and conservation of rare plants. Oxford Univ. Press, pp. 3–30.

Barton, N. H. and Whitlock, M. C. 1997. The evolution of metapopulations. – In: Hanski, I. A. and Gilpin, M. E. (eds), Metapopulation biology. Academic Press, pp. 183–210.

Belovsky, G. E. 1987. Extinction models and mammalian persistence. – In: Soulé, M. E. (ed.), Viable populations for conservation. Cambridge Univ. Press, pp. 35–57.

Boecklen, W. J. 1986. Optimal reserve design of nature reserves: consequences of genetic drift. – Biol. Conserv. 38: 323–338.

Boyce, M. S. 1992. Population viability analysis. – Annu. Rev. Ecol. Syst. 23: 481–506.

Brewer, B. A. et al. 1990. Inbreeding depression in insular and central populations of *Peromyscus* mice. – J. Heredity 81: 257–266.

Brook, B. W. et al. 1997. Does population viability analysis software predict the behaviour of real populations? A retrospective study on the Lord Howe Island woodhen *Tricholimnas sylvestris* (Sclater). – Biol. Conserv. 82: 119–128.

Brook, B. W. et al. 1999. Comparison of the population viability analysis packages GAPPS, INMAT, RAMAS and VORTEX for the whooping crane (*Grus americana*). – Anim. Conserv. 2: 23–31.

Brook, B. W. et al. 2000. Predictive accuracy of population viability analysis in conservation biology. – Nature 404: 385–387.

Brown, J. H. and Kodric-Brown, A. 1977. Turnover rates in insular biogeography: effect of immigration on extinction. – Ecology 58: 445–449.

Caro, T. M. and Laurenson, M. K. 1994. Ecological and genetic factors in conservation: a cautionary tale. – Science 263: 485–486.

Caughley, G. 1994. Directions in conservation biology. – J. Anim. Ecol. 63: 215–244.

Charlesworth, D. and Charlesworth, B. 1987. Inbreeding depression and its evolutionary consequences. – Annu. Rev. Ecol. Syst. 18: 237–268.

Clark, T. W. 1989. Conservation biology of the black-footed ferret. – Wildlife Preservation Trust International, Philadelphia.

Courchamp, F., Clutton-Brock, T. and Grenfell, B. 1999. Inverse density dependence and the Allee effect. – Trends Ecol. Evol. 14: 405–410.

Dietz, J. M., Baker, A. J. and Ballou, J. D. 2000. Demographic evidence of inbreeding depression in golden lion tamarins. – In: Young, A. and Clarke, G. (eds), Genetics, demography and population viability. Cambridge Univ. Press, in press.

Ellis, S. et al. 1992. Alala, akohekohe, and palila population and habitat viability assessment reports. – IUCN SSC Captive Breeding Specialist Group, Apple Valley, Minnesota.

Fahrig, L. and Merriam, G. 1985. Habitat patch connectivity and population survival. – Ecology 66: 1762–1768.

Fahrig, L. and Merriam, G. 1994. Conservation of fragmented populations. – Conserv. Biol. 8: 50–59.

Fisher, R. A. 1958. The genetical theory of natural selection. – Dover, New York.

Foley, P. 1994. Predicting extinction times from environmental stochasticity and carrying capacity. – Conserv. Biol. 8: 124–137.

Forman, L. et al. 1986. Genetic variation within and among lion tamarins. – Am. J. Phys. Anthropol. 71: 1–11.

Frankel, O. H. and Soulé, M. E. 1981. Conservation and evolution. – Cambridge Univ. Press.

Frankham, R. 1995a. Conservation genetics. – Annu. Rev. Genet. 29: 305–327.

Frankham, R. 1995b. Inbreeding and extinction: a threshold effect. – Conserv. Biol. 9: 792–799.

Franklin, I. R. 1980. Evolutionary change in small populations. – In: Soulé, M. E. and Wilcox, B. A. (eds), Conservation biology. An evolutionary-ecological perspective. Sinauer, pp. 135–149.

Gilpin, M. E. and Soulé, M. E. 1986. Minimum viable populations: the processes of species extinction. – In: Soulé, M. E. (ed.), Conservation biology: the science of scarcity and diversity. Sinauer, pp. 13–34.

Goodman, D. 1987. The demography of chance extinction. – In: Soulé, M. E. (ed.), Viable populations for conservation. Cambridge Univ. Press, pp. 11–34.

Gyllenberg, M. and Hanski, I. 1992. Single-species metapopulation dynamics: a structured model. – Theor. Popul. Biol. 42: 35–61.

Hedrick, P. W. 1994. Purging inbreeding depression and the probability of extinction: full-sib mating. – Heredity 73: 363–372.

Hedrick, P. W. and Miller, P. S. 1992. Conservation genetics: techniques and fundamentals. – Ecol. Appl. 2: 30–46.

Hedrick, P. W. et al. 1996. Directions in conservation biology: comments on Caughley. – Conserv. Biol. 10: 1312–1320.

Jiménez, J. A. et al. 1994. An experimental study of inbreeding depression in a natural habitat. – Science 266: 271–273.

Keane, B. 1990. The effect of relatedness on reproductive success and mate choice in the white-footed mouse, *Peromyscus leucopus*. – Anim. Behav. 39: 264–273.

Keller, L. F. et al. 1994. Selection against inbred song sparrows during a natural population bottleneck. – Nature 372: 356–357.

Kindvall, O. 1996. Habitat heterogeneity and survival of the bush cricket, *Metrioptera bicolor*. – Ecology 77: 207–214.

Lacy, R. C. 1987. Loss of genetic diversity from managed populations: interacting effects of drift, mutation, immigration, selection, and population subdivision. – Conserv. Biol. 1: 143–158.

Lacy, R. C. 1993/1994. What is population (and habitat) viability analysis? – Primate Conserv. 14/15: 27–33.

Lacy, R. 1996. Further population modelling of northern white rhinoceros under various management scenarios. – In: Foose, T. J. (ed.), Summary, Appendix 3 – Northern White Rhinoceros Conservation Strategy Workshop. International Rhino Foundation, Cumberland, Ohio, pp. 1–15.

Lacy, R. C. 1997. Importance of genetic variation to the viability of mammalian populations. – J. Mammal. 78: 320–335.

Lacy, R. C. 2000. Structure of the VORTEX simulation model for population viability analysis. – Ecol. Bull. 48: 191–203.

Lacy, R. C. and Lindenmayer, D. B. 1995. A simulation study of the impacts of population subdivision on the mountain brushtail possum, *Trichosurus caninus* Ogilby (Phalangeridae: Marsupialia), in south-eastern Australia. II. Loss of genetic variation within and between subpopulations. – Biol. Conserv. 73: 131–142.

Lacy, R. C. and Ballou, J. D. 1998. Effectiveness of selection in reducing the genetic load in populations of *Peromyscus polionotus* during generations of inbreeding. – Evolution 52: 900–909.

Lacy, R. C. and Miller, P. S. 2001. Managing the human animal: incorporating human populations and activities into PVA for wildlife conservation. – In: Beissinger, S. and McCullough, D. R. (eds), Population viability analysis. Univ. of Chicago Press, in press.

Lacy, R. C., Alaks, G. and Walsh, A. 1996. Hierarchical analysis of inbreeding depression in *Peromyscus polionotus*. – Evolution 50: 2187–2200.

Lande, R. 1988. Genetics and demography in biological conservation. – Science 241: 1455–1460.

Lande, R. 1995. Mutation and conservation. – Conserv. Biol. 9: 782–791.

Lindenmayer, D. B. and Lacy, R. C. 1995. Metapopulation viability of Leadbeater's possum, *Gymnobelideus leadbeateri*, in fragmented old-growth forests. – Ecol. Appl. 5: 164–182.

Lindenmayer, D. B. et al. 1995. A review of the generic computer programs ALEX, RAMAS/space and VORTEX for modelling the viability of wildlife populations. – Ecol. Model. 82: 161–174.

Lindenmayer, D. B., McCarthy, M. A. and Pope, M. L. 1999. Arboreal marsupial incidence in eucalypt patches in south-eastern Australia: a test of Hanski's incidence function meta-population model for patch occupancy. – Oikos 84: 99–109.

Lindenmayer, D. B., Lacy, R. C. and Pope, M. L. 2000. Testing a simulation model for population viability analysis. – Ecol. Appl. 10: 580–597.

Lynch, M. and Walsh, B. 1998. Genetics and analysis of quantitative traits. – Sinauer.

Margolis, H. 1996. Dealing with risk. Why the public and the experts disagree on environmental issues. – Univ. of Chicago Press.

Menges, E. S. 2000. Applications of population viability analyses in plant conservation. – Ecol. Bull. 48: 73–84.

Miller, P. S. 1994. Is inbreeding depression more severe in a stressful environment? – Zoo Biol. 13: 195–208.

Mills, L. S. and Smouse, P. E. 1994. Demographic consequences of inbreeding in remnant populations. – Am. Nat. 114: 412–431.

Mills, L. S. et al. 1996. Factors leading to different viability predictions for a grizzly bear data set. – Conserv. Biol. 10: 863–873.

Mirande, C., Lacy, R. and Seal, U. 1991. Whooping crane (*Grus americana*) conservation viability assessment workshop report. – IUCN SSC Captive Breeding Specialist Group, Apple Valley, Minnesota.

Pettersson, B. 1985. Extinction of an isolated population of the middle spotted woodpecker *Dendrocopos medius* (L.) in Sweden and its relation to general theories on extinction. – Biol. Conserv. 32: 335–353.

Piattelli-Palmarini, M. 1994. Inevitable illusions. How mistakes of reason rule our minds. – Wiley.

Pounds, J. A. and Crump, M. L. 1994. Amphibian declines and climate disturbance: the case of the golden toad and the harlequin frog. – Conserv. Biol. 8: 72–85.

Rabenold, K. N. 1990. *Campylorhynchus* wrens: the ecology of delayed dispersal and cooperation in the Venezuelan savanna. – In: Stacey, P. B. and Koenig, W. D. (eds), Cooperative breeding in birds. Cambridge Univ. Press, pp. 159–196.

Rabenold, P. P. et al. 1991. Density-dependent dispersal in social wrens: genetic analysis using novel matriline markers. – Anim. Behav. 42: 144–146.

Ralls, K., Ballou, J. D. and Templeton, A. 1988. Estimates of lethal equivalents and the cost of inbreeding in mammals. – Conserv. Biol. 2: 185–193.

Ronce, O., Perret, F. and Olivieri, I. 2000. Evolutionarily stable dispersal rates do not always increase with local extinction rates. – Am. Nat. 155: 485–496.

Ryan, K. K. 2000. Causes and consequences of male mate choice in a monogamous oldfield mouse, *Peromyscus polionotus*. – Ph.D. thesis, Univ. of Chicago.

Saccheri, I. et al. 1998. Inbreeding and extinction in a butterfly metapopulation. – Nature 392: 491–494.

Schonewald-Cox, C. M. et al. (eds) 1983. Genetics and conservation. A reference for managing wild animal and plant populations. – Benjamin/Cummings, Menlo Park, California.

Seal, U. S. and Lacy, R. C. 1989. Florida panther population viability analysis. – Report to the U.S. Fish and Wildlife Service. IUCN SSC Captive Breeding Specialist Group, Apple Valley, Minnesota.

Seal, U. S. et al. 1992. Genetic management strategies and population viability of the Florida panther (*Felis concolor coryi*). – Report to the U.S. Fish and Wildlife Service. IUCN SSC Captive Breeding Specialist Group, Apple Valley, Minnesota.

Shaffer, M. L. 1981. Minimum population sizes for species conservation. – BioScience 31: 131–134.

Shaffer, M. 1987. Minimum viable populations: coping with uncertainty. – In: Soulé, M. E. (ed.), Viable populations for conservation. Cambridge Univ. Press, pp. 69–86.

Sjögren-Gulve, P. 1994. Distribution and extinction patterns within a northern metapopulation of the pool frog, *Rana lessonae*. – Ecology 75: 1357–1367.

Sjögren-Gulve, P. and Ray, C. 1996. Using logistic regression to model metapopulation dynamics: Large-scale forestry extirpates the pool frog. – In: McCullough, D. R. (ed.), Metapopulations and wildlife conservation. Island Press, Washington, D.C., pp. 111–137.

Soulé, M. E. 1980. Thresholds for survival: maintaining fitness and evolutionary potential. – In: Soulé, M. E. and Wilcox, B. A. (eds), Conservation biology. An evolutionary-ecological perspective. Sinauer, pp. 151–169.

Soulé, M. E. 1987. Viable populations for conservation. – Cambridge Univ. Press.

Starfield, A. M. 1997. A pragmatic approach to modeling. – J. Wildl. Manage. 61: 261–270.

Starfield, A. M. and Bleloch, A. L. 1986. Building models for conservation and wildlife management. – MacMillan, New York.

Stephens, P. A. and Sutherland, W. J. 1999. Consequences of the Allee effect for behaviour, ecology and conservation. – Trends Ecol. Evol. 14: 401–405.

Varvio, S. L., Chakraborty, R. and Nei, M. 1986. Genetic variation in subdivided populations and conservation genetics. – Heredity 57: 189–198.

Vucetich, J. A. and Creel, S. 1999. Ecological interactions, social organization, and extinction risk in African wild dogs. – Conserv. Biol. 13: 1172–1182.

Vucetich, J. A., Peterson, R. O. and Waite, T. A. 1997. Effects of social structure and prey dynamics on extinction risk in gray wolves. – Conserv. Biol. 11: 957–965.

Westemeier, R. L. et al. 1998. Tracking the long-term decline and recovery of an isolated population. – Science 282: 1695–1698.

Ecological Bulletins 48: 53–71. Copenhagen 2000

Metapopulation viability analysis using occupancy models

Per Sjögren-Gulve and Ilkka Hanski

Sjögren-Gulve, P. and Hanski, I. 2000. Metapopulation viability analysis using occupancy models. – Ecol. Bull. 48: 53–71.

We review the use of patch occupancy metapopulation models for viability analysis of spatially structured populations. These models are parameterized using species presence-absence data from one or more surveys of a habitat patch network. Occupancy models do not include a description of local population dynamics within habitat patches. On the other hand, occupancy models such as the incidence function model and the logistic regression model can incorporate information on the sizes, qualities and spatial locations of habitat patches, and how these patch attributes influence local extinctions and colonizations. For metapopulations in markedly patchy habitats, occupancy models have been used successfully to predict patterns of local extinction and colonization and the persistence of species in patch networks. In such situations, it is doubtful whether more data-intensive and complex demographic models would yield any more reliable predictions than the relatively simple occupancy models. We review the Levins model, the incidence function model and the logistic regression model – their assumptions, data requirements and types of predictions – and discuss their advantages and limitations. We also review selected case studies in which these models have been used in conservation assessments.

P. Sjögren-Gulve (per.sjogren-gulve@environ.se), Dept of Conservation Biology and Genetics, Evolutionary Biology Centre, Uppsala Univ., Norbyvägen 18D, SE-752 36 Uppsala, Sweden; (present address: Swedish Environmental Protection Agency, SE-106 48 Stockholm, Sweden). – I. Hanski, Metapopulation Research Group, Dept of Ecology and Systematics, FIN-00014 Univ. of Helsinki, Helsinki, Finland.

Spatially subdivided populations, or metapopulations in a broad sense (see Hanski and Simberloff 1997), have attracted much attention in conservation biology and conservation planning. The term "metapopulation" was originally coined by Levins (1970) to denote a "population" of (local) populations of a species, interconnected to some degree by dispersal (for a historical overview, see Hanski 1999). Today, many species have greatly declined and their distribution has become fragmented due to anthropogenic habitat loss. Many threatened species fall into this category. But many species also exhibit a patchy distribution in a landscape for natural reasons, as they depend on a patchily distributed habitat or resource (e.g., host plant-specific insects, amphibians in ponds, mosses on coarse woody debris) or their distribution is constrained by low dispersal ability.

Along with the gradual acceptance of a broad view of metapopulations, a paradigm shift has occurred in conservation biology from the widespread use of the island biogeography theory to metapopulation approaches (Hanski and Simberloff 1997). At the same time, the applicability of metapopulation theory to practical conservation has been questioned and discussed, primarily because of unsuccessful application of simple models to real-world situations and because some researchers believe that classical metapopulation structures are not common in nature (e.g., Harrison 1994, Scheiner and Rey-Benayas 1997; but see Hanski 1999).

In this paper, we outline the situations in which a particular class of metapopulation models, called patch occupancy models, can be used for population viability analysis (PVA) in conservation assessments and planning. We start by reviewing the seminal metapopulation model of Levins (1969), then focus on two particular models that have been applied to real-world situations, namely state-transition models as exemplified by the logistic regression metapopulation model (LRM; Sjögren-Gulve and Ray 1996) and the incidence function model (IFM; Hanski 1994, 1999, Hanski et al. 1996b, Moilanen 1999). Both classes of models have substantial promise for conservation planning. We review the general features, assumptions and predictions of these models. We also provide some guidelines in the use of the models, and present some illustrative case studies that may further help managers and conservation biologists in choosing a model for their focal system and questions.

Occupancy models

General features, assumptions and predictions

Occupancy models describe the number (N) or proportion (P) of discrete habitat patches in a landscape or geographic region that are inhabited by the focal species. These models can be expanded to several species (reviewed in Hanski 1999), but most conservation applications are likely to be focused on single species (but see Fleishman et al. 2000). A general requirement for the use of these models is that the species' habitat can be readily identified and that distinct habitat patches in the landscape can be delimited. Contrary to demographically structured metapopulation models (see Burgman et al. 1993, Hanski 1999, Akcakaya 2000, Lacy 2000a, Menges 2000, Akçakaya and Sjögren-Gulve 2000 for reviews), occupancy models do not track variation in local population size and demography. Occupancy models simply score the habitat patches as either occupied or unoccupied by the focal species, and model local extinction and (re)colonization events that may be functions of some environmental variables. The occupancy models are most applicable to systems with a reasonably large number of relatively small habitat patches (see Choosing a model).

The classic model of Levins (1969, 1970) predicts the temporal change in the number or proportion of occupied patches in a patch network, with one parameter describing the extinction rate of existing local populations and another parameter describing the (re)colonization rate of unoccupied but suitable patches (see Levins' original model). Variants of Levins' (1969, 1970) model attempt to capture various aspects of metapopulation dynamics in relation to environmental and/or demographic factors that influence population turnover. Such models are reviewed and exemplified in Gotelli and Kelley

(1993), McCullough (1996), Hanski and Gilpin (1997) and Hanski (1999); some of these models have been developed for particular cases or problems whereas others are more general.

The Levins model assumes that all habitat patches are identical. For conservation and planning purposes, occupancy models that include variation in habitat quality and variation in extinction and colonization rates or probabilities among individual patches are more realistic and useful (Harrison 1994). To achieve such realism, spatially realistic metapopulation models (Hanski 1999) that specify the geographical location of each patch are needed (see Hanski and Simberloff [1997] for model terminology). The IFM and LRM are both spatially realistic metapopulation models. For the purpose of numerical predictions, they allow one to calculate, for each time-step, colonization and extinction probabilities for individual habitat patches as functions of their quality (e.g., patch size) and spatial position in relation to other patches (explained in detail below). In addition to the patch-specific colonization and extinction probabilities that can be calculated using various explanatory variables, Monte-Carlo simulations of the IFM and LRM predict the species' presence/absence pattern in each time step and produce trajectories over time of the proportion (or number) of occupied patches (e.g., Sjögren-Gulve and Ray 1996, Hanski et al. 1996b, Hanski 1999). Based on simulations, one can obtain the temporal pattern of occupancy for each patch in a network.

Occupancy models are appropriate for ranking habitat management alternatives. By explicit inclusion of habitat variables, the models facilitate simulation of metapopulation response to changes in the number, density and/or quality of habitat patches. In contrast, using demographic models to predict metapopulation response to habitat change may be problematic if the habitat effects are only implicit in demographic rates. In the applications of occupancy models so far, patch quality has generally been kept constant over time during simulation whereas distances to occupied patches are updated for all patches in each time step. Effects of habitat change have instead been modeled by using scenarios (e.g., the pool frog example in Selected case studies). Similar to vital rates in demographic simulation models (e.g., Akçakaya 2000, Lacy 2000b), patch quality could be allowed to vary temporally in a deterministic or stochastic manner also in occupancy models.

The LRM (Sjögren-Gulve and Ray 1996), the Levins (1969) model and its variants (Gotelli and Kelley 1993) all require that local extinctions and/or colonizations have been observed for the purpose of estimating model parameters. In contrast, the IFM can be parameterized with survey data from just one year and does not require that extinctions or colonizations have been observed, though more reliable parameter estimates can be obtained with more extensive data (Hanski 1994, 1999, Moilanen 1999).

Levins' original model

Levins (1969) outlined a deterministic model that demonstrates how a species may persist at the regional level despite repeated local extinctions. The key to regional persistence is that local population dynamics and thereby extinctions do not occur synchronously in space, and that extinctions are balanced by recolonization of vacant habitat patches ("metapopulation equilibrium"; Box 1). For this model to be useful, the patches must be easy to identify, homogeneous in quality, all within the dispersal range of the species, and local extinctions and colonizations must be fairly frequent events.

Using a stochastic version of the Levins model for a finite number of patches, Nisbet and Gurney (1982) presented an approximate condition for long-term metapopulation persistence given that $\hat{N}$ of the T suitable habitat patches are occupied at metapopulation equilibrium (stochastic steady state conditioned on metapopulation extinction not having occurred), viz.

$$\hat{N} \geq 3\sqrt{T}. \qquad (3)$$

In summary, the Levins (1969) model assumes that the number of patches is infinitely large and that all patches are identical and equally connected. These assumptions are convenient for mathematical analysis but too restrictive to make the model really useful for practical applications. In principle, it is possible to parameterize the model with empirical data (see Declining edible frog populations in southern Sweden). However, we advocate that spatially realistic models are used to incorporate the effects of landscape structure on metapopulation processes (see also Harrison 1994). The risk of extinction is typically different for different local populations, and not all patches are equally favorable nor equally accessible for the species. Nonetheless, the Levins model is helpful in providing insight into the processes that underpin the long-term persistence of classical metapopulations.

Recently, Hanski and Ovaskainen (2000, Ovaskainen and Hanski unpubl.) have developed a spatially realistic version of the Levins model. The new model is deterministic like the original model but applies to networks with a finite number of patches and spatial variation in patch area and connectivity. This model is closely related to the stochastic IFM in its basic assumptions.

State Transition Models: the Logistic Regression Model (LRM)

State transition models can be applied to metapopulations for which a sufficient number of observations are available on local extinctions and colonizations. The basic modeling approach is to find the main environmental correlates of the observed population turnover. Because a statistical model (multiple logistic regression) is used to analyze the colonization and extinction events, > 5 local extinction

---

Box 1

Levins' metapopulation model

Levins (1969) envisioned a total number of T patches of suitable habitat, of which N patches are currently occupied by the focal species. All local populations have the same extinction rate e, and the species disperses to and colonizes vacant patches with a rate set by parameter c. The change in the number of occupied patches with time (t) in a large network is given as

$$\frac{dN}{dt} = cN\left(1 - \frac{N}{T}\right) - eN. \qquad (1a)$$

Dividing by the total number of patches (T) we obtain an equation for the rate of change in the proportion of occupied patches (P = N/T),

$$\frac{dP}{dt} = cP(1 - P) - eP \qquad (1b)$$

(Levins 1970). The equilibrium number of occupied patches $\hat{N}$ is given by

$$\hat{N} = T\left(1 - \frac{e}{c}\right). \qquad (2)$$

Thus, provided that the ratio e/c < 1, the model predicts that the metapopulation will persist in a very large (infinite) network.

and colonization events, respectively, should be available for the analysis (see Selected case studies). This type of modeling is useful when the observed colonization and extinction events can be explained by some measurable environmental variable(s), as exemplified by the study on the Swedish pool frog by Sjögren-Gulve and Ray (1996). The LRM can include the effects of multiple variables on extinction and colonization, and it can be used to model spatially correlated extinctions. The latter makes it possible to model metapopulation dynamics in systems where "local populations" inhabit more than one patch, or systems with spatially correlated local extinctions due to "regional stochasticity" (Hanski 1991) such as, for example, spatially correlated weather conditions (Solbreck 1991, Sjögren-Gulve 1994).

In the LRM, patch colonization and extinction events are viewed as binary responses (yes or no, quantified as 1 or 0) to environmental conditions (Box 2). For instance, the habitat quality of a patch was so low that local extinction occurred. Multiple logistic regression (e.g., Hosmer and Lemeshow 1989, Engelman 1990, Sjögren-Gulve and Ray 1996) can be used to select environmental variables that together best discriminate between the response groups (e.g., populations that went extinct vs those that did not). The environmental variables can be either continuous or categorical. For example, the probability of colonization (vs continued vacancy) of a vacant patch in a metapopulation might be affected by its distance to the closest occupied patch (continuous variable) and by the presence or absence of a road barrier in intervening areas (categorical

Box 2

The logistic regression metapopulation model (LRM)

In logistic regression analysis, the predicted proportion of positive responses (response = 1; e.g. local extinction or colonization) in the total sample is assumed to follow the logistic model

$$\frac{\exp(u)}{1+\exp(u)}, \tag{4}$$

which ranges from 0 to 1. The logit u is a linear equation including any explanatory variables (X_i) and a constant (ε) that contribute to the discrimination between the two response groups (yes = 1 and no = 0):

$$u = \varepsilon + \beta_1 \cdot X_1 + \beta_2 \cdot X_2 + ... + \beta_n \cdot X_n \tag{5}$$

where β_n is the regression coefficient for X_n. For instance, a logistic regression analysis might select patch size (Area) and presence or absence of a predator (Pred = 1 or 0) as two environmental variables that differ significantly between patches in which local populations went extinct and those where populations persisted. The fitted regression constant is ε and the probability of extinction for a particular patch (i) at time (t) can be calculated from its area and the presence or absence of the predator using the logit

$$u_e(i,t) = \varepsilon - \beta_1 \cdot Area_i + \beta_2 \cdot Pred_i. \tag{6}$$

Negative β_1 indicates that the extinction probability $E_i(t)$ is lower for larger patches, and positive β_2 shows that the presence of the predator increases $E_i(t)$. The general expression for the conditional extinction probability of patch i is given by substituting u_e into eq. 4,

$$E_i(t) = \frac{\exp\big(u_e(i,t)\big)}{1+\exp\big(u_e(i,t)\big)}. \tag{7}$$

Colonization probabilities for unoccupied patches, $C_i(t)$, are calculated in a similar way by allowing multiple logistic regression to select the environmental variables that best discriminate among the patches that became colonized versus those that remained vacant.

The expected change in the number of occupied patches per time interval ($E[\Delta N/\Delta t]$) can be written as the sum of the colonization probabilities for the currently vacant patches (numbering $T - N$) minus the sum of the extinction probabilities for the N occupied patches (cf. Hosmer and Lemeshow 1989, p. 10),

$$E\left(\frac{\Delta N}{\Delta t}\right) = \sum_{i=1}^{T-N} C_i(t) - \sum_{i=1}^{N} E_i(t). \tag{8}$$

variable). Logistic regressions of colonizations and extinctions provide standardized models that quantify the colonization and extinction probabilities of the individual patches given their current values of the significant variables.

Using the equations for extinction and colonization events parameterized with empirical data, one may simulate metapopulation dynamics with the LRM by calculating, in each time interval, the extinction and colonization probabilities for all the T patches in the network. In the simulation, an extinction or colonization event occurs in patch i at time t if a uniformly distributed random number from the interval [0, 1) is smaller than or equal to the model-predicted probability. This modeling procedure is used in the simulation program METAPOP III (Ray et al. 1999) and is further described and exemplified by Sjögren-Gulve and Ray (1996) and in the section Selected case studies below.

The Incidence Function Model (IFM)

The incidence function model (Hanski 1994, 1999) was originally developed with a different approach than the LRM. Given distinct patches of habitat in a landscape and stochastic (steady state) dynamics under undisturbed conditions, the IFM models metapopulation dynamics and effects of habitat loss with focus on patch area and position in relation to occupied patches. More recent extensions (Hanski 1999, Moilanen 1999, Vos et al. 2000) also consider effects of additional environmental variables on local extinction and colonization (e.g., Moilanen and Hanski 1998). In the IFM, extinction probability is related to patch area for convenience, because data on areas are easy to obtain. The variable of fundamental interest is local population size, but it is often reasonable to assume that there exists a linear (Kindvall and Ahlén 1992) or some other simple relationship (Hanski et al. 1996b) between patch area and local population size, hence patch area can be used instead.

The IFM is based on a linear first-order Markov chain model for a single habitat patch, in which the patch has constant transition probabilities between the states of being vacant and being occupied. Thus if patch i is presently vacant it becomes recolonized with a patch-specific probability C_i in unit time. If patch i is presently occupied, the population goes extinct with a patch-specific probability E_i in unit time. With these assumptions, the stationary probability of patch i being occupied, which is called the incidence J_i of the species in patch i, is given by

$$J_i = \frac{C_i}{C_i + E_i}. \tag{9}$$

Equation (9) is valid on the condition that the metapopulation has not gone extinct. Including the rescue effect into this model, in other words taking into account

the reduction in extinction rate due to (past) immigration, leads to the expression (justified in Hanski 1997a, 1999)

$$J_i = \frac{C_i}{C_i + E_i(1 - C_i)}. \tag{10}$$

From here the model construction proceeds in three steps, of which steps 1–2 are presented in Box 3 (details in Hanski 1994, 1999). Step 3, model simulation is discussed below.

The method of parameter estimation described in Box 3 does not take into account spatial and temporal correlations in patch occupancies, and hence the parameter estimates are not true maximum likelihoods. Moilanen (1999) has developed an alternative approach to parameter estimation, based on a Monte Carlo method, which does not suffer from this problem (appropriate software can be downloaded from http://www.helsinki.fi/science/metapop/softa.html). The basic idea is to consider patterns of patch occupancy in the network, rather than occupancy of individual patches, and to calculate transition probabilities between the empirically observed occupancy patterns at the network level. This approach allows one to include both "spatial" data (occupancy patterns) and "temporal" data (extinction and colonization events) into parameter estimation. Generally, however, the simpler method described in Box 3 gives reasonably good estimates of the parameter values as shown by simulation studies (Hanski 1994, Ter Braak et al. 1998, Moilanen 1999).

Simulate metapopulation dynamics with the estimated parameter values

Having estimated the model parameters (Box 3), one may proceed to numerically simulate metapopulation dynamics in the same or in some other patch network to generate quantitative predictions about non-equilibrium (transient) dynamics and about the stochastic steady state (Hanski 1994, 1999). This step is similar to simulating the LRM. For applications of the IFM, see Selected case studies.

MVM and MASH

Hanski et al. (1996a) suggested two measures that are helpful for population viability analyses with occupancy models, namely the Minimum Viable Metapopulation (MVM) Size and the Minimum Amount of Suitable Habitat (MASH). MVM was defined as "the minimum number of interacting local populations necessary for long-term persistence of a metapopulation in a balance between local extinctions and recolonizations". MASH was defined as "the minimum density (or number) of suitable habitat patches necessary for metapopulation persistence".

Hanski et al. (1996a) used a modified Levins (1970)

Box 3

The incidence function metapopulation model (IFM)

Specific assumptions are made about the effects of landscape structure on the colonization and extinction probabilities

Often it is realistic to assume that the extinction probability depends on patch area (because extinction probability depends on expected population size which depends on patch area) but not much on isolation. A convenient functional form for this relationship is the following,

$$E_i = \min\left[\frac{e}{Area_i^x}, 1\right], \tag{11}$$

where $Area_i$ is the area of patch i, and e and x are two parameters. In this formulation, there is a minimum patch area $Area_0$ such that the extinction probability equals 1 for patches smaller or equal to $Area_0$.

The colonization probability C_i is an increasing function of the numbers of immigrants M_i arriving at patch i in unit time. In the case of mainland-island metapopulations (Hanski and Gyllenberg 1993, Hanski and Simberloff 1997), with a permanent "mainland" population as the sole or main source of colonists, a reasonable simple functional form is

$$C_i = \beta \exp(-\alpha D_i), \tag{12}$$

where D_i is the distance of patch (or island) i from the mainland, and α and β are two parameters. For common species, which recolonize a little-isolated patch (D_i close to zero) without delay, eq. (12) may be simplified by setting $\beta = 1$. In the case of metapopulations without a mainland, M_i is the sum of individuals originating from the surrounding local populations. Taking into account the sizes and spatial positions of these populations, we may assume that

$$M_i = \beta S_i = \beta \sum_{j \neq i}^{T} \exp(-\alpha D_{ij}) p_j Area_j^b, \tag{13}$$

where p_j equals 1 for occupied and 0 for vacant patches, D_{ij} is the distance between patches i and j, α and β are two parameters as in eq. (12), and b scales emigration from patch j in relation to its area ($Area_j$). The sum in eq. (13) is denoted by S_i for convenience. If there are no interactions among the immigrants in the establishment of a new population, C_i would increase exponentially with M_i. Often, however, the probability of successful establishment of a new population depends on the number of colonists arriving to the vacant patch (propagule size; Ebenhard 1991), in which case an s-shaped increase in C_i with increasing M_i is better justified,

$$C_i = \frac{M_i^2}{M_i^2 + y^2}, \tag{14}$$

where y is an extra parameter (notice that when eq. 13 is substituted into eq. 14, only the parameter combination y/β remains, which we denote simply by y below). The colonization probabilities do not remain constant when the pattern of patch occupancy (the p_j values) changes, but this violation of the assumption of eq. (9) is generally of little importance when the metapopulation is at the steady state (Hanski 1994).

Plugging equations (11) and (14) for E_i and C_i into eq. (10) leads to the expression

$$J_i = \frac{1}{1 + \dfrac{ey^2}{S_i^2 Area_i^x}}. \tag{15}$$

Estimate the model parameters α, x, e and y using non-linear maximum-likelihood regression or some other technique

In parameter estimation, the observed occupancies p_i are regressed against the incidences J_i (Hanski 1994). Minimally, one needs the following data from one metapopulation at a stochastic steady state: patch areas $Area_i$ and their spatial coordinates (the latter to calculate the pairwise distances D_{ij}), and the state of the patches at one point in time (the p_j values). It is desirable, however, to have censused the metapopulation a few times for more reliable parameter estimates (Hanski et al. 1996b, Moilanen 1999). The critical assumption at this stage is that the metapopulation from which the parameter values are estimated is at a stochastic steady state, that is, that there is no long-term increasing nor decreasing trend in patch occupancy.

The values of e and y cannot be estimated independently with eq. (15). To tease apart their values one may use either information on population turnover rate between two or more years, as explained in Hanski (1994; see also Moilanen 1999); or one may estimate (or guess) the minimum patch area $Area_0$ for which $e/Area_0 = 1$. Hence, $e = Area_0^x$ and the value of y can be solved from the known value of ey^2. The exact value of $Area_0$ will affect the predicted rates of extinction and colonization, but not the J_i-values nor metapopulation size (average number of occupied patches) at steady state.

model to examine how the reduced amount of suitable habitat and the rescue effect, respectively, might affect the criterion for long-term persistence as given by eq. (3). They also used the IFM applied to the Glanville fritillary butterfly to examine the influence of regional stochasticity on MVM, and they studied the response of the metapopulation to a gradual decline in the amount of habitat in the landscape.

Hanski et al. (1996a) found that, provided the metapopulation exists in a stochastic steady state, eq. (3) gives a reasonably good approximation of the number of local populations required for long-term persistence (MVM) in relation to the number of suitable patches (T). However, in cases where there is a time lag in the species' regional decline in response to habitat fragmentation, the extinction rate will be underestimated, inflating the estimate of the equilibrium number of local populations and making eq. (3) unreliable as a MVM criterion. A similar problem occurs if the metapopulation is affected by strong regional stochasticity. In cases with great variation in patch areas, in which case some local populations are large and buffer against metapopulation extinction, Hanski et al. (1996a) found that eq. (3) is liable to give a conservative criterion.

Hanski et al. (1996a) concluded that simulations of metapopulation dynamics with spatially realistic models may help detect non-equilibrium situations. Using the IFM for the Glanville fritillary butterfly, the MVM was of the order of 10 extant populations at any one time in a network of 20 or more suitable patches (MASH), in agreement with eq. (3). Although Hanski et al. (1996a) cautioned against putting too much trust into such simplified models, they found that empirical results from other butterfly studies, reviewed in Thomas and Hanski (1997), were in broad agreement with these MVM and MASH estimates.

Choosing a model

The choice of a PVA model should always be made in collaboration with ecologists knowledgeable about the ecology of the focal species and conservation biologists with some experience in PVA. Akçakaya and Sjögren-Gulve (2000) present guidelines that can be used to choose between occupancy models and structured (demographic) metapopulation models in the PVA planning process. Here, we make some further comments pertaining to the occupancy models.

When is an occupancy model appropriate?

Occupancy models require that the type of habitat that the focal species uses can be readily identified and that the patches are clearly delimited geographically. Because the models ignore local dynamics, they are best applied to systems of relatively small habitat patches and hence to "highly fragmented" landscapes, in which the habitat patches cover only maximally some 20% of the landscape (Hanski 1999). Naturally, the models also require that the presence or absence of the focal species in the patches can be reliably scored. Most occupancy models assume independent dynamics in different patches, but this assumption can be relaxed (e.g., modeling of correlated extinctions in the LRM). Occupancy models primarily apply to large networks of patches with moderate or high population turnover, or to moderate-sized systems with high turnover. They are inappropriate for systems with very few habitat patches and local populations (T < 20). The reasons for these recommendations are that relatively high turnover assures a fast approach to equilibrium (important for the IFM) and more observed turnover (greater statistical power in the LRM). In some applications, occupancy models have been applied to territories of breeding pairs interpreted as habitat patches (Lande 1987).

Occupancy models do not track temporal changes in local population size and they therefore cannot be applied when asking questions about trends in the numbers of individuals within habitat patches. Instead the basic variable that is modeled is the number of occupied patches projected over time, which may be examined under alternative environmental conditions and management regimes. As mentioned in General features, assumptions and predictions, most occupancy models regard patch quality variables (e.g., patch area, local climate, and even presence or absence of patch-associated predators) as static and only update distances to occupied patches (connectivity) in each time step during simulation. It is straightforward to allow for stochastic variation or deterministic changes in patch quality variables during simulations. Nonetheless, one must bear in mind the assumptions about patch quality while designing and interpreting simulations. Predictions should not be made over a period much longer than patch quality is expected to remain static, or the assumption about possible changes in patch quality should be made explicit.

It may be questioned whether species presence/absence is a sufficient measure for PVA assessments. Model choice must depend on the question(s) asked, but the decision is often constrained by limited information and financial resources. Demographically structured models require much data (Akçakaya and Sjögren-Gulve 2000), and measurements of abundance and vital rates are subject to observer bias as well as statistical and temporal sampling errors (Link and Sauer 1998, Droege et al. 1998, Akçakaya 2000). Similar errors apply to occupancy data (i.e. "pseudoturnover") but the effects of such errors are easier to control when only one type of data are needed (see also discussion in Vos et al. 2000). Like Akçakaya and Sjögren-Gulve (2000), we maintain that demographically structured models and occupancy models complement each other. Spatially realistic models may predict species presence-absence (and local extinction and colonization) with an accuracy > 80% that is equal to, and sometimes even better than, that of demographic models (Kindvall 2000; cf. Sjögren-Gulve and Ray [1996] and Brook et al. [2000]). In general, occupancy models are useful for metapopulations in which demographic rates and their temporal and spatial variation (Akçakaya 2000) are hard to quantify. Simulations of parameterized IFM and LRM (or other spatially realistic occupancy models) may help discern individual patches or conservation measures that particularly promote species persistence. Such models that explicitly incorporate habitat structure and quality to predict effects of habitat change on metapopulation size (N) may also allow a more straightforward and reliable viability assessment than structured demographic models for which the effects of habitat change on individual survival, fecundity and patch carrying capacity must first be quantified.

Which kind of occupancy model should I choose ?

If the only data available are reliable observations of species presence or absence in easily delimited habitat patches from one survey, then only the IFM can be used. However, this approach assumes that the metapopulation as a whole is in a stochastic steady state, and hence some additional information should be used to support (or reject) this assumption. One source of such information is repeated surveys of the metapopulation, which also provides a much more solid basis for parameterizing the IFM (Moilanen 1999). The LRM requires that repeated surveys have been carried out, minimally in two different years, and that local extinctions and (re)colonizations have been observed. Multiple surveys are very useful also in allowing tests of model predictions, though a more stringent test would require data for parameter estimation and model testing to be collected from spatially separate parts of the metapopulation (e.g. Hanski and Thomas 1994, Wahlberg et al. 1996, Kindvall 2000).

The Levins (1969) model provides results only at the metapopulation level. Usually, managers and conservation biologists wish to assess the viability of the focal metapopulation under management actions that affect only certain parts (patches) of it. Frequently the question is asked which number of interconnected patches needs to be maintained, protected and/or managed, and in which spatial configuration, to achieve or maintain a viable metapopulation? This sort of question calls for a spatially realistic model, such as the IFM and LRM, or a spatially realistic version of the Levins model (Hanski and Ovaskainen 2000). Some studies have nonetheless employed the original Levins model, and attempted to parameterize it for real cases (see the edible frog example under Selected case studies). We advise against such applications because of the excessive simplicity of the model and because the additional information needed to parameterize the IFM (patch areas and spatial locations) is usually easy to collect.

If data are available to parameterize both the IFM and the LRM, a number of factors need to be taken into account when choosing between these two models. Perhaps the first advice should be to use both models and compare their results. Second, one should ask whether it appears that the IFM assumption of metapopulation stochastic steady state is valid? If not, the LRM may be more appropriate (Kindvall 2000), though one should be aware of the risk of inaccurate predictions by the LRM if it is parameterized with data from a few (or an unrepresentative set of) turnover events (Hanski 1999, p. 97, Moilanen 2000). Third, prior to parameter estimation one has to decide whether to use information in the patch occupancy pattern (can be done only with the IFM) or in the population turnover patterns (can be done with both models), or both (can be done only with the IFM; see Vos et al. 2000). Generally, the more turnover has been observed, the less additional information is contained in the occupancy pattern, which reflects past turnover events. Finally, it is more straightforward to incorporate additional environmental variables apart from patch area and connectivity in the LRM, although this can be done in the IFM by letting them affect the effective area and connectivity of the patches (Moilanen and Hanski 1998, Moilanen 1999, Vos et al. 2000).

Selected case studies

In this section we review some case studies illustrating how the Levins (1969) model, the LRM and the IFM have been applied to empirical data for conservation purposes. These case studies reflect the sorts of problems and questions that have been considered with the different models and how the results have been interpreted.

Declining edible frog populations in southern Sweden (Levins' model)

The edible frog *Rana esculenta* is a hybridogenetic water frog species which is widely distributed in central Europe and which also occurs in southern Scandinavia (Denmark, southern Sweden) (Ebendal 1979). It inhabits permanent waters, primarily ponds, and is larger and more aquatic than its close congener, the pool frog (*Rana lessonae*; Sjögren 1991). The edible frog is vulnerable to predatory fish and benefits from a warm local climate (Kindvall 1996) just like the pool frog (Sjögren-Gulve 1994). In Sweden, it is red-listed as "near threatened" and its northernmost natural populations are found in the county of Scania.

Until 1989, local populations of the edible frog *Rana esculenta* occurred in a geographically distinct area in eastern Scania called Österlen (Ebendal 1979, Kindvall 1996). The frogs were not discovered by scientists until 1973 and their origin is unknown. In contrast to the majority of the Scanian populations in the west, the Österlen populations showed a decline from 1976 onwards (Table 1). A total of 26 ponds suitable for the species occurred in the area. The ponds were made for peat breaking and many of them were created in the mid-1900s (Kindvall 1996). Distances between unoccupied ponds and the closest inhabited pond in this area were all less than the maximum colonization distance that has been empirically observed (2.1 km; Kindvall 1996).

Kindvall (1996) parameterized Levins' (1969) model (eq. 1a) with turnover data for 1973–79 (t = 2 frog generations [3-yr periods]) (Table 1). Excluding one extinction that occurred due to introduction of predatory fish (Table 1), and assuming that all extinction events were correctly observed even though the surveys were not annual, the value of the extinction parameter e in eq. (1a) was roughly calculated as $9/(9 \times 2) = 0.50$ per 3-yr period. Similarly, the colonization parameter (c) became $7 \times 26/(9 \times 17 \times 2) = 0.59$ per 3-yr period. (These calculations are very approximate, because they do not take into account possible extinctions followed by colonizations, and vice versa, during the 3-yr periods, and the numbers of populations and patches are small.) With these values, the equilibrium number of occupied patches (eq. 2) was calculated to be 4, and by eq. (3) this does not satisfy the condition for long-term persistence $(4 < 3\sqrt{26}=15)$. The conclusion of this very rough calculation therefore is that, given the observed extinction rates in 1973–79, the metapopulation was prone to rapid stochastic (though not to deterministic) extinction.

Kindvall (1996) argued that, although a more detailed analysis might be desirable, the above modeling approach would have highlighted the imminent risk of extinction of this metapopulation already in 1979. If the modeling results had been used to initiate a monitoring and habitat management program in 1980, the regional extinction that occurred 10 yr later might have been prevented. Kindvall (1996) discussed alternative management actions that could have been adopted in the early 1980s, and he found that decreasing e (extinction rate) by at least 50% (by en-

Table 1. The number of suitable permanent ponds (lacking predatory fish) in the Österlen region with and without the edible frog (*Rana esculenta*) in 1973–1991, and the observed population turnover (from Kindvall 1996).

Year	Total (T)	Suitable ponds without predatory fish Occupied (N)	Vacant	Observed turnover since previous census Extinctions	Colonizations
1973	27	9	18	–	–
1976	26	12	14	3*	6
1979	26	6	20	7	1
1988	26	2**	24	4	0
1991	26	0	26	2	0

* Introduction of predatory fish in pond no. 27 caused a local extinction.
** In 1987, only three adult males were found in the entire metapopulation. Two of them were observed at yearly censuses until 1989.

61

hancing reproductive output and habitat quality) would have been most efficient. Artificially increasing c (colonization rate) by transplantations would carry the risk of increasing e (of donor patches; Akçakaya 2000, Lacy 2000a); a substantial increase of T (number of habitat patches) would require that dozens of new suitable ponds were created. On the premise that the system met the threshold condition of Levins' (1969) model, Kindvall (1996) suggested that e ≤ 0.25 per 3-yr interval would mean that the equilibrium number of occupied patches could be 15 or more, which according to eq. (3) implies long-term persistence, or at least a significantly reduced extinction risk, for the Österlen metapopulation.

Field voles in the Tvärminne archipelago (Levins' model)

A contrasting example in which Levins' (1969) original model was used to assess metapopulation persistence is due to Ås et al. (1997), who used population turnover data on field voles *Microtus agrestis* collected during 1972–1977 by Pokki (1981) on 71 islands in the Tvärminne archipelago, southwestern Finland. The field vole is a common rodent

in Scandinavia; it is a generalist, produces multiple litters per year averaging 4 – 5 young each, and it can both swim and run on the ice in winter between islands (Ås et al. 1997).

Of the 71 islands (T in eq. 1a), 69 were inhabited by the field vole in at least one year though the average number of occupied islands was only 39. During 1972–1977 (t = 5 years), a total of 81 local extinctions and 84 colonizations were observed, giving rough estimates of e = 0.42 and c = 0.95 in eq. (1a). Based on eq. (2), Ås et al. (1997) found the predicted equilibrium number of occupied islands to be 40, which is close to the observed average (= 39). Using eq. (3) to assess whether or not this metapopulation satisfied Nisbet and Gurney's (1982) criterion for long-term persistence despite the high turnover rate, they obtained the inequality 39>√71=25, which indicates a persistent metapopulation.

The pool frog at the Baltic coast of Sweden (LRM)

Sjögren-Gulve (1994) used multiple logistic regression to analyze the occupancy pattern of the pool frog (*Rana*

Fig. 1. a) Aerial photograph of a pool frog *Rana lessonae* pond in mixed coniferous/deciduous forest close to the Baltic sea of east-central Sweden. (Photo: Göran Hansson/N). b) Typical pool frog pond (Area = 0.36 ha) with warm local climate. c) Basking male pool frog during breeding season. (Photo b+c: Per Sjögren-Gulve.)

lessonae; red-listed as "vulnerable") at ponds in east-central Sweden. Sjögren (1991) and Sjögren-Gulve and Ray (1996) used the same method to examine how various environmental variables correlated with the observed patterns of local extinction and colonization. Using the obtained significant variables, metapopulation dynamics were simulated under different scenarios using eqs (7) and (8). In addition to evaluating the predictive power of the LRM, Sjögren-Gulve and Ray (1996) explored the effect of large-scale forestry on local and regional persistence of the Swedish pool frogs. Large-scale forestry includes clearcuts and/or ditching of areas ≥ 5 ha that evidently inhibit pool frog dispersal (Sjögren-Gulve and Ray 1996).

In the study region, the pool frog typically inhabits permanent relatively warm ponds (Fig. 1) that are situated close to other occupied ponds. Local climate is not spatially autocorrelated, which together with isolation-dependent extinctions suggests a significant rescue effect in the metapopulation dynamics (Sjögren-Gulve 1994).

Sjögren-Gulve and Ray (1996) compared the characteristics of 25 ponds where the pool frog went extinct prior to 1990 with 54 ponds with persisting populations. They found that the probability of local extinction decreased with warmer local climate (high Texp) and increased with pond area (Area; large ponds have predatory fish) and the presence of large-scale forestry in the nearby landscape (Ditch = 1). The logit describing local extinction was given as

$$u_e(i,t) = 26.92 - 2.179\text{Texp}_i + 0.4906\ln \text{Area}_i + 3.553\text{Ditch}_i$$

(16)

and the calculated probabilities of extinction using eqs (16) and (7) for the 79 ponds in 1987 are shown in Fig. 2.

In the analysis of colonization events (9 colonized vs 57 non-colonized ponds), no (re)colonization was observed at ponds with large-scale forestry nearby (Ditch = 1), and the colonization probability of vacant ponds decreased with increasing distance to the closest occupied pond (D_{ij}). The colonization logit was given as

$$u_c(i,t) = -0.001181D_{ij} - 99.83\text{Ditch}_i .$$ (17)

When simulating metapopulation dynamics from 1987 to 1994 (two frog generations) using eqs (16) and (17), the LRM correctly predicted a slight increase in metapopulation occupancy during the 7-yr period.

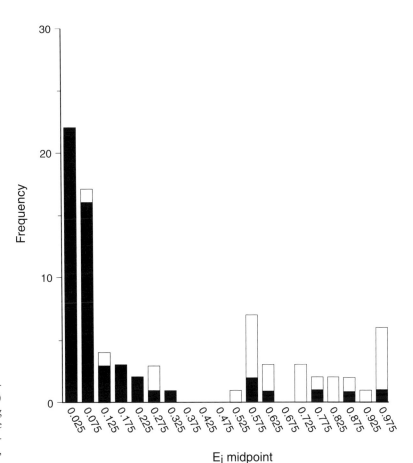

Fig. 2. Distribution of conditional extinction probabilities E_i based on eqs (7) and (16) for sites where the pool frog went extinct (open bars) and sites where it persisted (black bars) in 1987 (modified from Sjögren-Gulve and Ray 1996, © Island Press).

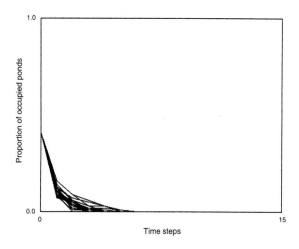

Fig. 3. Fraction of 102 ponds occupied by pool frogs during the course of a subset of 15 simulations × 15 census intervals (53 yr) of a LRM based on eqs (16) and (17) with large-scale forestry being omnipresent in the region (i.e. Ditch = 1 for all ponds) (reprinted from Sjögren-Gulve and Ray 1996, © Island Press).

Analyzing the accuracy of model predictions at the pond level, Sjögren-Gulve and Ray (1996) found that the model correctly classified 82 of the 102 ponds (80.4%) as either occupied or unoccupied in 1994.

Subsequent LRM simulations (Sjögren-Gulve and Ray 1996) explored two management alternatives. In one scenario, large-scale forestry would continue only in the areas affected through 1990. Under these conditions the model predicted that metapopulation occupancy would remain in a stochastic steady state with 44 ponds occupied on average. With T = 102 ponds in eq. (3), 44>√102=30, which indicates a persistent system. However, under an alternative scenario in which large-scale forestry expanded to affect all the 102 ponds in the study area, occupancy declined rapidly and the metapopulation went extinct with a probability of 0.99 within 53 yr (15 census intervals; Fig. 3). These simulation results indicate that large-scale forestry poses a severe threat to the pool frog. Large-scale forestry inhibits dispersal and colonization, even at close distances; setting Ditch = 1 in eq. (17) makes C_i = 0, regardless of close proximity to occupied ponds (see Sjögren-Gulve and Ray [1996] for discussion). Furthermore, presence of such forestry increases local extinction probability profoundly; Fig. 4 shows the distribution of extinction probabilities (E_i,

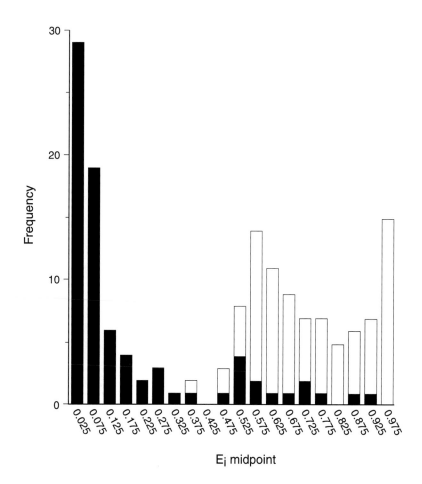

Fig. 4. Distribution of conditional extinction probabilities E_i for pool frog extinction and persistence sites in 1990 based on eqs (7) and (16) under two alternative scenarios where no sites have experienced large-scaled forestry (Ditch = 0 for all ponds; black bars) and all sites are affected (Ditch =1; open bars), respectively.

calculated from eqs 16 and 7) of ponds formerly or currently inhabited by pool frogs under the scenarios Ditch = 0 and Ditch = 1, respectively, for all ponds. The combined effect of this increase and the inhibition of dispersal and rescue effect destabilizes the metapopulation completely.

Recovery of the silver-spotted skipper butterfly in southern Britain (simplified LRM)

The silver-spotted skipper butterfly *Hesperia comma* inhabits heavily grazed calcareous grasslands in Britain. It has a 1-yr life cycle, and eggs are laid on sheltered *Festuca ovina* grass that is 1–5 cm tall and partially surrounded by bare ground. Before the 1950s, the species inhabited most of the grasslands of the southern British chalk hills (Thomas et al. 1986). Due to the spread of myxomatosis in the mid-1950s, there was wide-spread disappearance of rabbits, many of the grasslands became overgrown and the butterfly declined to occupy 46 or fewer sites in ten refuge regions (Thomas and Jones 1993). By 1982, at least 144 former butterfly sites became again suitable for the butterfly due to rabbit recovery and/or habitat restoration. However, by 1991, 109 of them had still not been recolonized.

Thomas and Jones (1993) used logistic regression to analyze the pattern of habitat occupancy, extinction and (re)colonization of the skipper butterfly based on data from two surveys conducted in 1982 and 1991. Using patch-specific colonization and extinction probabilities derived from logistic regression analysis, they examined whether the observed pattern of extinctions and recolonizations during 1982–1991 could explain why the skipper butterfly had recovered only partially, and why many restored habitat patches remained vacant.

Thomas and Jones (1993) found that the occupied patches in 1991 were larger and less isolated from neighboring local populations than the unoccupied patches. Variation in habitat quality was not a significant determinant of patch occupancy, but since patches were identified and delimited based on habitat-quality criteria this was not very surprising. Patches that had been recolonized during 1982–1991 (n = 29) were significantly larger and less isolated from the occupied patches than the 109 patches that remained unoccupied. The resulting logit of recolonization (u_c) for a vacant patch i, of size $Area_i$ ha, at a distance D_{ij} m from the closest occupied patch j, was given as

$$u_c(i,t) = 11.41 + 2.56\log_{10} Area_i - 3.68\log_{10} D_{ij}. \quad (18)$$

This logit expression plugged into eq. (4) explained 69% of the observed variation in recolonization. Turning to local extinctions, the ten sites at which populations went extinct were significantly smaller and more isolated from neighboring populations than the 50 sites where populations persisted. The resulting logit of extinction (u_e) for an occupied patch i was given as

$$u_e(i,t) = -7.67 - 1.52\log_{10} Area_i + 2.71\log_{10} D_{ij}. \quad (19)$$

The extinction model combining eqs (19) and (7) explained 29% of the observed variation in local extinction. The colonization and extinction events of the skipper butterfly in relation to patch area and isolation are shown in Fig. 5.

On the basis of the observed rate of spread (0.73 km yr^{-1} in areas with abundant habitat) Thomas and Jones (1993) predicted that the presently vacant areas would become repopulated within 55 – 70 yr. However, simulations of eqs (18) and (19) for 100 yr from 1991 onwards predicted that further spread of the species would occur only in East Sussex. In the other areas, > 10 km wide zones of unsuitable habitat acted as barriers that prevented recolonization of suitable patches, in agreement with the situation that had already been observed (see also Hanski and Thomas 1994). In summary, this analysis highlighted the importance of both the proximity to extant local populations and large patch area for the skipper butterfly to successfully recolonize restored habitats.

Fig. 5. a) Colonizations and b) extinctions of the butterfly *Hesperia comma* in southern Britain from 1982 to 1991 in relation to patch area and isolation from the nearest populated patch. Solid symbols show patches that were (a) colonized and (b) still populated in 1991. Open symbols show patches that (a) remained vacant in 1991 and (b) became extinct. Fitted lines give 90%, 50% and 10% probabilities of (a) colonization and (b) extinction based on eqs (18) and (19) respectively (reprinted from Thomas and Jones 1993, © Blackwell Publ.).

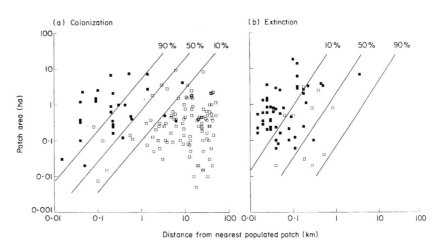

The Glanville fritillary in the Åland Islands, Finland (IFM)

The Glanville fritillary *Melitaea cinxia* (Fig. 6) is another butterfly species that has suffered from habitat loss and fragmentation in northern Europe over the past decades (Hanski and Kuussaari 1995). In Finland, the Glanville fritillary went extinct on the mainland in the 1970s and it now survives only in the Åland Islands in the northern Baltic. In the Åland, there is still a very large network of suitable habitat patches, dry meadows with the larval host plants (Fig. 6). Of ca 1700 patches surveyed in 1993–99, 20–30% have been occupied in any one year, with a high rate of population turnover, a reflection of the very small size of most local populations (Hanski et al. 1996b, Hanski 1999).

The IFM was parameterized for a metapopulation of the Glanville fritillary using data collected in two years from a network of 50 patches (Hanski 1994), a small subset of the entire study area. Hanski et al. (1996b) tested the IFM by using the parameter values estimated from this subset to predict the pattern of occupancy across all patches in the Åland Islands (Fig. 7). Predictions are generally expected to be worse for networks with a small number of patches, because the dynamics in such networks are greatly influenced by extinction-colonization stochasticity (Hanski 1991, 1999). For most networks with >15 patches, observed and predicted occupancy were in good agreement (Hanski et al. 1996b). Predictions failed in one region, which may have been far from equilibrium occupancy (Hanski et al. 1996b, Moilanen and Hanski 1998).

Wahlberg et al. (1996) applied the IFM parameterized for the Glanville fritillary to another related checkerspot butterfly, the false heath fritillary *Melitaea diamina*, which is highly endangered in Finland. *Melitaea diamina* has only two remaining metapopulations in Finland, of which one is located in one small area in south-central Finland. The caterpillars feed exclusively on *Valeriana sambucifolia* and the butterflies occur only on the moist meadows where this plant species grows (Wahlberg 1997). Using the previously estimated parameter values for the Glanville fritillary, as well as data on the patch areas and spatial locations in the *M. diamina* network, Wahlberg et al. (1996) used the IFM to predict the pattern of patch occupancy in the latter species. This prediction was subsequently tested with empirical data on the actual pattern of patch occupancy in the *M. diamina* metapopulation. The match between the IFM-predicted incidences (long-term probability of patch occupancy) and the observed snapshot of patch occupancy was good (Fig. 8), suggesting that if the opportunity exists it may be worthwhile to consider using data for a common related species to parameterize a model for an endangered species. However, this approach must be used with great caution, as it is generally impossible to rule out differences in the biology of the two species that would affect their metapopulation dynamics.

The tree frog in the Netherlands (IFM)

Vos et al. (2000) studied the tree frog (*Hyla arborea*; redlisted as "endangered") within an area of 250 km^2 in the western part of the Zealand Flanders in the Netherlands. In this agricultural landscape, the tree frogs inhabit habitat patches of dissimilar sizes and shapes including cattle drinking ponds and their surroundings with shrubs, bushes and vegetation of high herbs. The suitable habitat covers ca 1.5% of the total landscape. Vos et al. (2000) studied a network of ca 500 patches (ponds), of which ca 10% were occupied in any one year. They examined whether empirical evidence for landscape effects (habitat patch area and connectivity) on tree frog extinctions and colonizations could be found, whether dispersal was observed indicating patch connectivity and a functioning metapopulation, and whether an IFM could successfully model future patch occupancy.

Tree frogs were surveyed in three successive years 1981–83 and in 1986. For the occupancy analysis of 1981, 187 ponds were chosen for which good data on habitat quality were available (Vos and Stumpel 1996). Individual frogs dispersed up to 12.6 km (mean = 1.5 km) with a preference for occupied target ponds. During the period 1981–1983, 10 local populations went extinct. Extinction probability increased with decreasing habitat patch area but also with increasing water conductivity. A total of 16 colonization events were observed. Colonization probability increased with proximity to other occupied patches and patch size. Pond vegetation cover and year also affected colonizations.

To model the tree frog metapopulation, Vos et al. (2000) combined information from the spatial pattern of patch occupancy and from the pattern of extinctions and colonizations. The data of 1981–83 were used to parameterize an "extended IFM", which incorporated also the effects of water conductivity and vegetation cover. This model was then used to predict the pattern of patch occupancy in 1986. The model predicted well the pattern of occupancy at the network level (59% of the variation) and at the pond level (87% of the ponds correctly classified as vacant or occupied). Although water conductivity was not as strong as patch area in explaining local extinctions, vegetation cover strongly affected pond colonization emphasizing the need for an extended IFM.

The method used by Vos et al. (2000) allowed them to use information in both the spatial pattern of patch occupancy and the observed turnover events. Their study showed that occupancy modeling and a metapopulation approach are useful for tree frog conservation in the Netherlands, emphasizing the importance of both habitat patch quality and landscape connectivity for population persistence. They used binary non-linear regression in parameter estimation as originally suggested by Hanski (1994). Because spatial and temporal correlations in patch occupancy can cause spurious results, future applications of this approach should follow the method of Moilanen (1999).

Fig. 6. a) Dry meadow representing suitable habitat for the Glanville fritillary *Melitaea cinxia*. b) Glanville fritillary. (Photo: Taipo Gustafsson).

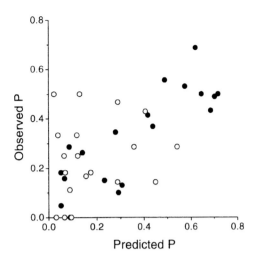

Fig. 7. Comparison of the predicted and observed fractions of patches (P) occupied by the Glanville fritillary *Melitaea cinxia* in 4 × 4 km² areas in western Åland. Black dots represent squares with > 15 habitat patches, open symbols are for squares with fewer patches. Statistical tests and further details in Hanski et al. (1996b).

The bush cricket *Metrioptera bicolor* in southern Sweden (IFM, LRM, RAMAS/GIS)

Kindvall (2000) used metapopulation occupancy data from 1989 to 1994, supplemented with demographic census data collected during the same time period, to model the dynamics of a metapopulation of the bush cricket *Metrioptera bicolor*. This species is red-listed as "vulnerable"

in Sweden, and the study metapopulation was modeled using the IFM, the LRM and the RAMAS/GIS program (see Akçakaya 2000). The habitat patch network inhabited by the metapopulation consists of a western and an eastern half, separated by distances ≥ 550 m, which exceed the maximum dispersal distance (350 m) of the bush cricket observed during Kindvall's study. The western half consisted of 66 habitat patches with an average area of 6.2 ha, and the eastern half of 50 patches with an average area of 1.6 ha. Although the western patches were on average larger than the eastern ones, inter-patch distances were similar (median = 30 m). Kindvall (2000) used data from one half of the metapopulation to parameterize the models, and he tested model predictions using data from the other half.

In both halves of the patch network, probability of local extinction was negatively correlated with increasing patch area and habitat heterogeneity (of which the IFM was parameterized using area only), and colonization was negatively correlated with distance to the nearest occupied patch. The model predictions that were tested included the proportion of occupied patches (P) in each half, the numbers of colonizations and extinctions, and the proportion of time each patch was occupied during 1990-1994 (p_i). The spatial configuration of five patches changed during 1990-94; these changes were modeled as having occurred prior to 1989 in the simulations.

Kindvall (2000) found that, in general, the predictions of the simpler models (IFM and LRM) were not significantly less accurate than those of the demographic RAMAS/GIS model. Only in two out of 18 comparisons was an observed value outside the range of values predicted by the models: the patch occupancy of the eastern half was higher

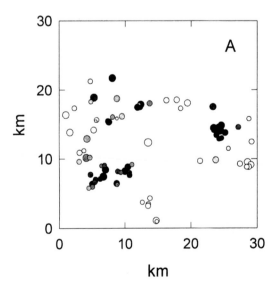

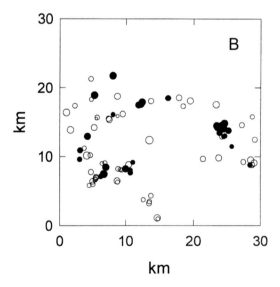

Fig. 8. A) A map of the patch network occupied by the butterfly *Melitaea diamina* showing the relative patch sizes, their spatial locations and the predicted incidence of occupancy based on parameter values estimated for the congeneric Glanville fritillary *Melitaea cinxia*. Higher incidences are shown by darker shading. B) A snapshot of patch occupancy in 1995 (based on Wahlberg et al. 1996).

than predicted by the IFM, and the number of local extinctions observed in the eastern half exceeded that predicted by RAMAS/GIS. Among the models, the IFM on average made the best predictions of number of local extinctions in the eastern half, the RAMAS/GIS predictions were best regarding P of the western half and p_i in the eastern half, and the LRM predictions were best regarding P of the eastern half, p_i in the western half, number of colonizations in both metapopulation halves, and number of extinctions in the western half. Kindvall (2000) concluded that all three types of models are worth considering for the PVA of the bush cricket. He emphasized the great potential of more advanced occupancy modeling packages for PVA purposes that can include multiple predictive variables and their temporal trends and variation.

Conclusions

"The most obvious question a metapopulation biologist may expect to be asked is whether some species X is likely to persist, as a metapopulation, in some particular set of habitat patches Y. In the context of conservation biology, the set of patches Y is often a subset of some larger number of larger patches, and the ecologist is asked to predict whether species X, present in the current patch network, would still persist if some patches were removed or their areas were reduced" (Hanski 1994). We have attempted to show in this paper that in the case of highly fragmented habitats one may successfully use patch occupancy models such as the IFM and the LRM to make such predictions. Both models are spatially realistic and they can be extended to deal with non-spatial patch quality variables influencing patch occupancy and persistence (Sjögren-Gulve and Ray 1996, Moilanen 1999, Hanski 1999 section 12.1).

We add two important remarks to this conclusion. First, the patch occupancy models are most appropriate for metapopulations with high or relatively high population turnover rate and inhabiting reasonably large networks of generally small habitat patches. Because extinctions and colonizations are modeled as explicit functions of environmental variables, which can be subject to management, the IFM and LRM constitute complementary, and in many cases preferable (e.g. Kindvall 2000) PVA approaches to the traditional demographic models. However, it is clear that not all species of conservation concern satisfy the assumptions of patch occupancy models. Exactly when it is appropriate to use these models must be decided case by case; we have given some guidelines in the section Choosing a model, and others are given by Akçakaya and Sjögren-Gulve (2000). Second, we echo the point made by many others (Hanski 1997b, Ralls and Taylor 1997, Beissinger and Westphal 1998, Akçakaya and Sjögren-Gulve 2000) that the most useful application of any PVA model is in comparisons of alternative landscape scenarios and management actions rather than in making predictions about the absolute risk of extinction of particular metapopulations. To estimate absolute risk, we would need a very large amount of data to parameterize the model and accurate information about how the landscape will change in the future. Especially the latter is hardly ever available. In contrast, while making comparisons it is reasonable to assume that our ranking of different scenarios is robust to many uncertainties that will affect the prediction of the absolute risk of extinction. We believe that it is in this comparative context that occupancy models can make a constructive contribution to landscape-level planning and conservation.

Acknowledgements – We thank Susan Harrison, Linda Hedlund, Oskar Kindvall and Chris Ray for valuable comments on a draft of this paper. IH acknowledges the support for project 44887 (Finnish Centre of Excellence Programme, 2000–2005).

References

Akçakaya, H. R. 2000. Population viability analyses with demographically and spatially structured models. – Ecol. Bull. 48: 23–38.

Akçakaya, H. R. and Sjögren-Gulve, P. 2000. Population viability analyses in conservation planning: an overview. – Ecol. Bull. 48: 9–21.

Ås, S., Bengtsson, J. and Ebenhard, T. 1997. Archipelagoes and theories of insularity. – Ecol. Bull. 46: 88–116.

Beissinger, S. R. and Westphal, M. I. 1998. On the use of demographic models of population viability in endangered species management. – J. Wildl. Manage. 62: 821–841.

Brook, B. W. et al. 2000. Predictive accuracy of population viability analysis in conservation biology. – Nature 404: 385–387.

Burgman, M. A., Ferson, S. and Akçakaya, H. R. 1993. Risk assessment in conservation biology. – Chapman and Hall.

Droege, S., Cyr, A. and Larivée, J. 1998. Checklists: an underused tool for the inventory and monitoring of plants and animals. – Conserv. Biol. 12: 1134–1138.

Ebendal, T. 1979. Distribution, morphology and taxonomy of the Swedish green frogs (*Rana esculenta* complex). – Mitt. Zool. Mus. Berlin 55: 143–152.

Ebenhard, T. 1991. Colonization in metapopulations: a review of theory and observations. – Biol. J. Linn. Soc. 42: 105–121.

Engelman, L. 1990. Stepwise logistic regression. – In: Dixon, W. J. (ed.), BMDP statistical software manual, Vol. 2. Univ. of California Press, pp. 1013–1046.

Fleishman, E., Jonsson, B. G. and Sjögren-Gulve, P. 2000. Focal species modeling for biodiversity conservation. – Ecol. Bull. 48: 85–99.

Gotelli, N. J. and Kelley, W. G. 1993. A general model of metapopulation dynamics. – Oikos 68: 36–44.

Hanski, I. 1991. Single-species metapopulation dynamics: concepts, models and observations. – Biol. J. Linn. Soc. 42: 17–38.

Hanski, I. 1994. A practical model of metapopulation dynamics. – J. Anim. Ecol. 63: 151–162.

Hanski, I. 1997a. Predictive and practical metapopulation models: the incidence function approach. – In: Tilman, D. and Kareiva, P. (eds), Spatial ecology. Princeton Univ. Press, pp. 21–45.

Hanski, I. 1997b. Habitat destruction and metapopulation dynamics. – In: Pickett, S. T. A. et al. (eds), The ecological basis of conservation: heterogeneity, ecosystems and biodiversity. Chapman and Hall, pp. 217–227.

Hanski, I. 1999. Metapopulation ecology. – Oxford Univ. Press.

Hanski, I. and Gyllenberg, M. 1993. Two general metapopulation models and the core-satellite species hypothesis. – Am. Nat. 142: 17–41.

Hanski, I. and Thomas, C. D. 1994. Metapopulation dynamics and conservation: a spatially explicit model applied to butterflies. – Biol. Conserv. 68: 167–180.

Hanski, I. and Kuussaari, M. 1995. Butterfly metapopulation dynamics. – In: Cappucino, N. and Price, P. W. (eds), Population dynamics: new approaches and synthesis. Academic Press, pp. 149–171.

Hanski, I. and Gilpin, M. E. (eds) 1997. Metapopulation biology; ecology, genetics and evolution. – Academic Press.

Hanski, I. and Simberloff, D. 1997. The metapopulation approach, its history, conceptual domain, and applicability to conservation. – In: Hanski, I. and Gilpin, M. E. (eds), Metapopulation biology – ecology, genetics and evolution. Academic Press, pp. 5–26.

Hanski, I. and Ovaskainen, O. 2000. The metapopulation capacity of a fragmented landscape. – Nature 404: 755–758.

Hanski, I., Moilanen, A. and Gyllenberg, M. 1996a. Minimum viable metapopulation size. – Am. Nat. 147: 527–541.

Hanski, I. et al. 1996b. The quantitative incidence function model and persistence of an endangered butterfly metapopulation. – Conserv. Biol. 10: 578–590.

Harrison, S. 1994. Metapopulations and conservation. – In: Edwards, P. J., Webb, N. R. and May R. M. (eds), Large-scale ecology and conservation biology. Blackwell, pp. 111–128.

Hosmer Jr, D. W. and Lemeshow, S. 1989. Applied logistic regression. – Wiley.

Kindvall, O. 1996. En naturvårdsbiologisk betraktelse över de österlenska gröngrodornas undergång. – In: Gärdenfors, U. and Carlson, A. (eds), Med huvudet före – Festskrift till Ingemar Ahléns 60-årsdag. Rapport 33, Dept of Wildlife Ecology, Swedish Univ. of Agricult. Sci., Uppsala, pp. 69–80, in Swedish.

Kindvall, O. 2000. Comparative precision of three spatially realistic simulation models of metapopulation dynamics. – Ecol. Bull. 48: 101–110.

Kindvall, O. and Ahlén, I. 1992. Geometrical factors and metapopulation dynamics of the bush cricket, *Metrioptera bicolor* Philippi (Orthoptera: Tettigoniidae). – Conserv. Biol. 6: 520–529.

Lacy, R. C. 2000a. Considering threats to the viability of small populations using individual-based models. – Ecol. Bull. 48: 39–51.

Lacy, R. C. 2000b. Structure of the VORTEX simulation model for population viability analysis. – Ecol. Bull. 48: 191–203.

Lande, R. 1987. Extinction thresholds in demographic models of territorial populations. – Am. Nat. 130: 624–635.

Levins, R. 1969. Some demographic and genetic consequences of environmental heterogeneity for biological control. – Bull. Entomol. Soc. Am. 15: 237–240.

Levins, R. 1970. Extinction. – In: Gerstenhaber, M. (ed.), Some mathematical problems in biology. Am. Mathemat. Soc., Providence, RI, pp. 75–107.

Link, W. A. and Sauer, J. R. 1998. Estimating population change from count data: application to the North American Breeding Bird Survey. – Ecol. Appl. 8: 258–268.

McCullough, D. R. (ed.) 1996. Metapopulations and wildlife conservation. – Island Press.

Menges, E. S. 2000. Applications of population viability analyses in plant conservation. – Ecol. Bull. 48: 73–84.

Moilanen, A. 1999. Patch occupancy models of metapopulation dynamics: efficient parameter estimation using implicit statistical inference. – Ecology 80: 1031–1043.

Moilanen, A. 2000. The equilibrium assumption in estimating the parameters of metapopulation models. – J. Anim. Ecol. 69: 143–153.

Moilanen, A. and Hanski, I. 1998. Metapopulation dynamics: effects of habitat quality and landscape structure. – Ecology 79: 2503–2515.

Nisbet, R. M. and Gurney, W. S. C. 1982. Modelling fluctuating populations. – Wiley.

Pokki, J. 1981. Distribution, demography and dispersal of the field vole, *Microtus agrestis* (L.), in the Tvärminne archipelago, Finland. – Acta Zool. Fenn. 164: 1–48.

Ralls, K. and Taylor, B. L. 1997. How viable is population viability analysis? – In: Pickett, S. T. A. et al. (eds), The ecological basis of conservation: heterogeneity, ecosystems and biodiversity. Chapman and Hall, pp. 228–235.

Ray, C., Sjögren-Gulve, P. and Gilpin, M. E. 1999. METAPOP, ver. 3.3. – Unpubl. simulation software.

Scheiner, S. M. and Rey-Benayas, J. M. 1997. Placing empirical limits on metapopulation models for terrestrial plants. – Evol. Ecol. 11: 275–288.

Sjögren, P. 1991. Extinction and isolation gradients in metapopulations: the case of the pool frog (*Rana lessonae*). – Biol. J. Linn. Soc. 42: 135–147.

Sjögren-Gulve, P. 1994. Distribution and extinction patterns in a northern metapopulation of the pool frog, *Rana lessonae*. – Ecology 75: 1357–1367.

Sjögren-Gulve, P. and Ray, C. 1996. Using logistic regression to model metapopulation dynamics: large-scale forestry extirpates the pool frog. – In: McCullough, D. R. (ed.), Metapopulations and wildlife conservation. Island Press, pp. 111–137.

Solbreck, C. 1991. Unusual weather and insect population dynamics: *Lygaeus equestris* during an extinction and recovery period. – Oikos 60: 343–350.

Ter Braak, C. J. F., Hanski, I. A. and Verboom, J. 1998. The incidence function approach to modeling of metapopulation dynamics. – In: Bascompte, J. and Solé, R. V. (eds), Modeling spatiotemporal dynamics in ecology. Springer, pp. 167–188.

Thomas, C. D. and Jones, T. M. 1993. Partial recovery of a skipper butterfly (*Hesperia comma*) from population refuges: lessons for conservation in a fragmented landscape. – J. Anim. Ecol. 62: 472–481.

Thomas, C. D. and Hanski, I. 1997. Butterfly metapopulations. – In: Hanski, I. and Gilpin, M. E. (eds), Metapopulation biology – ecology, genetics and evolution. Academic Press, pp. 359–386.

Thomas, J. A. et al. 1986. Ecology and declining status of the silver-spotted skipper butterfly (*Hesperia comma*) in Britain. – J. Appl. Ecol. 23: 365–380.

Vos, C. C. and Stumpel, A. H. P. 1996. Comparison of habitat-isolation parameters in relation to fragmented distribution patterns in the tree frog (*Hyla arborea*). – Landscape Ecol. 11: 203–214.

Vos, C. C., Ter Braak, C. J. F. and Nieuwenhuizen, N. 2000. Incidence function modelling and conservation of the tree frog *Hyla arborea* in the Netherlands. – Ecol. Bull. 48: 165–180.

Wahlberg, N. 1997. The life history and ecology of *Melitaea diamina* (Nymphalidae) in Finland. – Nota Lepid. 20: 70–81.

Wahlberg, N., Moilanen, A. and Hanski, I. 1996. Predicting the occurrence of endangered species in fragmented landscapes. – Science 273: 1536–1538.

Ecological Bulletins 48: 73–84. Copenhagen 2000

Applications of population viability analyses in plant conservation

Eric S. Menges

Menges, E. S. 2000. Applications of population viability analyses in plant conservation. – Ecol. Bull. 48: 73–84.

A review of population viability analyses (PVAs) for plants shows that studies are usually of short-duration and consider few populations and species, limiting their applicability in conservation. Some aspects of plant life history require extensive data or special approaches in modeling PVAs. Plant dormancy, seed dormancy, periodic recruitment and flowering, and clonal growth require additional data, experiments, or alternative modeling approaches. PVAs have been applied to conservation issues such as the effects of grazing, pollinator limitation of exotics, and minimum viable population sizes. Models incorporating disturbance, which can often be manipulated by land managers, can be used to pinpoint disturbance regimes that increase population viability. These approaches have been applied to trampling, grazing, fire, and sod-cutting disturbances using a variety of approaches. Metapopulation modeling, including the use of incidence function models, and other kinds of spatially explicit modeling can suggest appropriate management tactics, but have been little-used for plants. The most realistic and conservative uses of PVA for plants involve comparative analyses of various management regimes without misleadingly precise projections of exact outcomes. Continued feedback between management and PVAs will benefit both endeavors.

Eric S. Menges (emenges@archbold-station.org), Archbold Biological Station, P.O. Box 2057, Lake Placid, FL 33862, USA.

Conservation and management of endangered species often must proceed with incomplete knowledge of their requirements. However, as data accumulate, they can be organized into population viability analyses (PVAs). PVAs usually consist of demographic, environmental, and genetic data together with simulation models, and can provide assessments of population persistence (or extinction risk). While PVAs are anchored in basic science, they have been influential in conservation and management (Beissinger and Westphal 1998).

Plant studies are not prominent in most recent reviews of population viability analysis. For example, Beissinger and Westphal (1998) concentrated on animal studies with only 2 of 136 references focusing on plants. They identified some key problems with PVAs (e.g. measuring survival rates) that are secondary for plant studies. Boyce (1992) cited 166 papers in his review, but none were PVA studies of plants. Groom and Pascual (1997) reviewed PVA studies from four journals, but only 3 of 58 studies that they found involved plants. This paper updates a recent review of plant PVAs (Menges 2000) and discusses how plant PVAs can be useful in conservation.

Review of PVA studies for plants

Methods

I considered a study as a PVA if it included empirical data on the entire life cycle of a wild population, and if it projected future populations (e.g. the finite rate of increase (λ), extinction probability, time to extinction, or future population size or structure). Literature searches (through 1998) were conducted using UnCover, Current Contents, and the author's files which have been updated continously since the mid-1970s. The review coverage is extensive, but not complete. Many studies that meet the criteria for a PVA (as outlined above) do not use keywords that lead searches to them, but were identified from direct readings. Most studies are not self-referenced as population viability studies, have disparate goals and approaches, and did not have a major slant toward conservation biology or habitat management. I did not include studies without projected futures, although many detailed demographic studies included enough life history detail so that projection matrices were later derived for meta-analyses (e.g. Silvertown et al. 1996). For statistics on PVAs, I included only the most recent of multiple, overlapping articles on the same species at the same site (this occurred four times).

Statistics and characteristics of PVAs

Altogether, I identified 99 population viability analyses (Table 1). Only five studies referred to themselves as PVAs. Plant PVAs have become more common in recent years (Fig. 1). Most of the PVAs in this review were found in major ecological journals such as The Journal of Ecology (24%), Ecology (11%), Conservation Biology (8%) and the Australian Journal of Ecology (7%). In recent years (1995–98), The Journal of Ecology and Conservation Biology have published many plant PVAs (23% and 16% respectively). PVAs have been published based on species from many plant families and ecosystems, and from every continent except Antarctica. However, most studies have taken place in North America and western Europe.

Most PVAs are based on demographically structured models, in which individuals are grouped into classes (Akçakaya 2000). Unlike animal studies (e.g., reviewed in Groom and Pascual 1997), most plant PVAs are based on stage or size classified matrices (80%), rather than on populations classified by age (8%), or by both age and stage (4%). Some studies have contrasted projections based on populations classified by stage and age independently (e.g. Werner and Caswell 1977, Enright and Watson 1992). With the methods of Cochrane and Ellner (1992) to estimate age-based parameters from stage-based matrices, more plant studies will integrate stage and age approaches (Boucher and Mallona 1997, Morris and Doak 1998). The breakpoints for size classes are usually made subjec-

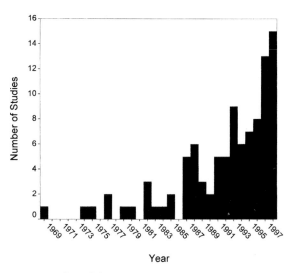

Fig. 1. Number of plant PVA articles by year.

tively or on the basis of minimum reproductive size, although there is some use of the algorithms of van der Meer (1978) and Moloney (1986) to minimize the sampling and distribution errors.

Most plant PVA studies are of short duration, with a mean of 5 yr, and median and mode of 4 yr each (similar to typical dissertation and grant study periods). Only four studies in this review were based on more than ten years of field data. Such short periods of study may not result in PVAs that capture the essence of a species' population dynamics (Fiedler et al.1998), especially demographic and environmental variation. Even qualitative predictions may be suspect. Bierzychudek (1999) returned 15 yr later to field populations in order to check predictions based on three years of data. In one of two populations, the projected population increase was contradicted by an actual decrease over the intervening years. The ability of projections based on short-term studies to adequately anticipate longer-term results needs more evaluation in species with a range of life histories and ecological contexts.

Most plant PVA studies consider a single species (83%). Fiedler (1987) was among the most ambitious in this regard, comparing population viability among four species of *Calochortus* lilies. O'Connor (1993) worked on 6 species of African savanna grasses in relation to grazing and climate; finding that population growth rates among species were correlated because of the positive effects of rainfall. Comparative approaches can provide context that may aid in conservation. For example, Byers and Meagher (1997) showed that while short-term population viability was favorable in the rare *Eupatorium resinosum* relative to its more common congener, lower seed production meant that colonization of new habitats was less frequent. One might use this information as an argument to manage for higher seed production or for establishment of new populations from seeds.

Most studies also consider just a few populations (mean 4.1, median 3, mode 1; only 8% consider > 10 populations). Since populations vary widely in demographic parameters, studies based on only a few populations would seem incomplete at best. Demographic parameters often follow different temporal patterns among multiple populations. Therefore, trends from only a few populations may not be generalizable to unstudied populations. No organized analyses have explored how many populations should be studied, but probably practical limitations will keep that number low in most future studies.

Most assessments of plant PVA used deterministic approaches, and 84% calculated λ (the finite rate of increase), a measure of the population's growth rate and an indicator of population viability. In contrast, few studies calculated a stochastic growth rate (Nakaoka 1996). Population structure was usually modeled (54%), but population size alone (without structure) was projected in 21% of studies. Many other deterministic parameters were calculated, with elasticities (57%) being among the most common. Elasticities within species (Akçakaya 2000) vary across space and time (e.g. Horvitz and Schemske 1995, Silvertown et al. 1996, Oostermeijer et al. 1996).

In contrast to animal PVAs, fewer plant studies (11%) modeled density dependence (e.g. de Kroon et al. 1987, Solbrig et al. 1988, Alvarez-Buylla 1994, Crone and Gehring 1998). In many of the endangered species for which PVAs are needed, density dependence is unlikely or will be difficult to parameterize. Similarly, few plant studies included demographic stochasticity (3%), disturbances or catastrophes (16%), genetics (3%), spatial models (5%) or metapopulation dynamics (3%).

Environmental stochasticity was used to predict extinction in 22% of the studies surveyed. The time periods over which extinction risk was modeled varied from 25 to 1000 yr, with 50, 100, and 200 yr being most frequently used. Full distributions of times to extinction are recommended as less misleading than single results (Beissinger and Westphal 1998). The probability of population decline has also been recommended in comparing management scenarios (Akçakaya and Sjögren-Gulve 2000).

Challenges to plant PVAs

Plant dormancy

Many perennial plants with underground storage organs (e.g. geophytes) can remain alive, but dormant and missing aboveground, for several years. This prolonged dormancy presents a difficult challenge for plant population studies (Lesica and Steele 1994). It is a logistical challenge to demonstrate plant dormancy and keep track of dormant individuals, since some "missing" plants are likely even in species without plant dormancy. Short-term studies are likely to inflate mortality estimates if dormant plants at the

beginning and end of the study are not considered (Lesica and Steele 1994). Deleting or adjusting mortality data from the end of the study may be required. For species with plant dormancy, only longer-term studies can be expected to provide reliable statistics on mortality and plant dormancy.

Modeling plant dormancy can utilize stage classes for dormant individuals, perhaps divided by size or duration. Dormant orchids were studied by Waite and Hutchings (1991), who divided plants into first and second year dormant classes. The breaking of dormancy in this species increased with a shift in the grazing regime during the study. Guerrant (1996) modeled dormant plants broken down by the size class when they were last seen. Menges and Quintana-Ascencio (1998) divided dormant plants into several classes based on pre-dormancy size (which affected subsequent transitions) and length of dormancy, when modeling a prairie clover based on eleven years of data. The approach chosen should depend on whether pre-dormancy size and duration affect subsequent growth, mortality, and fecundity. However, such statistics will require both long-term studies and a large sample size of plants becoming dormant during the study.

Seed dormancy

Persistent seed banks are reserves of viable but ungerminated seeds that remain dormant until the second (or later) germination season. Seed banks may buffer environmental variation and reduce extinction risk (Kalisz and McPeek 1992) and conserve genetic variation. Because seed banks must be studied experimentally to obtain transitions for modeling, data on seed dormancy and seed banks are fragmentary in many species. Seed bank data may be obtained using experiments with seeds in soil microcosms or mesh envelopes, with seeds recovered and examined for survival at annual intervals (Kalisz and McPeek 1992, Quintana-Ascencio 1997). Even with seed bank data from only two years, it is possible to use simulations to examine what proportion of "good" years are required for a viable population (Kalisz and McPeek 1992). Even if seed bank transitions make minor contributions to a population's finite rates of increase, they may add to persistence time of populations until a favorable disturbance event makes conditions once again (Quintana-Ascencio 1997).

Periodic recruitment and flowering

Episodic seedling recruitment is common for many species. Modeling seedling recruitment rates as constants, as implied by the calculation of finite rates of increase, does not capture this variation. Stochastic modeling can be useful as a more realistic approximation to naturally varying seedling recruitment or fecundity. Menges and Dolan

Table 1. Information on plant PVAs included in this review. Y = year of publication (1900s or 2000); y = number of years of study; S = number of species; P = number of populations; M = type of model (S = stage; I = Size; A = Age; B = both age and stage; C = Cell; G = gap (usually with stage also); N = individual; U = unstructured); D = density-dependence; E = environmental stochasticity; d = demographic stochasticity; C = catastrophes or disturbance; G = genetics; m = metapopulation; s = spatial; V = minimum viable population; e = extinction modeling (0 = none; 1 = extinction probability; 2 = time to extinction; 3 = -both); t = time for extinction; λ = finite rate of increase; p = population characters (1 = size; 2 = structure); L = elasticities. Under Comments: CI = confidence intervals for λ.

Authors	Y	y	S	P	M	D	E	d	C	G	m	s	V	e	t	λ	p	L	Comments
Abe et al.	98	8	1	1	S	0	0	0	0	0	0	0	0	0	.		1	2	1	Shade vs gap populations
Aberg	96	4	1	2	I	0	1	0	1	0	0	0	0	3	300	0	2	0	Seaweed, ice effects
Allphin and Harper	97	5	1	6	I	0	0	0	0	0	0	0	0	0	.	1	0	0	Life table approach
Alvarez-Buylla	94	8	1	1	G	1	0	0	0	0	1	0	0	0	.	1	2	1	CI, 4 models
Aplet et al.	94	2	1	1	I	0	0	0	0	0	0	0	0	0	.	1	1	0	Risk scenarios
Batista et al.	98	14	1	1	I	0	0	0	0	0	0	0	0	0	.	1	2	1	Hurricane effects
Bengtsson	93	7	1	3	S	0	0	0	0	0	0	0	0	0	.	1	0	1	Temporal, site variation
Bernal	98	2	1	1	S	0	0	0	0	0	0	0	0	0	.	1	2	0	Harvest, loglinear
Bierzychudek	82	3	1	2	S	0	1	0	0	0	0	0	0	0	32	1	2	0	Environmental stochasticity
Boeken and Canham	95	2	1	9	S	0	0	0	0	0	0	0	0	0	.	1	2	1	Shrub stem demography
Boucher and Mallona	97	5	1	1	S	0	0	0	0	0	0	0	0	0	.	1	2	1	Hurricane, Cochrane meth.
Bradstock and O'Connell	88	4	2	7	U	0	0	0	0	0	0	0	0	0	.	0	1	0	Fire return interval
Bradstock et al.	96	.	1	.	C	0	0	0	1	0	0	1	0	1	200	0	0	0	Fire frequencies
Bradstock et al.	98	.	3	.	C	0	0	0	1	0	0	1	0	1	200	0	1	0	Fire strategies, spatial anal.
Bullock	80	2	1	1	S	0	0	0	0	0	0	0	0	0	.	1	0	0	Sensitivity analysis
Bullock et al.	94	7	1	16	S	0	0	0	0	0	0	0	0	0	.	1	0	1	Strategy, grazing
Burns and Ogden	85	2	1	1	S	0	0	0	0	0	0	0	0	0		1	2	0	Verify with aerials
Burgman and Lamont	92	.	1	1	A	0	1	1	1	1	0	0	0	1	50	0	1	0	Fire, drought
Byers and Meagher	97	2	2	3	S	0	0	0	0	0	0	0	0	0	.	1	1	1	Compare rare, common spp.
Calvo	93	3	1	1	I	0	0	0	0	0	0	0	0	0	.	1	0	0	Pollination
Canales et al.	94	1	1	2	A	1	0	0	0	0	0	0	0	0	.	1	2	0	Fire, annual
Carlsson and Callaghan	91	4	1	1	B	0	1	0	0	0	0	0	0	0	25	1	1	1	Tiller demography
Chapman et al.	89	10	1	4	A	0	1	0	0	0	0	0	0	0	.	0	1	0	Cyclic fire
Cipollini et al.	93	14	1	3	G	0	0	0	1	0	0	0	0	0		1	2	0	Canopy, patch dynamics
Cipollini et al.	94	2	1	1	I	0	0	0	1	0	1	1	0	0	.	1	2	1	Canopy, gender
Cochrane and Ellner	92	4	1	1	S	0	0	0	0	0	0	0	0	0	.	1	2	0	Age-related parameters
Crone and Gehring	98	6	1	1	U	1	1	0	0	0	1	1	0	2	100	1	1	0	Environmental correlations
Damman and Cain	98	7	1	2	S	0	1	1	0	0	0	1	3		100	1	2	1	Ramets/genets, MVP
DeKroon et al.	87	1	1	1	S	1	0	0	0	0	0	0	0	0	.	0	2	1	Mowing regimes
Desmet et al.	96	.	1	1	S	0	0	0	0	0	0	0	0	0	0	1	2	1	Harvesting
Ehrlén	95	4	1	6	S	0	1	0	0	0	0	0	1		100	1	2	1	Herbivory
Ehrlén and Eriksson	95	2	1	1	S	0	0	0	0	0	0	0	0	0	.	1	2	0	Pollinator limitation
Enright	82	3	2	4	S	0	0	0	0	0	0	0	0	0	.	1	2	0	Variation in λ
Enright and Ogden	79	2	2	4	S	0	0	0	0	0	0	0	0	0	.	1	2	0	Rainforest trees
Enright and Watson	91	7	1	1	S	0	0	0	0	0	0	0	0	0	.	1	2	1	Elasticities, strategies
Enright and Watson	92	6	1	1	B	0	0	0	0	0	0	0	0	0	.	1	2	1	Age/size in palm
Enright et al.	98	13	1	15	A	0	1	0	1	0	0	0	0	0	.	1	1	0	Fire return interval
Enright et al.	98	13	1	15	A	0	1	0	1	0	0	0	0	0	.	1	1	0	Serotiny, weather
Eriksson	88	4	1	1	S	0	0	0	0	0	0	0	0	0	.	1	2	1	Ramet demography
Eriksson	92	4	1	1	S	0	1	0	0	0	0	1	1		50	1	2	0	Stochastic ramet demog.
Eriksson	94	4	3	1	S	0	1	0	1	0	0	1	3		50	1	2	0	Ramets and genets
Fiedler	87	4	4	4	S	0	0	0	0	0	0	0	0	0	.	1	2	0	Four congeners
Fone	89	4	2	1	A	0	0	0	0	0	0	0	0	0	.	1	0	1	Modeled seed bank
Greenlee and Kaye	97	4	1	4	S	0	1	0	0	0	0	0	1		25	1	1	0	Matrix selection effects
Gross et al.	98	5	1	1	I	0	1	0	0	0	0	0	0	0	200	1	2	1	Fire, trampling
Guerrant	96	5	1	4	S	0	0	0	0	0	0	0	0	0	.	1	1	0	Clonal growth
Harcombe	87	.	1	.	S	0	0	0	0	0	0	0	0	0	.	1	0	0	Tree life tables
Hartshorn	75	2	1	1	S	0	0	0	0	0	0	0	0	0	.	1	2	0	Sensitivity analysis
Horvitz and Schemske	95	6	1	4	S	0	0	0	0	0	0	0	0	0	.	1	0	1	Spatial, temporal variation
Huenneke and Marks	87	3	1	2	I	0	0	0	0	0	0	0	0	0	.	1	2	0	Shrub stem dynamics

Table 1. Cont.

Authors	Y	y	S	P	M	D	E	d	C	G	m	s	V	e	t	λ	p	L	Comments
Kalisz and McPeek	92	2	1	1	A	0	1	0	0	0	0	0	0	3	250	1	2	0	Seed bank methodology
Kawano et al.	87	8	4		S	0	0	0	0	0	0	0	0	0		0	2	0	Comparative life histories
Kephart and Paladino	93	4	1	2	S	1	0	0	0	0	0	0	0	0	.	1	0	1	Habitat differences
Klemow and Raynal	85	5	2	1	S	0	0	0	0	0	0	0	0	0	.	1	0	0	Sparse vs dense pops
Kubota	97	4	3	1	A	1	0	0	0	0	0	0	0	0	.	0	2	0	Tree effects on understory
Lennartsson	97	5	1	1	S	0	1	0	0	0	0	0	1	1	50	1	0	1	Grazing, climate
Lesica	95	8	1	2	S	0	0	0	0	0	0	0	0	0	.	1	1	1	Herbivory
Lesica and Shelly	95	5	1	3	S	0	0	0	0	0	0	0	0	0	.	1	0	1	Reproductive mode
Lonsdale et al.	98	1	1	3	U	1	0	0	0	0	0	0	0	0	.	0	1	0	Postfire annual
Maschinsky et al	97	7	1	1	S	0	0	1	0	0	0	0	0	1	100	1	1	0	Trampling; before/after
McPherson and Williams	96	3	1	1	I	0	0	0	0	0	0	0	0	0	.	0	2	0	Stage duration
Meagher	82	5	1	3	S	0	0	0	0	0	0	0	0	0	.	1	2	0	Two sexes
Menges	90	4	1	14	S	0	1	0	1	0	0	0	0	1	100	1	0	0	Moisture, cover
Menges and Dolan	98	7	1	16	S	0	0	0	0	1	0	0	0	1	1000	1	0	0	Management, size,genetics
Moloney	88	3	1	5	S	0	0	0	0	0	0	0	0	0	.	1	2	1	Fine-scale demography
Morris and Doak	98	2	1	1	I	0	0	0	0	0	0	0	0	0	.	1	2	1	Cochrane methods
Nantel et al.	96	5	2	5	S	0	1	0	0	0	0	0	1	3	200	1	1	0	Harvesting
Nault and Gagnon	93	5	1	1	S	0	0	0	1	0	0	0	0	0	.	1	2	1	Harvest, ramet
O'Connor	93	4	6	12	I	0	0	0	0	0	0	0	0	0	.	1	2	1	Grazing
Olmsted and Alvarez-B.	95	2	2	5	S	0	0	0	0	0	0	0	0	0	.	1	2	1	Cl, harvest
Oostermeijer	00	7	1	1	S	1	1	0	1	1	0	0	0	3	50	0	1	1	Genetics, management
Oostermeijer et al.	96	7	1	6	S	0	0	0	0	0	0	0	0	0	.	1	0	1	Strategy, management
Parker	97	2	1	4	S	0	0	0	0	0	0	0	0	0	.	1	1	0	Pollination
Pascarella and Horvitz	98	2	1	4	S	0	0	0	1	0	0	0	0	0	.	1	2	1	Megamatrices, hurricanes
Pinard	93	1	1	5	I	0	0	0	0	0	0	0	0	0	.	1	2	1	Harvest, lgln
Piñero et al.	84	7	1	1	S	0	0	0	0	0	0	0	0	0	.	1	2	0	Sensitivity
Platt et al.	88	5	1	1	I	0	0	0	0	0	0	0	0	0	.	1	2	1	Integrative approach
Quintana-Ascencio	97	3	1	14	S	0	1	0	1	0	0	0	0	3	600	1	0	1	Fire and recovery models
Ratsivarson et al.	96	4	1	1	S	0	0	0	0	0	0	0	0	0	.	1	0	1	Harvesting palm leaves
Reader and Thomas	77	2	1	1	S	1	0	0	0	0	0	0	0	0	38	0	2	0	Patch dynamics
Santos	93	1	1	1	I	0	0	0	0	0	0	0	0	0	.	1	2	0	Two types of harvesting
Sarukhan and Gadgil	74	2	3	23	S	0	0	0	0	0	0	0	0	0	.	1	0	0	Crucial plant demography
Scandrett and Gimingham	89	.	1	.	S	0	0	0	1	0	0	0	0	0	.	0	1	0	Growth phases, dif
Schwartz et al.	00	9	1	4	N	0	0	0	0	0	0	0	0	3	100	0	1	0	Individual based model
Shimizu et al.	98	4	1	2	S	0	1	0	0	0	0	0	0	0	100	1	2	0	Seedling recruits
Silander	83	.	1	3	A	0	0	0	0	0	0	0	0	0	.	1	2	0	Grazing; Australian shrub
Silva et al.	91	6	1	2	I	0	0	0	0	0	0	0	0	0	.	1	2	1	Fire effects on grass species
Solbrig et al.	88	5	1	1	S	1	1	0	0	0	0	0	0	0	80	0	2	0	Seed bank, density-depend.
Svensson et al.	93	8	3	1	S	0	0	0	0	0	0	0	0	0	.	1	2	1	Loglinear analysis
Usher	66	6	1	1	S	0	0	0	0	0	0	0	0	0	.	1	0	0	Scots pine
Valverde and Silvertown	97	3	1	8	S	0	0	0	1	0	1	0	0	3	.	1	0	0	Succession,metapopulation
Valverde and Silvertown	98	3	1	8	S	0	0	0	0	0	0	0	0	0	.	1	2	1	Comparison of light envs.
Van Groenendael and Slim	88	3	1	2	B	0	1	0	0	0	0	0	0	2	200	1	2	1	Clonal, stochastic analyses
Vavrek et al.	97	1	1	1	I	0	0	0	0	0	0	0	0	0	.	1	0	1	Seasonal variation
Waite and Hutching	91	10	1	1	S	0	0	0	0	0	0	0	0	0	.	1	0	1	Grazing experiments
Wardle	98	4	1	2	S	0	0	0	0	0	0	0	0	0	.	1	0	0	Loop analysis
Watkinson	90	9	1	2	U	1	0	0	0	0	0	0	0	0	.	1	1	0	Observed extinction
Werner and Caswell	77	6	1	8	B	0	0	0	0	0	0	0	0	0	.	1	0	0	Groundbreaking research

(1998), working with the prairie herb *Silene regia*, modeled episodic seedling recruitment by constructing matrices representing years with and without recruitment. Different management practices created different proportions of years with seedling recruitment during the study. These were used as probabilities applied to the recruitment matrices, used in modeling extinction probabilities. The highest probabilities of seedling recruitment were found in management that included prescribed burning. Subsequent extinction risks were lowest in burned areas (Menges and Dolan 1998).

When flowering or seedling recruitment is episodic at unknown intervals, modeling can be used to define minimal intervals required for viable populations. For the periodically flowering tiller populations of *Carex bigelowii*, stable population growth rates resulted from flowering once every three years on average (Carlsson and Callaghan 1991).

Clonal growth

Although animal species subject to PVAs (mainly vertebrates and butterflies) have distinct genetic individuals, many plant genets consist of multiple physiologically independent individuals, or ramets. Modeling ramet demography is a common approach in plant PVAs of clonal species, especially when genets cannot be easily identified (e.g. Bierzychudek 1982, Huenneke and Marks 1987, Boeken and Canham 1995). New ramets can be considered offspring of individual ramets, or aggregate ramet production can be calculated from groups of plants.

New clonal ramets can be considered merely as additional plants of the appropriate size or can be modeled separately from other plants. Separating new clonal ramets is more realistic, since their demographic rates will usually differ from seedling-derived plants. In demographic modeling, new clonal ramets may be further subdivided by their size or stage (Guerrant 1996) or simultaneously by size and flowering status (Byers and Meagher 1997). Subsequent transitions of older clonal ramets will usually involve transitions to appropriate size or stage classes without respect to clonal origin, but with sufficient data, one can test for continuing demographic differences between clonal-derived and seed-derived plants.

When genets can be distinguished, both ramet and genet demography can be studied simultaneously. For example, in three species, genet extinctions occurred on the time scale of centuries, while ramet extinctions were possible for small populations within 50 yr (Eriksson 1994). For the clonal woodland herb wild ginger *Asarum canadense*, the minimum viable population size (considering both environmental and demographic stochasticity) was 25 for genets and 1000 for ramets (Damman and Cain 1998).

Consideration of ramet vs genet population viability will be important for wild populations of clonal plants, which may only rarely recruit new genets and may have complex patterns in ramet establishment and mortality (Eriksson 1992). In ex situ populations, managers will need to keep viable populations of both genets (considering genetic factors explicitly) and ramets. Multiplication of ramets in ex situ populations may buffer extinctions from demographic fluctuations.

Plant PVA applications to conservation

Plant-animal interactions

While most PVAs are concentrated on the demography of focal species, interactions with other species may be included in various ways. Careful observations or experiments can examine how pollinators, herbivores, or mutualists affect demography. For example, Parker (1997) included results from pollinator limitation experiments, calculating the costs of reproduction to examine pollinator limitation effects on the spread of an invasive exotic shrub. Ehrlén (1995) contrasted the effects of three different types of herbivory and their effects on different parts of the life cycle of *Lathyrus vernus*. Other studies predicted that rotational grazing would not seriously impact population growth in *Astragalus scaphoides* (Lesica 1995) and that grazing by sheep or rabbits would be likely to cause extinctions in local populations of an Australian desert shrub (Silander 1983). The results from these studies can provide management recommendations to promote population viability.

Minimum viable populations

Because larger populations of plants have a lower risk of extinction than smaller populations (e.g. Fischer and Stöcklin 1997), PVAs can be used to define a minimum viable population (MVP; Shaffer 1981). This MVP is contingent upon decisions of tolerable risk of extinction and a defined time period (e.g., what is the MVP to reduce extinction risk < 5% for 100 yr?). Most authors appear unwilling to calculate MVPs, perhaps because data uncertainty and modeling assumptions are not represented in a single number MVP. MVPs have been calculated in relation to harvesting pressure on two understory herbs (Nantel et al. 1996). These analyses were used to define tolerable harvest levels and were influential in the legislation of harvesting bans in one species. Other MVPs have been made for ramets of three clonal plant species (Eriksson 1994) and for ramet and genet MVPs of the clonal herb *Asarum canadense* (Damman and Cain 1998).

One of the reasons for caution in publishing MVPs

may be a general lack of evidence for thresholds in extinction risk in relation to population size in wild plants. Several particular scenerios could result in extinction thresholds, including loss of pollinator services in small and isolated populations (Ågren 1996).

Genetic problems in very small populations can also limit fecundity and may cause thresholds in extinction risk in smaller populations. Some plants are unable to produce offspring except in crosses with plants of other mating types. In extreme cases, this may require crosses from faraway populations (DeMauro 1993). In Florida ziziphus *Ziziphus celata*, fruit production never occurs in most wild populations. Limited numbers of compatible genotypes that can produce offspring even in more genetically diverse populations may limit fecundity and population viability (Weekley et al. 1999). This complicates both ex situ approaches and planned re-introductions of plants to protected areas.

Reintroduction of endangered plants can make use of population viability approaches to determine the number and type of propagules likely to result in a viable population. For example, introduction of later life history stages is likely to require fewer plants to reduce extinction risk because of higher survival of those stages (Guerrant and Pavlik 1998).

PVAs and managing disturbance

Because population growth and extinction risks vary with disturbance regimes, and because disturbance regimes can often be managed and manipulated, PVAs considering the effects of disturbance regimes can be crucial in designing effective management. Disturbances include outside, unusual events that are often minimized, as well as natural disturbances that must be managed or initiated in a fragmented landscape.

Many plants may be threatened by harvest or damage caused by humans. Harvesting of plant parts will tend to reduce finite rates of increase and increase extinction risks, but some harvesting may be sustainable. A typical approach in modeling is to determine the demographic effects of harvesting and then model increasing levels or frequencies of harvesting until finite rates of increase drop to one (e.g. Nault and Gagnon 1993, Nantel et al. 1996, Bernal 1998) or until all adults disappear (Olmsted and Alvarez-Buylla 1995). Approaches incorporating environmental variation and harvesting pressure will provide more conservative results.

Human trampling effects were studied by Maschinski et al. (1997), based on seven years of data before and after protection from trampling. Population growth was predicted to be more stable after protection from trampling. Finite rates of increase were also correlated with weather, both before and after protection, and both favorable and unfavorable weather years occurred before and after protection. Therefore, despite the lack of a true control, the authors felt confident that protection from trampling was necessary to minimize extinction risks.

Population modeling can be used to make the most of demographic data by modeling management options and combinations. The combination of prescribed burning at 6–16 yr intervals and protection from trampling was necessary for viable populations of *Hudsonia montana* (Gross et al. 1998). Traditional European grassland management (summer mowing with autumn grazing) predicted lower extinction risk for two grassland indicator species than the current practice of grazing only (Lennartsson 1997). Grazing was also the treatment in analyses of 16 populations of *Cirsium vulgare*; in this case the treatments were imposed in an experimental design that allowed statistical analyses of treatment effects on finite rates of increase (Bullock et al. 1994).

Fire is one of the most commonly manipulated disturbances. Many plants are dependent on periodic fire, but disturbances fire frequencies, seasons, and intensities may vary widely and affect population viability. Some plant PVAs consider whether fire management, in general, is favorable for population viability. For example, finite rates of increase were more strongly related to a regular fire regime in *Silene regia* than they were to population size, isolation, or genetic factors (Menges and Dolan 1998). Population growth rates were four-fold higher in burned than unburned savanna for an annual grass (Canales et al. 1994).

More explicitly dynamic modeling approaches can address subtle elements of disturbance regimes. Demographic rates in species of ecosystems subject to periodic disturbances do not vary randomly, but instead cyclically. The effects of periodic disturbance and subsequent recovery can be modeled by a combination of information about the disturbance/recovery cycle, the frequency and intensity of disturbance, and demographic responses to disturbances. A megamatrix approach, with individual matrices linked by transitions representing the probabilities of going from one environmental state to another, has been used to model the viability of a tropical understory shrub in the face of hurricane disturbance and resulting canopy dynamics (Pascarella and Horvitz 1998). Open, post-hurricane environments supported the highest population growth rates and elasticities, although such environments are less common than more closed canopy patches. Similar approaches were used to model gap dynamics and demography of the rain forest tree *Cecropia obtusifolia* (Alvarez-Buylla 1994) and a forest understory shrub *Lindera benzoin* (Cipollini et al. 1994).

An alternative approach is built around the time since the last disturbance and chances of a new disturbance. For example, several authors have used demographic models to identify disturbance cycles that would minimize extinction risk. Modeling on *Gentiana pneumonanthe* (Oostermeijer et al. 1996, Oostermeijer 2000) showed that a sod-cutting interval of ca 13 yr was optimal to forestall local extinction due to demographic causes, but that inbreeding depression reduced the optimal disturbance frequency, especially for small populations. Fire return intervals promoting high stochastic rates of increase have been modeled for both non-sprouting and sprouting species of the Australian shrub genus *Banksia* (Enright et al. 1998 a, b). For *Banksia attenuata*, a resprouting, seroninous shrub, fire return intervals of 7–40 yr promoted long-term population growth (Enright et al. 1998b). Model runs suggested that serotiny was favored by stochasticity in the fire regime. These studies suggest not only the range of fire return intervals within which these species will persist, but the importance of variation in fire return times.

Finally, Burgman and Lamont (1992) considered fire frequency effects, environmental stochasticity, and potential inbreeding in studying population viability of *Banksia cuneata*. The fire frequency that maximized population size did not minimize extinction because it exposed vulnerable seedlings to catastrophic mortality during droughts. It is likely that minimizing extinction and maximizing population size will not require the same conditions in other species.

Metapopulation dynamics and spatially-explicit modeling

Metapopulations are assemblages of local populations linked by movements of individuals among populations (Hanski and Gilpin 1997). In plants, movements are usually either via pollen (and perhaps pollinators) or via seed dispersal. Pollen movements provide gene flow and often occur over larger distances than seed dispersal, which is often strongly skewed to very local dispersal. However, only seed dispersal and subsequent seedling establishment (or long-distance movements of small vegetative parts) are capable of recolonizing empty patches after local extinctions.

To understand persistence in plants, one may often need to consider the metapopulation level, for a number of reasons. Many plant species have patchy distributions, often occurring on specialized sites that can be identified and censused for species occupancy. Some additional patchiness is created by disturbances and disturbance-specialized or disturbance-avoiding species. Because dispersal of many species is limited, suitable unoccupied patches may remain. The concepts of minimum viable metapopulation and minimum available suitable habitat (Hanski et al. 1996b, Sjögren-Gulve and Hanski 2000) are applicable to many plant species.

Data on species presence/absence in suitable habitat patches can be used to infer colonization and extinction rates using incidence-based metapopulation models (e.g. Hanski et al. 1996a). Incidence-based metapopulation modeling has been used to examine distributions of Florida scrub plants (Quintana-Ascencio and Menges 1996). Extinction risks as a function of patch sizes were predicted to differ between shrubs and herbs.

More explicitly dynamic metapopulation models have been formulated for *Primula vulgaris* by linking individual population demography with transition probabilities among canopy stages and estimating colonization and extinction rates (Valverde and Silvertown 1997). Both seed dispersal and forest disturbance rates increased metapopulation growth rates.

These explicit metapopulation approaches have great potential for plant studies, but will be difficult because of the additional data requirements. In Valverde and Silvertown (1997), colonization was not estimated directly, but modeled as varying fractions of seeds being dispersed outside of individual populations, combined with data on plant densities and fecundities. While direct estimates of colonization and extinction are difficult to obtain, demographic and genetic approaches can be used to obtain these data in long-term studies. However, plants with seed banks provide an additional challenge, since observed (aboveground) local extinctions can mask the presence of persistent seed banks.

Spatially-explicit PVAs have proven useful in predicting how animal species may be affected by human-caused changes in landscapes (reviewed in Reed et al. unpubl.) and the importance of individual patches in a landscape (Lindenmayer and Possingham 1995). Such patch-level models have potential in plants to evaluate the importance of metapopulation dynamics (Harrison and Ray 2001). Spatially explicit models have been used to explore fire effects on demography in Australian shrublands (Bradstock et. al. 1998). Model runs varied burn size, randomness of ignitions, and fire frequency and looked at effects on extinction of plant functional groups. Extinction risks for resprouters were unaffected by fire, but obligate seeders were threatened by large random fires (Bradstock et al. 1998).

Relating genetics and demography

Most studies of plant PVAs do not incorporate genetic information, despite a general call for such integration (e.g. Reed et al. 1998, Reed et al. unpubl., but see Beissinger and Westphal 1998). Even with data on genetic variation across species ranges or in relation to population sizes, it is rare to have data on the effects of genetic variation on demographic rates in wild populations. Observed demographic rates in wild populations may already incorporate some of the deleterious effects of inbreeding on past and current demography, but will be unable to predict differential future rates.

For plants, there are few examples integrating genetics and demography. Genetic variation was used as one of many predictors of population viability by Menges and Dolan (1998), but was less significant than the fire regime in affecting population growth or extinction. Genetics is included with many other demographic and environmental factors in exploring population viability in an Australian shrub (Burgman and Lamont 1992). Oostermeijer (2000) incorporated population size, selfing rate, and their effects on inbreeding depression of fecundity into demographic simulations involving disturbance cycles. Inbreeding had significant effects on the regeneration capacity of small populations, although larger populations were not strongly affected.

Assessing management regimes and conservation strategies using comparative PVAs

More realistic PVAs provide detailed results under a range of scenarios and assumptions (e.g. Crone and Gehring 1998). PVAs that merely provide single results give the impression of superiority to more qualitative approaches because of the "fallacy of illusory precision" (Reed et al. 1998). A range of results also allows comparisons among conservation strategies, management regimes, and sites (Akçakaya and Sjögren-Gulve 2000). This makes it possible to examine relative rather than absolute rates of extinction. Presumably, this comparative approach is more robust to the limitations of the "few-populations/few years" data that plague plant PVAs.

Comparative analyses can be particularly useful in contrasting alternative management regimes. In the shrub *Fumana procumbens*, for example, sparse vegetation maintained by light (but not heavy) grazing created safe sites for seedling recruitment, which was associated with relatively high finite rates of increase during a seven year study (Bengtsson 1993). In the perennial herb *Silene regia,* higher finite rates of increase and lower extinction probabilities were predicted for burned populations, largely due to increased opportunities for seedling recruitment (Menges and Dolan 1998). Although there are many uncertainties that might caution against believing specific projections of population futures, comparative approaches can distinguish between alternative management strategies in a way that is more comprehensive than considering single life history responses.

Plant PVAs seem to be in their infancy. Few studies that could be identified as plant PVAs are labeled as such, leading general reviews to include little mention of plants. Most of the studies include conservation issues only tangentially, but PVAs still have much to offer plant conservation. Limitations such as short study duration, few populations being studied, and challenging life history stages can be overcome, albeit with considerable time, energy, and expertise. However, PVAs can offer crucial feedback to support, modify, or expand useful management regimes, which can continue to provide data for improved and updated PVAs.

Acknowledgements – I thank Daniel Gagnon, Martha Groom, Ed Guerrant, Samara Hamzé, Christine Hawkes, Tom Kaye, Tommy Lennartsson, Thomas Nilsson, Gerard Oostermeijer, Pedro Quintana-Ascencio, Per Sjögren-Gulve and the participants in the 1997 Swedish workshop: "The use of population viability analyses in conservation planning", for valuable conversations and reviews.

References

Abe, S., Nakashizuka, T. and Tanaka, H. 1998. Effects of canopy gaps on the demography of the subcanopy tree *Styrax obassia*. – J. Veg. Sci. 9: 787–796.

Ågren, J. 1996. Population size, pollinator limitation, and seed set in the self-incompatible herb *Lythrum salicaria*. – Ecology 77: 1779–1790.

Akçakaya, H. R. 2000. Population viability analyses with demographically and spatially structured models. – Ecol. Bull. 48: 23–38.

Akçakaya, H. R. and Sjögren-Gulve, P. 2000. Population viability analyses in conservation planning: an overview. – Ecol. Bull. 48: 9–21.

Allphin, L. and Harper, K. T. 1997. Demography and life history characteristics of the rare Kachina daisy (*Erigeron kachinensis*, Asteraceae). – Am. Midl. Nat. 138: 109–120.

Alvarez-Buylla, E. R. 1994. Density dependence and patch dynamics in tropical rain forests: matrix models and applications to a tree species. – Am. Nat. 143: 155–191.

Aplet, G. H., Laven, R. D. and Shaw R. B. 1994. Application of transition matrix models to the recovery of the rare Hawaiian shrub, *Tetramolium arenarium* (Asteraceae). – Natural Areas J. 14: 99–106.

Batista, W. B., Platt, W. J. and Macchiavelli, R. E. 1998. Demography of a shade-tolerant tree (*Fagus grandifolia*) in a hurricane-disturbed forest. – Ecology 79: 38–53.

Beissinger, S. R. and Westphal, M. I. 1998. On the use of demographic models of population viability in endangered species management. – J. Wildl. Manage. 62: 821–841.

Bengtsson, K. 1993. *Fumana procumbens* on Öland – population dynamics of a disjunct species at the northern limit of its range. – J. Ecol. 81: 745–758.

Bernal, R. 1998. Demography of the vegetable ivory palm *Phytelephas seemannii* in Columbia, and the impact of seed harvesting. – J. Appl. Ecol. 35: 64–74.

Bierzychudek, P. 1982. The demography of jack-in-the-pulpit, a forest perennial that changes sex. – Ecol. Monogr. 52: 335–351.

Bierzychudek, P. 1999. Looking backwards: assessing the projections of a transition matrix model. – Ecol. Appl. 9: 1278–1287.

Boeken, B. and Canham, C. D. 1995. Biotic and abiotic control of the dynamics of gray dogwood (*Cornus racemosa* Lam.) shrub thickets. – J. Ecol. 83: 569–580.

Boucher, D. H. and Mallona, M. A. 1997. Recovery of the rain forest tree *Vochysia ferruginea* over five years following hurricane Joan in Nicaragua: a preliminary population projection matrix. – For. Ecol. Manage. 91: 195–204.

Boyce, M. 1992. Population viability analysis. – Annu. Rev. Ecol. Syst. 23: 481–506.

Bradstock, R. A. and O'Connell, M. A. 1988. Demography of woody plants in relation to fire: *Banksia ericifolia* L.f. and *Petrophile pulchella* (Schrad) R.Br. – Aust. J. Ecol. 13: 505–518.

Bradstock, R. A. et al. 1996. Simulation of the effect of spatial and temporal variation in fire regimes on the population viability of a *Banksia* species. – Conserv. Biol. 10: 776–784.

Bradstock, R. A. et al. 1998. Spatially-explicit simulation of the effect of prescribed burning on fire regimes and plant extinctions in shrublands typical of south-eastern Australia. – Biol. Conserv. 86: 83–95.

Bullock, S. H. 1980. Demography of an undergrowth palm in littoral Cameroon. – Biotropica 12: 247–255.

Bullock, J. M., Clear Hill, B. and Silvertown, J. 1994. Demography of *Cirsium vulgare* in a grazing experiment. – J. Ecol. 82: 101–111.

Burgman, M. A. and Lamont, B. B. 1992. A stochastic model for the viability of *Banksia cuneata* populations: environmental, demographic, and genetic effects. – J. Appl. Ecol. 29: 719–727.

Burns, B. R. and Ogden, J. 1985. The demography of a temperate mangrove [*Avicennia marina* (Forsk.) Vierh.] at its southern limit in New Zealand. – Aust. J. Ecol. 10: 125–133.

Byers, D. L. and Meagher, T. R. 1997. A comparison of demographic characteristics in a rare and a common species of *Eupatorium*. – Ecol. Appl. 7: 519–530.

Calvo, R. N. 1993. Evolutionary demography of orchids: intensity and frequency of pollination and the cost of fruiting. – Ecology 74: 1033–1042.

Canales, J. et al. 1994. A demographic study of an annual grass (*Andropogon brevifolius* Schwarz) in burnt and unburnt savanna. –Acta Oecol. 15: 261–273.

Carlsson, B. Å. and Callaghan, T. V. 1991. Simulation of fluctuating populations of *Carex bigelowii* tillers classified by type, age, and size. – Oikos 60: 231–240.

Chapman, S. B., Rose, R. J. and Clarke, R. T. 1989. The behavior of populations of the marsh gentian (*Gentiana pheumonanthe*): a modeling approach. – J. Appl. Ecol. 26: 1059–1072.

Cipollini, M. L., Whigham, D. F. and O'Neill, J. 1993. Population growth, structure, and seed dispersal in the understory herb *Cynoglossum virginianum*: a population and patch dynamics model. - Plant Species Biol. 8: 117–129.

Cipollini, M. L., Wallace-Senft, D. A. and Whigham, D. F. 1994. A model of patch dynamics, seed dispersal, and sex ratio in the dioecious shrub *Lindera benzoin* (Lauraceae). – J. Ecol. 82: 621–633.

Cochrane, M. E. and Ellner, S. 1992. Simple methods for calculating age-based life history parameters for stage-structured populations. – Ecol. Monogr. 63: 345–364.

Crone, E. E. and Gehring, J. L. 1998. Population viability of *Rorippa columbiae*: multiple models and spatial trend data. – Conserv. Biol. 12: 1054–1065.

Damman, H. and Cain, M. L. 1998. Population growth and viability analyses of the clonal woodland shrub, *Asarum canadense*. – J. Ecol. 86: 13–26.

deKroon, H., Plaiser, A. and van Groenendael, J. 1987. Density-dependent simulation of the population dynamics of a perennial grassland species, *Hypochaeris radicata*. – Oikos 50: 3–12.

DeMauro, M. M. 1993. Relationship of breeding system to rarity in the Lakeside Daisy (*Hymenoxys acaulis* var. *glabra*). – Conserv. Biol. 7: 542–550.

Desmet, P. G., Shackleton, C. M. and Robinson, C. R. 1996. The population dynamics and life-history attributes of a *Pterocarpus angolensis* DC population in the northern Province, South Africa. – S. Afr. J. Bot. 62: 160–166.

Ehrlén, J. 1995. Demography of the perennial herb *Lathyrus vernus*. II. Herbivory and population dynamics. – J. Ecol. 83: 297–308.

Ehrlén, J. and Eriksson, O. 1995. Pollen limitation and population growth in a herbaceous perennial legume. – Ecology 76: 652–656.

Enright, N. J. 1982. The ecology of Araucaria species in New Guinea. III. Population dynamics of sample stands. – Aust. J. Ecol. 16: 507–520.

Enright, N. J. and Ogden, J. 1979. Applications of transition matrix models in forest dynamics: *Araucaria* in Papua New Guinea and *Nothofagus* in New Zealand. – Aust. J. Ecol. 4: 3–23.

Enright, N. J. and Watson, A. D. 1991. A matrix population model analysis for the tropical tree, *Araucaria cunninghamii*. – Aust. J. Ecol. 16: 507–520.

Enright, N. J. and Watson, A. D. 1992. Population dynamics of the nikau palm, *Rhopalostylis sapida* (Wendl. et Drude), in a temperate forest remnant near Auckland, New Zealand. – N. Z. J. Bot. 30: 29–43.

Enright, N. J. et al. 1998a. The ecological significance of canopy seed storage in fire-prone environments: a model for non-sprouting shrubs. – J. Ecol. 86: 946–959.

Enright, N. J. et al. 1998b. The ecological significance of canopy seed storage in fire-prone environments: a model for re-sprouting shrubs. – J. Ecol. 86: 960–973.

Eriksson, O. 1988. Ramet behavior and population growth in the clonal herb *Potentilla anserina*. – J. Ecol. 76: 522–536.

Eriksson, O. 1992. Population structure and dynamics of the clonal dwarf-shrub *Linnaea borealis*. – J. Veg. Sci. 3: 61–68.

Eriksson, O. 1994. Stochastic population dynamics of clonal plants: numerical experiments with ramet and genet models. – Ecol. Res. 9: 257–268.

Fiedler, P. L. 1987. Life history and population dynamics of rare and common mariposa lilies (*Calochortus* Pursh: Liliaceae) – J. Ecol. 75: 977–995.

Fiedler, P. L., Knapp, B. E. and Fredricks, N. 1998. Rare plant demography: lessons from the Mariposa lilies (*Calochortus*: Liliaceae). – In: Fiedler, P. L. and Kareiva, P. M. (eds), Conservation biology for the coming decade. Chapman and Hall, pp. 28–48.

Fischer, M. and Stöcklin, J. 1997. Local extinctions of plants in remnants of extensively used calcareous grasslands 1950–1985. – Conserv. Biol. 11: 727–737.

Fone, A. L. 1989. A comparative demographic study of annual and perennial *Hypochoeris* (Asteraceae). – J. Ecol. 77: 495–508.

Groom, M. J. and Pascual, M. A. 1997. The analysis of population persistence: an outlook on the practice of viability analysis. – In: Fiedler, P. L. and Kareiva, P. M. (eds), Conservation biology for the coming decade. Chapman and Hall, pp. 4–27.

Gross, K. et al. 1998. Modeling controlled burning and trampling reduction for conservation of *Hudsonia montana*. – Conserv. Biol. 12: 1291–1301.

Greenlee, J. and Kaye, T. N. 1997. Stochastic matrix projection: a comparison of the effect of element and matrix selection methods on quasi-extinction risk for *Haplopappus radiatus* (Asteraceae). – In: Kaye, T. N. et al. (eds), Conservation and management of native plants and fungi. Native Plant Soc. of Oregon. Corvallis, OR, pp. 66–71.

Guerrant, E. O. Jr 1996. Comparative demography of *Erythronium elegans* in two populations: one thought to be in decline (Lost Prairie) and one presumably healthy (Mt. Hebo): interim report on four transitions, or five years of data. – Bureau of Land Manage., Salem, Oregon.

Guerrant, E. O. Jr and Pavlik, B. 1998. Reintroduction of rare plants: genetics, demography, and the role of ex situ conservation methods. – In: Fiedler, P. L. and Kareiva, P. M. (eds), Conservation biology for the coming decade. Chapman and Hall, pp. 80–108.

Hanski, I. and Gilpin, M. E. 1997. Metapopulation biology: ecology, genetics, and evolution. – Academic Press.

Hanski, I. et al. 1996a. The quantitative incidence function model and persistence of an endangered butterfly metapopulation. – Conserv. Biol. 10: 578–590.

Hanski, I., Moilanen, A. and Gyllenberg, M. 1996b. Minimum viable metapopulation size. – Am. Nat. 147: 527–541.

Harcombe, P. A. 1987. Tree life tables. – BioScience 37: 557–568.

Harrison, S. and Ray, C. 2001. Plant population viability and metapopulation processes. – In: Beissinger, S. and McCulloch, D. (eds), Population viability analysis. Univ. of Chicago Press, in press.

Hartshorn, G. S. 1975. A matrix model of tree population dynamics. – In: Golley, F. B. and Medina, E. (eds), Tropical ecological systems. Springer, pp. 41–51.

Horvitz, C. C. and Schemske, D. W. 1995. Spatiotemporal variation in demographic transitions of a tropical understory herb: projection matrix analysis. – Ecol. Monogr. 65: 155–192.

Huenneke, L. F. and Marks, P. L. 1987. Stem dynamics of the shrub *Alnus incana* ssp. *rugosa*: transition matrix models. – Ecology 68: 1234–1242.

Kalisz, S. and McPeek, M. A. 1992. Demography of an age-structured annual: resampled projection matrices, elasticity analyses, and seed bank effects. – Ecology 73: 1082–1093.

Kawano, S. et al. 1987. Demographic differentiation and life-history evolution in temperate woodland plants. – In: Urbanska, K. M. (ed.), Differentiation patterns in higher plants. Academic Press, pp. 153–180.

Kephart, S. R. and Paladino, C. 1993. Demographic change and microhabitat variability in a grassland endemic, *Silene douglasii* var. *oraria* (Caryophyllaceae). – Am. J. Bot. 84: 179–189.

Klemow, K. M. and Raynal, D. J. 1985. Demography of two facultative biennial plant species in an unproductive habitat. – J. Ecol. 73: 147–167.

Kubota, Y. 1997. Demographic traits of understory trees and population dynamics of a *Picea-Abies* forest in Taisetsuzan National Park, northern Japan. – Ecol. Res. 12: 1–9.

Lennartsson, T. 1997. Demography, reproductive biology and adaptive traits in *Gentiana campestris* and *G. amarella*: evaluating grassland management for conservation by using indicator plant species. – Ph.D. thesis, Swedish Univ. of Agricult. Sci., Uppsala.

Lesica, P. 1995. Demography of *Astragalus scaphoides* and effects of herbivory on population growth. – Great Basin Nat. 55: 142–150.

Lesica, P. and Steele, B. M. 1994. Prolonged dormancy in vascular plants and implications for monitoring studies. – Nat. Areas J. 14: 209–212.

Lesica, P. and Shelley, J. S. 1995. Effects of reproductive mode on demography and life history in *Arabis fecunda* (Brassicaceae). – Am. J. Bot. 82: 752–762.

Lindenmayer, D. B. and Possingham, H. P. 1995. Modelling the viability of metapopulations of the endangered Leadbetter's possum in southeastern Australia. – Biodiv. Conserv. 4: 984–1018.

Lonsdale, W. M. et al. 1998. Modelling the recovery of an annual savanna grass following a fire-induced crash. – Aust. J. Ecol. 23: 509–513.

Maschinski, J., Frye, R. and Rutman, S. 1997. Demography and population viability of an endangered plant species before and after protection from trampling. – Conserv. Biol. 11: 990–999.

McPherson, K. and Williams, K. 1996. Establishment growth of cabbage palm, *Sabal palmetto* (Arecaceae). – Am. J. Bot. 83: 1566–1570.

Meagher, T. R. 1982. The population biology of *Chamaelirium luteum*, a dioecious member of the lily family: two-sex population projections and stable population structure. – Ecology 63: 1701–1711.

Menges, E. S. 1990. Population viability analysis for an endangered plant. – Conserv. Biol. 4: 52–62.

Menges, E. S. 2000. Population viability analyses in plants: challenges and opportunities. – Trends Ecol. Evol. 15: 51–56.

Menges, E. S. and Dolan, R. W. 1998. Demographic viability of populations of *Silene regia* in midwestern prairies: relationships with fire management, genetic variation, geographic location, population size, and isolation. – J. Ecol. 86: 63–78.

Menges, E. S. and Quintana-Ascencio, P. F. 1998. Population modeling for the prairie bush clover, *Lespedeza leptostachya*. – Final report to the Minnesota Dept of Natural Resources.

Moloney, K. A. 1986. A generalized algorithm for determining canopy size. – Oecologia 69: 176–180.

Moloney, K. A. 1988. Fine-scale spatial and temporal variation in the demography of a perennial bunchgrass. – Ecology 69: 1588–1598.

Morris, W. F. and Doak, D. F. 1998. Life history of the long-lived gynodioecious cushion plant *Silene acaulis* (Caryophyllaceae), inferred from size-based population projection matrices. – Am. J. Bot. 85: 784–793.

Nakaoka, M. 1996. Dynamics of age- and size-structured populations in fluctuating environments: applications of stochastic matrix models to natural populations. – Res. Pop. Ecol. 38: 141–152.

Nantel, P., Gagnon, D. and Nault, A. 1996. Population viability analysis of American ginseng and wild leek harvested in stochastic environments. – Conserv. Biol. 10: 608–621.

Nault, A. and Gagnon, D. 1993. Ramet demography of *Allium tricoccum*, a spring ephemeral, perennial forest herb. – J. Ecol. 81: 101–119.

O'Connor, T. G. 1993. The influence of rainfall and grazing on the demography of some African savanna grasses: a matrix modelling approach. – J. Appl. Ecol. 30: 119–132.

Olmsted, I. and Alvarez-Buylla, E. R. 1995. Sustainable harvesting of tropical trees: demography and matrix models of two palm species in Mexico. – Ecol. Appl. 5: 484–500.

Oostermeijer, J. G. B. 2000. Population viability analysis of the rare *Gentiana pneumonanthe*: importance of demography, genetics, and reproductive biology. – In: Young, A. and Clarke, G. (eds), Genetics, demography, and viability of fragmented populations, in press.

Oostermeijer, J. G. B. et al. 1996. Temporal and spatial variation in the demography of *Gentiana pneumonanthe*, a rare perennial herb. – J. Ecol. 84: 153–166.

Parker, I. M. 1997. Pollinator limitation of *Cytisus scoparius* (scotch broom), an invasive exotic shrub. – Ecology 78: 1457–1470.

Pascarella, J. B. and Horvitz, C. C. 1998. Hurricane disturbance and the population dynamics of a tropical, understory shrub: megamatrix elasticity analysis. – Ecology 79: 547–563.

Pinard, M. 1993. Impacts of stem harvesting on populations of *Iriartea deltoidea* (Palmae) in an extractive reserve in Acre, Brazil. – Biotropica 25: 2–14.

Piñero, D., Martinez-Ramos, M. and Sarukhán, J. 1984. A population model of *Astrocaryum mexicanum* and a sensitivity analysis of its finite rate of increase. – J. Ecol. 72: 977–991.

Platt, W. J., Evans, G. W. and Rathbun, S. L. 1988. The population dynamics of a long-lived conifer (*Pinus palustris*). – Am. Nat. 131: 491–525.

Quintana-Ascencio, P. F. 1997. Population viability analysis of a rare plant species in patchy habitats with sporadic fire. – Ph.D. thesis, State Univ. of New York at Stony Brook.

Quintana-Ascencio, P. F. and Menges, E. S. 1996. Inferring metapopulation dynamics from patch level incidence of Florida scrub plants. – Conserv. Biol. 10: 1210–1219.

Ratsirarson, J., Silander, Jr J. A. and Richard, A. F. 1996. Conservation and management of a threatened Madagascar palm species, *Neodypsis decaryi*, Jumelle. – Conserv. Biol. 10: 40–52.

Reader, R. J. and Thomas, A. G. 1977. Stochastic simulation of patch formation by *Hieracium floribundum* (Compositae) in abandoned pastureland. – Can. J. Bot. 55: 3075–3079.

Reed, J. M., Murphy, D. D. and Brussard, P. F. 1998. Efficacy of population viability analysis. – Wildl. Soc. Bull. 26: 244–251.

Santos, R. 1993. Plucking or cutting *Gelidium sesquipedale*? A demographic simulation of harvest impact using a population projection matrix model. – Hydrobiologia 260/261: 269–276.

Sarukhan, J. and Gadgil. M. 1974. Studies on plant demography: *Ranunculus repens* L., *R. bulbosus* L. and *R. acris* L. III. A mathematical model incorporating multiple modes of reproduction. – J. Ecol. 62: 921–936.

Scandrett, E. and Gimingham, C. H. 1989. A model of *Calluna* population dynamics: the effects of varying seed and vegetative regeneration. – Vegetatio 84: 143–152.

Schwartz, M. W., Hermann, S. M. and van Mantgem, P. J. 2000. Population persistence in Florida torreya (*Torreya taxifolia* Arn.) – slowly declining toward extinction. – Conserv. Biol., in press.

Shaffer, M. L. 1981. Minimum population sizes for species conservation. – BioScience 31: 131–134.

Shimizu, T. et al. 1998. The role of sexual and clonal reproduction in maintaining population in *Fritillaria camtschatcensis* (L.) Ker-Gawl. (Liliaceae). – Ecol. Res. 13: 27–39.

Silander, J. A. Jr 1983. Demographic variation in the Australian desert cassia under grazing pressure. – Oecologia 60: 227–233.

Silva, J. F. et al. 1991. Population responses to fire in a tropical savanna grass, *Andropogon semiberbis*: a matrix model approach. – J. Ecol. 79: 345–356.

Silvertown, J., Franco, M. and Menges, E. 1996. Interpretation of elasticity matrices as an aid to the management of plant populations for conservation. – Conserv. Biol. 10: 591–597.

Solbrig, O. T., Sarandón, R. and Bossert, W. 1988. A density-dependent growth model of a perennial herb, *Viola fimbriatula*. – Am. Nat. 131: 385–400.

Sjögren-Gulve, P. and Hanski, I. 2000. Metapopulation viability analysis using occupancy models. – Ecol. Bull. 48: 53–71.

Svensson, B. M. et al. 1993. Comparative long-term demography of three species of *Pinguicula*. – J. Ecol. 81: 635–645.

Usher, M. B. 1966. A matrix model for forest management. – Biometrika 25: 309–315.

Valverde, T. and Silvertown, J. 1997. A metapopulation model for *Primula vulgaris*, a temperate forest understory herb. – J. Ecol. 85: 193–210.

Valverde, T. and Silvertown, J. 1998. Variation in the demography of a woodland understorey herb (*Primula vulgaris*) along the forest regeneration cycle: projection matrix analysis. – J. Ecol. 86: 545–562.

Van der Meer, J. 1978. Choosing category size in a stage projection matrix. – Oecologia 32: 79–84.

Van Groenendael, J. M. and Slim, P. 1988. The contrasting dynamics of two populations of *Plantago lanceolata* classified by age and size. – J. Ecol. 76: 585–599.

Vavrek, M. C., McGraw, J. B. and Yang, H. S. 1997. Within-population variation in demography of *Taraxacum officinale*: season- and size-dependent survival, growth, and reproduction. – J. Ecol. 85: 277–287.

Waite, S. and Hutchings, M. J. 1991. The effects of different management regimes on the population dynamics of *Ophrys sphegodes*: analysis and description using matrix models. – In: Wells, T. C. E. and Willems, J. H. (eds), Population ecology of terrestrial orchids. SPB Academic Publ., pp. 161–175.

Wardle, G. M. 1998. A graph theory approach to demographic loop analysis. – Ecology 79: 2539–2549.

Watkinson, A. R. 1990. The population dynamics of *Vulpia fasciculata*: a ten year study. – J. Ecol. 78: 196–209.

Weekley, C., Race, T. and Hardin, D. 1999. Saving Florida ziziphus: recovery of a rare Lake Wales Ridge endemic. – Palmetto 19: 9–10, 20.

Werner, P. A. and Caswell, H. 1977. Population growth rates and age versus stage-distribution models for teasel (*Dipsacus sylvestris* Huds.). – Ecology 58: 1103–1111.

Ecological Bulletins 48: 85–99. Copenhagen 2000

Focal species modeling for biodiversity conservation

Erica Fleishman, Bengt Gunnar Jonsson, Per Sjögren-Gulve

Fleishman, E., Jonsson, B. G. and Sjögren-Gulve, P. 2000. Focal species modeling for biodiversity conservation. – Ecol. Bull. 48: 85–99.

Managers in diverse landscapes seek to detect and conserve localities with relatively rich or intact assemblages of native species. We present a four-step procedure to select and screen potential focal species for conservation planning and monitoring of species richness. Species inventories and nestedness analysis are used to identify potential indicators of high species richness. Next, linear regression is employed to predict species richness as a function of environmental parameters that are tractable to quantify or manage. Simultaneously, occupancy modeling tests whether the presence of each focal species mainly correlates with high species richness or tends to be explained by other environmental variables. Finally, repeated inventories empirically validate the use of the focal taxa and parameterize further analyses of their population dynamics and viability. We illustrate this general approach with case studies of cryptogams in boreal forests of northern Sweden and butterflies in mountain ranges in the western United States. Our results demonstrate that the roles that focal species realistically can play in monitoring and management vary between ecosystems and taxonomic groups. Regardless of whether individual focal species can serve as dependable indicators of species richness, our methods will assist managers in assessing the potential effects of alternative management strategies on species distributions.

E. Fleishman (efleish@leland.stanford.edu), Center for Conservation Biology, Dept of Biological Sciences, Stanford Univ., Stanford, CA 94305, USA. – B. G. Jonsson, Dept of Ecology and Environmental Science, Umeå Univ., SE-901 87 Umeå, Sweden. – P. Sjögren-Gulve, Dept of Conservation Biology and Genetics, Evolutionary Biology Centre, Uppsala Univ., Norbyv. 18D, SE-752 36 Uppsala, Sweden, (present address: Swedish Environmental Protection Agency, SE-106 48 Stockholm, Sweden).

Conservation and monitoring of biotic diversity is a major task for countries around the world (e.g. UNCED 1992). Because time and funding to conduct scientific research inevitably are limited, ecologists and resource managers seek scientifically defensible tools to evaluate the status of the biota. These evaluations help guide conservation planning and land use decisions (Stohlgren et al. 1995, Oliver and Beattie 1996, Longino and Colwell 1997, Niemi et al. 1997, Simberloff 1998). In this context, use of species whose measurement provides a scientifically reliable and cost-effective surrogate measure of another ecological parameter that is difficult to assess directly holds tremendous

appeal. Because these species will be the focus of many exercises in conservation planning, we refer to them as "focal species." Thus defined, the concept of focal species is particularly attractive for assessing taxonomic groups that are relatively challenging to sample and identify in terms of logistics, money, or technical expertise. It usually is faster and cheaper to monitor a subset of the species in such a taxonomic group than to perpetually track all its species.

We define a focal species as any taxon that receives considerable attention from conservation biologists and practitioners (see also Holt 1997, Thomas and Hanski 1997). This definition is quite broad, and encompasses species that fall into several loose categories. For example, there are "indicator species," defined here as taxa that exhibit distributions, abundances, or population dynamics that can serve as substitute measures of the status of other species or environmental attributes. Other groups of focal species include "umbrella species," taxa whose conservation confers a protective umbrella to numerous co-occurring species, and "keystone species," taxa with a disproportionate impact on the dynamics of their ecosystems. These examples are not exhaustive (Simberloff 1998). Our definition of focal species differs somewhat from that of Lambeck (1997), who characterized focal species as those whose requirements for persistence outline the ecological attributes that are necessary to meet the requirements of the species that occur in a given landscape. In his framework, focal species are employed to develop "explicit guidelines regarding the composition, quantity, and configuration of habitat patches" (p. 851) and management regimes; different focal species are used to define different aspects of the landscape and management strategy.

Although the concept of focal species is intriguing, empirical evidence that focal species can function as valid conservation tools is limited. So far, few organisms have been identified that are dependable and affordable surrogate measures of community-level or ecosystem-level variables (Scott 1998). The difficulties inherent in selecting appropriate focal species for particular situations often are exacerbated by fuzzy nomenclature. For example, the term "indicator species" has been applied to species whose richness is believed to correlate with that of other taxonomic groups (Kremen 1994, Carroll and Pearson 1998), that typically are associated with a specific habitat type or land use (Erhardt 1985, Hooson 1995, Blair and Launer 1997), or that might respond to anthropogenic ecological change (Temple and Wiens 1989, Noss 1990, Kremen et al. 1993, New et al. 1995, Hamer et al. 1997). As a result, there is little consensus on how focal species should be selected, what information their distribution or status might provide, and what their management might accomplish.

It is absolutely vital to outline the objectives of employing focal species in a given conservation or management scenario. In this paper, we assume that the management objective is to reliably detect and conserve localities with relatively rich or intact assemblages of native species. An associated goal is to identify any other localities that must be maintained to support the continued presence of particular (e.g. red-listed) species across the landscape. In light of resource constraints and confusion about terminology, it is also imperative that conservation biologists develop objective methods for selecting focal species. We outline here a methodology for selection and more detailed screening of potential focal species for conservation planning and monitoring of species richness. Our protocol can easily be applied to many existing data sets that have been compiled by researchers or practitioners, emphasizing that scientific defensibility and practicality are complementary rather than

mutually exclusive. We suggest that 1) species inventories at several localities, followed by 2) nestedness analysis (e.g. Patterson 1987), 3) occupancy modeling of selected taxa (with logistic regression and incidence function models; Sjögren-Gulve and Hanski 2000), and 4) empirical validation and further analysis constitute a powerful method for assessment of whether focal species can serve as trustworthy surrogates for monitoring of multiple species. We illustrate this procedure with case studies of cryptogams in boreal forests of northern Sweden and butterflies in mountain ranges in the western United States.

Working procedure

We present a stepwise working procedure, developed using two case studies as model systems, to analyze and validate the use of focal species for monitoring species richness. The procedure can be applied to diverse ecosystems and taxonomic groups despite their inherent variance in species richness and composition. The scientific basis for conservation planning and monitoring will be increased by completing even the first two or three of the four steps that we describe. Accordingly, even practitioners that are substantially constrained by time, money, or management alternatives can benefit from these methods. Because ecological and political situations are not static, successful conservation demands flexible approaches to biological analysis and monitoring; cooperation between researchers and managers dramatically improves the probability of developing such tools. Thus, development of a biologically sound monitoring scheme that could prove applicable to multiple management areas within a region is a valuable long-term investment.

Case study I: cryptogams

Species richness of vascular plants in northern boreal forests is relatively low, but species richness of bryophytes, lichens, and fungi often is comparatively high. During the past century, intensified forestry resulted in a loss of habitat for many species of cryptogams. These losses were pronounced for species that largely occur upon dead trees or coarse woody debris (CWD). Species on CWD are a well-delineated group, and most have fairly large distributional ranges within boreal forests (Fig. 1). The volume of CWD in logged stands tends to be smaller and more variable than in old-growth stands, and extremely large and highly decayed logs generally are quite rare in logged stands (Kruys et al. 1999). How much CWD is sufficient, and what other environmental conditions are necessary, to maintain viable populations of CWD-dependent taxa in managed forests are questions of great concern. Recent studies have focused on analyzing distributional patterns of CWD-dependent species in managed stands of mature Norway

Fig. 1. Granlandet, an old-growth boreal spruce forest in northern Sweden. These habitats are characterized by large volumes of coarse woody debris and the associated flora is generally rich and composed by many species, rare or threatened in managed forests.

spruce *Picea abies* (L.) Karst. and attempting to detect values of environmental variables associated with >50% probability of these species' presence.

Case study II: butterflies

The Great Basin of western North America is dominated by > 200 mountain ranges. After the Pleistocene, these ranges were isolated from the surrounding lower-elevation valleys as the regional climate became warmer and drier (Brown 1978, Grayson 1993). Individual mountain ranges function as discrete islands of habitat for numerous taxa that either are restricted to montane communities or have relatively low mobility, including butterflies (McDonald and Brown 1992, Murphy and Weiss 1992). The United States Forest Service, which manages some of the largest and most biologically diverse mountain ranges in the Great Basin, generally develops conservation plans for individual mountain ranges.

Most ranges have been incised deeply by canyons. For many resident butterflies, canyons within a mountain range also represent archipelagos of habitat islands. Land uses within mountain ranges commonly are delineated at the level of individual or several adjacent canyons. Therefore, canyons are an appropriate spatial scale for analysis of butterfly distributions.

Step 1: species inventories

Collection of data on species composition is the first step. An assumption of our working procedure is that relatively complete data on presences and absences of resident species within the studied taxonomic group have been compiled, using standard methods, for a representative sample of locations (preferably > 20) to be managed and monitored in the future. Our methods are designed to identify focal species within a discrete geographic area. In a different landscape, the same taxa may not function as focal species. Although data on abundance and its variance yield considerable insight into population viability, our methods only require presence–absence data. Spatial and temporal variation in abundance for many taxa is substantial, subject to observer bias, and difficult to attribute to discrete causes (Simberloff 1998, Link and Sauer 1998, Droege et al. 1998). Standardized inventory guidelines are readily available for numerous taxonomic groups, including

Inventory methods

Cryptogams

Twenty-five old-growth stands were randomly selected from > 30 000 included in the stand register of the forestry company SCA Forest and Timber and from inventories of small, privately-owned stands conducted by local forestry boards. All study stands are located in central Sweden between 62.5–64°N and 14–21°E, are on mesic ground, and are dominated by 115–135 yr old Norway spruce. In each study stand, all units of CWD > 5 cm in diameter and 30 cm in length were identified. Species composition of each CWD unit was inventoried. Bryophytes, lichens (fruticose, foliose, and calicioid species) and wood-fungi in the family Polyporaceae were included in the inventories. A few wood-fungi in the order Corticiaeae also were included because they frequently are targeted in inventories conducted by Swedish forestry boards (Norén et al. 1995). The final data set comprised 845 CWD units, 19 species of bryophytes, 12 species of lichens, and 19 species of wood-fungi.

Butterflies

From 1994 through 1996, comprehensive inventories of butterflies were conducted in 19 canyons in the Toiyabe Range, a large mountain range in the central Great Basin (Fig. 2), using standard methods (Kremen 1992, Pollard and Yates 1993, Harding et al. 1995). It is reasonable to interpret that a given butterfly species is absent from a location if the area has been inventoried with standard methods during the appropriate season and weather conditions (Pullin 1995, Reed 1996). Field methods are described in detail elsewhere (Fleishman 1997, Fleishman et al. 1998). A total of 68 resident species (taxa believed to complete their entire life cycle in the Toiyabe Range) was recorded in these inventories.

mammals, amphibians, and butterflies (e.g. Pollard and Yates 1993, Heyer et al. 1994, Wilson et al. 1996). Inventory methods for the cryptogam and butterfly case studies are summarized in Box 1. We further assume that species composition at each locality does not vary dramatically from year to year. Ramifications of violating the latter assumption are explored in more detail in the discussion.

Relatively complete presence–absence data have not been compiled for all landscapes of conservation interest, and may need to be collected at the outset. Although it may be tempting to overlook this requirement, our methods are of limited value without this information. Robust analyses cannot be conducted in a data vacuum. With our approach, we have tried to strike a balance between modeling ideals and management realities that remains scientifically defensible.

Step 2: nestedness analysis

After inventory data have been compiled, nestedness analysis is the next step in evaluating potential focal species. A nested biota is one in which the species present in comparatively depauperate locations represent statistically proper, and thus statistically predictable, subsets of the species present in locations with greater species richness (Patterson and Atmar 1986) (Table 1). Many taxonomic groups exhibit nested distributional patterns, and nested species assemblages have been reported from diverse biogeographic regions around the world (Wright et al. 1998). Nestedness analyses can provide information crucial for designing a monitoring program because they allow for both documentation of distributional patterns (e.g. Patterson and Atmar 1986, Patterson 1987, 1990, Hager 1998) and exploration of mechanisms that may produce those patterns (e.g. Silvertown and Wilson 1994, Cook and Quinn 1995, Lomolino 1996, Worthen and Rohde 1996). Not only natural processes but also human activities may produce nested species distributions (Hecnar and M'Closkey 1997). Among the factors most commonly believed to generate nested distributions are differences among species in extinction propensity or colonization ability and spatial nesting of critical habitat features or resources (Cody 1986, Cutler 1991, Simberloff and Martin 1991, Kadmon 1995, Cook and Quinn 1995, Lomolino 1996, Worthen and Rohde 1996).

Several aspects of nestedness analyses are appealing to conservation biologists and managers (Patterson 1987, Simberloff and Martin 1991, McDonald and Brown 1992, Doak and Mills 1994, Boecklen 1997). First, nestedness analyses facilitate prediction of the order in which species are likely to be extirpated from (or to colonize) a suite of localities. Second, it may be possible to identify particular species that correlate with the presence of other

Fig. 2. Toiyabe Range, east slope, from the crest of Stewart Canyon.

species with high statistical credibility. Third, Atmar and Patterson's nestedness temperature calculator (1993) can be used to identify individual "idiosyncratic" species with distributions that may be affected by biogeographic factors different from those that affected the assemblage as a whole, such as association with unique environmental features at a subset of locations (Atmar and Patterson 1993). Fourth, differences in relative nestedness among species with distinct life history characteristics can help identify factors that shape community structure, including anthropogenic phenomena. For example, variation in nestedness among comparatively sedentary and mobile taxa suggests that dispersal ability has influenced local patterns of species composition (Cook and Quinn 1995, Kadmon 1995, Hecnar and M'Closkey 1997). Similarly, differential nestedness among species that vary in their sensitivity to human disturbance may indicate that anthropogenic habitat modifications are causing local extirpations (Hecnar and

Table 1. A nested system. Locations are listed as columns in order of decreasing species richness (# spp.); species are listed as rows in order of decreasing ubiquity. Species presence is indicated by "+".

	1	2	3	4	5	6	7	8	9	10	# locations
A	+	+	+	+	+	+	+	+	+	+	10
B	+	+	+	+	+	+		+	+		8
C	+	+	+	+		+	+	+	+		8
D	+	+	+		+	+	+	+			7
E	+	+	+	+	+	+	+				7
F	+	+	+	+	+	+					6
G	+	+		+	+		+				5
H	+	+	+	+							4
I		+	+								2
J	+										1
# spp.	9	9	8	7	6	6	5	4	3	1	

Table 2. Partial species-by-location matrix of liverworts on coarse woody debris in old-growth forest stands in northern Sweden. Data are shown for the 12 sites with the greatest species richness and the 12 sites with the lowest species richness; row and column totals reflect data for the full matrix of 43 sites. Two liverworts, *Blepharostoma trichophyllum* (species O) and *Cephalozia loitlesbergii* (species T), appeared to be potential indicators of high species richness in the landscape as a whole. Species codes: A = *Ptilidium pulcherrimum*, B = *Lophozia ciliata*, C = *Lophozia longiflora*, D = *Lophozia silvicola*, E = *Cephalozia bicuspidata*, F = *Lophozia longidens*, G = *Anastrophyllum hellerianum*, H = *Cephalozia lunulifolia*, I = *Lophozia ventricosa*, J = *Barbilophozia attenuata*, K = *Riccardia latifrons*, L = *Calypogeia integristipula*, M = *Calypogeia neesiana*, N = *Riccardia palmata*, O = *Blepharostoma trichophylla*, P = *Lepidozia reptans*, Q = *Lophozia incisa*, R = *Cephalozia leucantha*, S = *Cephalozia pleniceps*, T = *Cephalozia loitlesbergii*, U = *Harpanthus flotovianus*, V = *Jungermannia leiantha*, W = *Geocalyx graveolens*, X = *Anastrophyllum minutum*.

	1	2	3	4	5	6	7	8	9	10	11	12	32	33	34	35	36	37	38	39	40	41	42	43	# sites
A	+	+	+	+	+	+	+	+	+	+	+	+	+	+	+	+	+	+	+	+	+	+	+	+	43
B	+	+	+	+	+	+	+	+	+	+	+	+	+	+	+	+	+	+		+		+	+	+	40
C	+	+	+	+	+	+	+	+	+	+	+	+	+	+	+	+	+	+	+		+	+	+	+	39
D	+	+	+	+	+	+	+	+	+	+	+	+	+	+	+	+		+				+	+	+	37
E	+	+	+	+	+	+	+	+	+	+	+	+			+		+	+	+					+	34
F	+		+	+	+	+		+	+	+	+	+			+			+	+			+	+		32
G	+	+	+	+	+	+	+	+	+	+	+	+	+	+	+			+							32
H	+	+	+	+	+	+	+	+	+	+		+			+			+			+	+			28
I	+			+			+	+	+		+	+				+									17
J		+	+			+	+	+			+	+			+	+	+						+		17
K	+	+	+	+	+		+	+		+	+														14
L	+	+	+	+	+		+								+		+		+	+					13
M	+	+	+			+	+	+				+					+			+					13
N		+	+	+	+	+			+		+														7
O	+	+	+	+	+	+																			6
P	+		+	+	+		+																		5
Q	+				+																				5
R	+	+						+				+													4
S	+	+				+																			3
T	+	+																							2
U				+																					1
V														+											1
W																									1
X																									1
# spp.	18	16	15	14	14	13	13	12	11	11	11	11	7	7	6	6	6	6	6	6	6	6	6	5	

M'Closkey 1997, Jonsson and Jonsell 1999).

There exist numerous methods for quantifying nestedness statistically (see Wright et al. 1998 for a review). Community-level metrics assess whether groups of species in relatively depauperate locations are statistically proper subsets of the species present in relatively rich locations. Most of these metrics are calculated with fairly straightforward computer software packages that are available upon request from their authors (e.g. RANDOM0 and RANDOM1, Patterson and Atmar 1986; NESTCALC, Wright and Reeves 1992; PN, Lomolino 1996; Nestedness Calculator, Atmar and Patterson 1993, http://aics-research.com/nestedness/tempcalc.html). Species-level nestedness analyses, by comparison, assess whether a given species' sequence of presences and absences is significantly different from random. In other words, a species with a nested distribution tends to be present in relatively species-rich locations (which also may be relatively large or non-isolated) and absent from relatively species-poor locations. A spe-

cies-by-location matrix of liverworts growing on coarse woody debris in old-growth forests in northern Sweden provides an excellent illustration (Table 2). Not only is the species assemblage strongly nested, but also two liverworts, *Blepharostoma trichophyllum* and *Cephalozia loitlesbergii*, appear to be potential focal species in the system. With the exception of *Ptilidium pulcherrimum*, which was ubiquitous, *B. trichophyllum* and *C. loitlesbergii* were the only two species with no "holes" in their distributional patterns – i.e. they were not absent from any location with greater species richness than another location in which they were present. The presence of the latter two species may suggest that > 12 or 15 additional species of liverworts, respectively, occur in the same locations.

At least four assumptions are made when conducting most of the common nestedness analyses (Patterson and Atmar 1986, Patterson 1987, Wright and Reeves 1992, Atmar and Patterson 1993, Wright et al. 1998). First, locations have a common biogeographic history. Second, loca-

Box 2

Nestedness of cryptogams

Nestedness of each species-by-stand matrix was computed with the model RANDOM1 (Patterson and Atmar 1986), which quantifies whether species present in relatively depauperate locations represent statistically proper subsets of the species present in locations that are richer in species. Overall species composition was significantly nested (Z = 3.57, p < 0.01). Bryophytes and lichens were significantly nested as groups (bryophytes: Z = 4.67, p < 0.01; lichens: Z = 1.61, p = 0.05). As a group, however, the wood-fungi did not have a significantly nested distributional pattern (Z = 0.90, p = 0.18). Wood-fungi indeed may have more random patterns of occurrence than either bryophytes or lichens; it also is possible that temporal variation in the presence of fruiting bodies, which are unique to the wood-fungi, may obscure the actual pattern of occurrence. Regardless of the mechanisms that influence the distribution of wood-fungi, results suggest that larger sample plots or repeated samples over several years may be necessary to define reliable indicators of species richness within this taxonomic group.

Three species seemed particularly well nested and thus were selected as potential focal species. Two of the species, *Anastrophyllum hellerianum* and *Lophozia ciliata*, are red-listed bryophytes confined to large and well decayed logs. Although small, they typically are found on conifer logs and their occurrence should be fairly easy to monitor. The third species, *Chaenotheca gracillima*, is a calicioid lichen (a lichen with stalked, pin-like fruiting bodies, apothecia) that grows on decayed snags in closed forest. The latter species is relatively large and also should be tractable to monitor.

Box 3

Nestedness of butterflies

Relative nestedness (C) of the species-by-canyons matrix was computed with the program NESTCALC (Wright et al. 1990). Cochran's Q statistic was used to test whether the matrix was significantly nested (Wright and Reeves 1992). Wilcoxon two-sample rank tests (Mann-Whitney U-tests) were used to test whether individual species had nested distributions (Simberloff and Martin 1991, Kadmon 1995, Hecnar and M'Closkey 1997).

The distributional pattern of all resident butterfly species in the Toiyabe Range was significantly nested (C = 0.521, p < 0.001). Wilcoxon tests demonstrated that > 80% of the butterfly species that inhabit the Toiyabe Range had nested patterns of occurrence (p < 0.5); these patterns were statistically significant (p < 0.05) for roughly one-third of the species (Fleishman and Murphy 1999).

tions shared an ancestral pool of species. Third, occurrences of species in the assemblage of interest are affected by similar environmental gradients. Fourth, relatively complete species inventories have been conducted. In addition, it generally is assumed that the short-term presence of species in each locality is independent (i.e. individuals are not shared among localities). As with all quantitative methods, interpretation of the results of nestedness analyses must reflect the extent to which the assumptions of the method were met. Systems in which the assumptions are violated are less likely to be strongly nested. Results from nestedness analyses of the cryptogam and butterfly systems are presented in Boxes 2 and 3, respectively.

Step 3: linear regression or occupancy modeling ?

The third step in selecting focal species focuses on assessment of environmental variables that may influence occupancy patterns. Regardless of whether the species assemblage is significantly nested, it may be possible to use stepwise multiple linear regression (MR) to predict species richness as a function of environmental parameters that are tractable to quantify or manage, such as precipitation, elevation, volume of CWD, or logging intensity. To illustrate, we used MR to build a predictive model of butterfly species richness in relation to environmental variables (Box 4).

Multiple linear regression of butterfly species richness

This analysis was based on species inventories and environmental variables for 102 canyon segments, each extending for ca 100 vertical meters ("elevational bands"), in 15 of the Toiyabe Range canyons included in the nestedness analysis described in Box 3 (see Fleishman et al. 1998 for details). Species richness of the elevational bands ranged from 7 to 53. Roughly 50 environmental attributes of each band, including area, elevation, slope, aspect, solar insolation, topographic position (valley, ridge, or open slope), and distance from water, were either measured in the field or derived digitally using a Geographic Information System.

Of the 48 independent variables included in the MR analysis, eight were significantly correlated with species richness of butterflies. Species richness tended to increase with increasing band length, topographic heterogeneity, solar insolation, area, and proximity to water. The MR model was highly significant, and explained 77% of the variation in species richness ($r^2 = 0.77$). For example, the model correctly classified 88% of the 49 elevational bands with > 25 butterfly species, and 46% of the 13 elevational bands with > 40 species. The risk that the model incorrectly would predict that > 25 species occurred in an elevational band actually occupied by < 25 species was 19%; in no case did the model incorrectly predict that an elevational band actually occupied by < 40 species contained > 40 species. By comparison, the single environmental variable that explained the greatest percentage of the variance in species richness had an r^2 of 0.47.

A compelling reason to employ occupancy modeling in conjunction with nestedness analysis is that selective protection of species-rich localities may not guarantee the viability of all taxa in those areas. Often, populations in diverse areas may be augmented by immigrants from less diverse localities.

If a species assemblage is highly nested, the presence of focal species may discern relatively species-rich locations with higher precision than a MR model. For example, linear regression analysis demonstrated that variation in species richness of cryptogams in managed boreal forest stands (see Box 1) was best explained by number of CWD units per ha. This variable, easily measured in the field, explained 44% of the variation in species richness (species richness ranged from 1 to 30 species). However, one of the focal species, the liverwort *Anastrophyllum hellerianum*, was present in all 12 of the 25 stands with 16 or more species (100% precision), and only occurred in one stand with < 16 species (8% risk of misidentification).

We strongly advocate analysis of the occupancy patterns of potential focal species with stepwise logistic regression (LR; see Sjögren-Gulve and Hanski 2000) to assess their reliability as indicators of species richness. Because LR simultaneously can analyze associations between the occurrence pattern of a given taxon, environmental variables, and species richness, it can elucidate whether the presence of a focal species primarily correlates with high species richness or largely is explained by other variables. If presence of the focal species indeed is strongly correlated with high species richness, then LR may identify environmental factors that explain additional variance in its occurrence. For example, a focal species that is a relatively poor disperser may be absent from species-rich sites that are relatively isolated. In addition, recognition of variables that

are significantly associated with presence of a focal species may suggest how natural or anthropogenic ecological changes might affect its distribution and viability. Examples from the cryptogam and butterfly systems are shown in Boxes 5 and 6.

The relative advantages of using the MR model as opposed to the presence of the focal species *Satyrium sylvinum* (Fig. 3) to predict species richness of butterflies depend upon 1) what managers seek to measure and 2) what mismeasurements are more likely to result in erroneous – and potentially costly – management decisions. Presence of *S. sylvinum* indicated elevational bands with > 40 butterfly species with 54% precision (Table 3), while the precision of the MR model was 46%. However, the risk that presence of *S. sylvinum* erroneously would suggest that an elevational band was occupied by > 40 species was 30%; the corresponding misindication risk of the MR model was 0%. The MR model classified elevational bands with more 25 species with 88% precision and a 19% misindication risk; the presence of *S. sylvinum* classified bands with > 25 species with 20% precision and 0% risk of misindication (Table 3).

Step 4: empirical validation and further analysis

The contention that the presence of a particular species functions as an effective substitute measure of species richness must be treated as a hypothesis (Landres et al. 1988, Swengel and Swengel 1997, Carroll and Pearson 1998). The fourth step of our method establishes a program of empirical validation and further analysis to test this hypothesis. Periodic repetition of species inventories is inte-

Fig. 3. *Satyrium sylvinum.*

gral to validating the use of focal taxa. Iterative data on species composition may facilitate detection of population turnover, evaluation of the temporal stability of nested distributional patterns, and examination of colonization and extinction patterns of focal species. In the butterfly system, for example, inventories in subsequent years showed that turnover does not always affect the relative nestedness of the entire species assemblage, but it may affect the degree of nestedness of individual species (Fleishman unpubl.).

Building upon the LR analysis of occupancy patterns in Step 3, a preliminary population viability analysis of focal species can be performed with an incidence function model (IFM; Hanski 1994, Hanski et al. 1996). The IFM can be parameterized with data on the species' occupancy pattern from a single time step, and yields location-specific extinction and colonization probabilities for PVA (see Sjögren-Gulve and Hanski 2000). However, repeated inventories are imperative to ascertain whether the proportion of localities occupied by the focal species changes significantly over time. If occupancy is fairly constant, it is valid to base the PVA upon an IFM; if occupancy fluctuates, it may be more appropriate to model extinction and colonization patterns with LR (Sjögren-Gulve and Hanski 2000). Either approach may provide additional important information on how ecological perturbations may affect the species' distribution and viability. An illustrative example in-

volving multiple taxa comes from the Baltic coast of Sweden, where ponds occupied by pool frogs *Rana lessonae* had higher species richness of semi-aquatic plants and dragonflies than unoccupied ponds (Fig. 4). Occupancy modeling of the pool frog (Sjögren-Gulve and Ray 1996) showed that large-scale forestry had detrimental effects on population viability. The potential value of the pool frog as an indicator of high species richness was corroborated by Sahlén (1999) who showed that ponds with nearby clearcuts, performed 5–15 yr ago, had lower species richness of dragonflies than other ponds.

Discussion

Our results demonstrate that the roles that focal species realistically can play in monitoring and management vary among ecosystems and taxonomic groups. For example, presence of focal species of boreal cryptogams appeared to indicate locations with high species richness more reliably than critical levels of some environmental variables. In the Great Basin, presence of the butterfly *Satyrium sylvinum* likewise seemed to be a promising indicator of high species richness. In the latter system, however, a MR model also generated fairly accurate predictions of specified levels of species richness. Moreover, the MR model yielded consid-

Box 5

Logistic regression of focal cryptogams

Occupancy patterns of the three potential focal species of cryptogams (see Box 2) further were examined with LR. Environmental variables used in the analyses included size and degree of decay (six stages) of each CWD unit; area; crown cover; canopy height; basal area of each tree species and across tree species; productivity (tree volume per year and per ha); size distributions of trees, cut stumps, and logs; number of CWD units per ha, volume of CWD per ha; and number of uprooted trees, snapped trees, snags, and cut logs.

When species richness was included as an independent variable in the LR models, no other environmental variable significantly improved the model for any of the focal taxa (p > 0.15). Tight correlations between occurrences of the three focal cryptograms and species richness suggested that they may function as dependable indicators of species richness.

When species richness was excluded from the LR analyses, highly statistically significant models were obtained for all three species (see table below). The percent of stands for which the LR models correctly predicted whether the focal species was present or absent was 88% for *Anastrophyllum hellerianum*, 92% for *Chaenotheca gracillima*, and 88% for *Lophozia ciliata*. These LR models not only identified environmental factors that explained significant variation in species occupancy patterns, but also suggested thresholds beyond which those variables were correlated with a > 50% probability of species occurrence. Threshold values for *A. hellerianum* were > 35 logs and 100 trees > 20 cm DBH. For *C. gracillima*, there was a threshold value of 16 snags. Threshold values for *L. ciliata* were > 40 logs and a basal area > 20 m^2 ha^{-1}. Although correlation does not necessarily imply causality, the strength of the relationships between these environmental variables and occurrence of the focal cryptograms may serve as a basis for more detailed occupancy models (see Step 4).

Results from LR analysis of the occurrence of potential focal cryptograms on coarse woody debris in boreal forest. Threshold p value for variable to enter the model was 0.10.

Species	Significant variables, direction of correlation with occurrence, and model improvement statistics (all DF = 1)	Deviance goodness-of-fit statistics of final model
Anastrophyllum hellerianum	# logs (+) (χ^2 = 9.36, p < 0.005) # trees > 20 cm DBH (+) (χ^2 = 7.73, p = 0.005)	χ^2 = 15.27 DF = 23, p = 0.88
Chaenotheca gracillima	# snags (+) (χ^2 = 16.61, p < 0.0001)	χ^2 = 25.92 DF = 24, p = 0.36
Lophozia ciliata	basal area (+) (χ^2 = 5.76, p < 0.05) # logs (+) (χ^2 = 15.46, p < 0.0001)	χ^2 = 11.72 DF = 23, p = 0.97

erable insight into environmental variables that may influence butterfly species richness in the region. Such knowledge may help practitioners to evaluate the potential effects of alternative management strategies on species distributions.

In numerous management situations, species-based conservation assessments are preferable to assessments that are based exclusively upon environmental variables. Species presence and persistence is an ultimate verification that environmental conditions at present and in the recent past meet their requirements. Although it may be difficult to identify taxa that can provide an early warning of environmental change, certain species may be useful for quantifying the effects of known ecological changes on the biota. Tracking the effects of environmental perturbations fre-

quently is a high priority for practitioners, particularly if the changes can be modified by management. Under such circumstances, focusing inventory and monitoring efforts on a subset of species may be considerably cheaper than attempting to measure continuously the status of an entire ecological community. Use of focal species also may increase the speed at which management can be adapted in response to data. Furthermore, incorporating focal species into conservation assessments and monitoring provides a general means to link ecosystem management with conservation of individual species of concern (Lambeck 1997, Noss et al. 1997, Simberloff 1998).

Researchers and managers alike must keep in mind that need for a tool does not always imply its existence; the sci-

Box 6

Logistic regression of focal butterflies

Four species (*Lycaena helloides*, *L. rubidus*, *Satyrium sylvinum*, and *Speyeria nokomis*), all with nested patterns of occurrence, were selected for LR analysis on the basis of their potential to function as umbrella species for butterflies in the Toiyabe Range. None of these species is presently red-listed. To identify potential umbrella species, we used an index, described in detail elsewhere (Fleishman et al. 2000), that incorporates three parameters: median-rarity, percentage of co-occurring species, and disturbance sensitivity. We considered median-rarity because species with an intermediate degree of ubiquity may be most likely to function as effective umbrella species. Relatively rare species (e.g., those that occupy less than five percent of the area to be managed), by comparison, may not be distributed across enough of the landscape to ensure the viability of populations of many other species. The second parameter in the umbrella index, percentage of co-occurring species, measures the extent to which the presence of a given species overlaps with the presence of other species in its taxonomic group. Disturbance sensitivity was evaluated with an index that quantifies the likelihood that a butterfly species will decline as a result of human activities. LR analyses were based on the same set of data on species composition and environmental variables described in Box 4.

Species richness was included as an independent variable in the LR analyses. The analysis for *S. sylvinum* – the only taxon for which species richness entered the LR model – also was repeated excluding species richness. Highly statistically significant models were obtained for all four butterfly species (see table below). Three species, *L. helloides*, *L. rubidus*, and *S. nokomis*, were rejected as potential umbrella species because correlations between their presence and species richness were weaker than correlations between their presence and several environmental variables. Protection of *S. sylvinum*, by contrast, might serve to protect many co-occurring species. Of the four variables significantly correlated with the presence of *S. sylvinum*, species richness was the strongest correlate. The other three variables connected with the presence of this species – slope, precipitation, and spring solar insolation – are relatively easy to measure in the field or to derive digitally.

Results from LR analysis of the occurrence of potential focal butterfly species. Threshold p value for variable to enter the model was 0.10.

Species	Significant variables, direction of correlation with occurrence, and model improvement statistics (all DF = 1)	Deviance goodness-of-fit statistics of final model
Lycaena helloides	distance to nearest occupied location (–) ($\chi^2 = 98.17$, $p < 0.0001$) aspect (east-facing) (+) ($\chi^2 = 4.27$, $p < 0.05$)	$\chi^2 = 40.02$ DF = 100, p = 1.000
Lycaena rubidus	distance to nearest occupied location (–) ($\chi^2 = 45.66$, $p < 0.001$) minimum distance to water (–) ($\chi^2 = 6.26$, $p = 0.01$) maximum distance to water (+) ($\chi^2 = 3.13$, $p < 0.10$)	$\chi^2 = 58.28$ DF = 99, p = 1.000
Satyrium sylvinum	species richness (+) ($\chi^2 = 16.95$, $p < 0.0001$) minimum slope (–) ($\chi^2 = 6.00$, $p = 0.01$) precipitation (–) ($\chi^2 = 7.84$, $p = 0.005$) equinox insolation (–) ($\chi^2 = 4.06$, $p < 0.05$)	$\chi^2 = 31.13$ DF = 98, p = 1.000
Speyeria nokomis	distance to nearest occupied location (–) ($\chi^2 = 34.74$, $p < 0.0001$) minimum distance to water (–) ($\chi^2 = 23.38$, $p < 0.0001$) equinox insolation (+) ($\chi^2 = 11.57$, $p < 0.001$) aspect (east-facing) (–) ($\chi^2 = 4.95$, $p < 0.05$) gradient (–) ($\chi^2 = 3.43$, $p < 0.10$)	$\chi^2 = 22.09$ DF = 97, p = 1.000

Reanalysis excluding species richness:

Species		
Satyrium sylvinum	minimum slope (–) ($\chi^2 = 26.36$, $p < 0.0001$) minimum distance to water (–) ($\chi^2 = 6.10$, $p = 0.01$)	$\chi^2 = 42.72$ DF = 100, p = 1.000

Table 3. Observed species richness (S_{obs}), species richness predicted by the MR model (S_{pred}), and presence (+) of the butterfly *Satyrium sylvinum* in elevational bands occupied by > 25 species of butterflies.

Band	S_{obs}	S_{pred}	*Satyrium sylvinum*
KI04	53	47	+
BI03	48	37	+
BI01	47	43	+
BI04	47	41	+
SU03	46	38	
SU05	45	41	
BC01	44	33	
BI02	44	45	+
SU04	44	38	+
WS02	44	36	
KI05	42	42	
OP05	42	34	
WS03	42	39	+
BC02	40	37	
SU02	40	30	
BC03	39	37	
KI06	39	40	
SU01	39	33	
JE03	38	38	
WS01	38	37	
KI02	37	40	
SJ02	37	39	
SU06	37	39	
BI05	36	33	
KI03	36	39	
OP06	36	30	
BC04	35	30	
OP04	35	35	
JE04	34	40	
OP03	33	30	
SU07	33	30	
SJ01	32	39	+
WS04	32	35	+
JE02	31	32	
ST05	31	29	
BI06	30	30	
JE01	30	29	
OP01	30	32	
BC05	29	30	
KI01	28	28	
WS05	28	29	
CT02	27	25	
NT06	27	23	
OP02	27	25	
OP07	27	28	
OP08	27	21	
SJ05	27	25	
CT03	26	33	+
WA06	26	19	

entific defensibility of proposed focal species must be evaluated empirically. Nonetheless, our four-step procedure represents a powerful method for screening focal species for conservation planning. Few effective and efficient methods for identifying potential indicators of species richness currently exist. Two-way indicator species analysis (TWINSPAN, Hill 1979) and cluster analysis widely have been used to define ecological communities, especially vegetational communities, on the basis of characteristic species composition (e.g. Dufrêne and Legendre 1997). Assemblages that are relatively species-rich then can be identified, and taxa with high affinity for those communities may be used as "indicators" of species richness. However, the primary purpose of methods like TWINSPAN and cluster analysis is to detect and describe variation in species composition. If species distribution patterns within a well-defined ecological community are strongly nested with respect to species richness, then nestedness analyses may provide a more straightforward and intuitively comfortable approach to selecting potential indicators of species richness.

In addition to serving as a tool for identification of focal species, nestedness analyses may inform conservation efforts ranging from land use delineation to habitat restoration. The sequence in which species drop out from an array of habitat remnants, for instance, not only may suggest which taxa are most sensitive to fragmentation (Patterson 1987, Cutler 1991, Lomolino 1996), but also may provide managers with information on critical thresholds of habitat area. By analyzing whether systems are nested with respect to area, isolation, abiotic parameters, or environmental gradients of management concern (e.g. concentrations of pollutants, grazing intensity, or other forms of human disturbance), practitioners may gain insight into mechanisms driving distributional patterns in landscapes under their jurisdiction. A related approach to elucidating causal mechanisms is to examine species distributions among plots of similar size. This method controls for species-area relationships and suggests whether habitat quality may be generating nested distribution patterns. Studies of boreal cryptogams, for example, demonstrate that species composition in plots of similar area often may be nested (Berglund 1997, Kruys and Jonsson 1997, Skarstedt 1998, Jonsson and Jonsell 1999). Nested patterns also were detected when analyses were restricted to old-growth stands, implying that even fairly short environmental gradients may produce nested patterns.

Even if managers do not have the time or resources to execute all four steps that we have outlined, species inventories (Step 1) and nestedness analysis (Step 2) alone can facilitate identification of potential focal species. Regression modeling (Step 3), repeated inventories (Step 4), and more detailed population models (Step 4) offer incremental gains for conservation planning. Steps 3 and 4 validate the proposed use of focal taxa as indicators of species richness and may distinguish environmental variables associat-

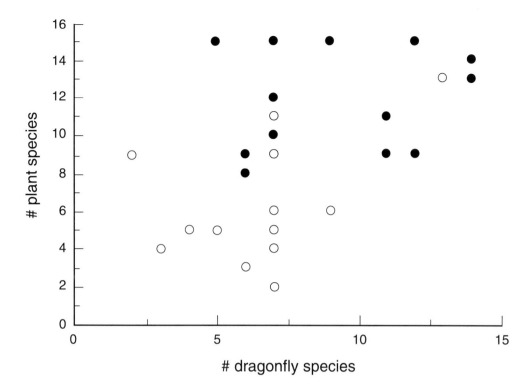

Fig. 4. Species richness of dragonflies and semi-aquatic vascular plants in ponds with (●) and without (○) pool frogs *Rana lessonae* along the Baltic coast of east-central Sweden (Sahlén and Ekestubbe unpubl.). Presence of the pool frog reliably indicated high species richness of both dragonflies and plants (F = 9.23, p < 0.005 and F = 25.08, p < 0.0001, respectively).

ed with their presence. Gradient analyses are an alternative method for examining how species richness varies in response to environmental variables such as elevation, precipitation, and soil chemistry. However, these analyses invoke assumptions concerning the statistical relationship (e.g. linear or exponential) between the selected gradients and species richness. Moreover, single environmental variables often do not explain a high proportion of the variance in species richness (see examples in Step 3). Occupancy modeling may prove more informative than regressing a certain range of species richness values on a suite of environmental variables. In addition, knowledge of the relative impact of diverse environmental factors on species distributions is critical for effective conservation planning regardless of whether species richness is a key correlate of species occurrence. Occupancy modeling also can shed light on environmental factors correlated with turnover events (Sjögren-Gulve and Ray 1996) that affect nestedness and species richness. As noted, locations with certain environmental characteristics, regardless of their relative degree of species diversity, may serve as key supports for populations in species-rich areas.

If it appears that a modest number of environmental variables are correlated with the presence of focal species, then repeated inventories (Step 4) may provide a strong foundation for parameterizing more detailed analyses of distribution or turnover patterns, including incidence function models or other population viability analyses (Akçakaya and Sjögren-Gulve 2000). This knowledge also may help improve management and conservation planning. We emphasize the tremendous value of close collaboration between ecologists and resource managers for the development of scientifically defensible tools for assessment of species richness and other ecosystem attributes.

Acknowledgements – We thank the Nevada Biodiversity Research and Conservation Initiative, the Swedish Environmental Protection Agency, and the Swedish Council for Forestry and Agricultural Research for financial support; Göran Sahlén for supplying unpubl. data, and George Austin, Nicholas Kruys, Dennis Murphy, and Dan Rubinoff for field assistance and numerous valuable discussions.

References

Akçakaya, H. R. and Sjögren-Gulve, P. 2000. Population viability analyses in conservation planning: an overview. – Ecol. Bull. 48: 9–21.

Atmar, W. and Patterson, B. D. 1993. The measure of order and disorder in the distribution of species in fragmented habitat. – Oecologia 96: 373–382.

Berglund, H. 1997. Är förekomsten av vedlevande svampar slumpmässig i ett mosaikartat skog-myr landskap? – M.S. thesis, Umeå Univ., Sweden.

Blair, R. B. and Launer, A. E. 1997. Butterfly diversity and human land use: species assemblages along an urban gradient. – Biol. Conserv. 80: 113–125.

Boecklen, W. J. 1997. Nestedness, biogeographic theory, and the design of nature reserves. – Oecologia 112: 123–142.

Brown, J. 1978. The theory of insular biogeography and the distribution of boreal mammals and birds. – Great Basin Nat. Mem. 2: 209–228.

Carroll, S. S. and Pearson, D. L. 1998. Spatial modeling of butterfly species richness using tiger beetles (Cicindelidae) as a bioindicator taxon. – Ecol. Appl. 8: 531–543.

Cody, M. L. 1986. Diversity, rarity, and conservation in Mediterranean-climate regions. – In: Soulé, M. E. (ed.), Conservation biology: the science of scarcity and diversity. Sinauer, pp. 122–152.

Cook, R. and Quinn, J. F. 1995. The influence of colonization in nested species subsets. – Oecologia 102: 413–424.

Cutler, A. 1991. Nested faunas and extinction in fragmented habitats. – Conserv. Biol. 5: 496–505.

Doak, D. F. and Mills, L. S. 1994. A useful role for theory in conservation. – Ecology 75: 615–626.

Droege, S., Cyr, A. and Larivée, J. 1998. Checklists: an underused tool for the inventory and monitoring of plants and animals. – Conserv. Biol. 12: 1134–1138.

Dufrêne, M. and Legendre, P. 1997. Species assemblages and indicator species: the need for a flexible asymmetrical approach. – Ecol. Monogr. 67: 345–366.

Erhardt, A. 1985. Diurnal lepidoptera: sensitive indicators of cultivated and abandoned grassland. – J. Appl. Ecol. 22: 849–861.

Fleishman, E. 1997. Mesoscale patterns in butterfly communities of the central Great Basin and their implications for conservation. – Ph.D. thesis, Univ. of Nevada, Reno, NV.

Fleishman, E. and Murphy, D. D. 1999. Patterns and processes of nestedness in a Great Basin butterfly community. – Oecologia 119: 133–139.

Fleishman, E., Austin, G. T. and Weiss, A. D. 1998. An empirical test of Rapoport's rule: elevational gradients in montane butterfly communities. – Ecology 79: 2482–2493.

Fleishman, E., Murphy, D. D. and Brussard, P. F. 2000. A new method for selection of umbrella species for conservation planning. – Ecol. Appl. 10: 569–579.

Grayson, D. K. 1993. The desert's past: a natural prehistory of the Great Basin. – Smithsonian Inst. Press, Washington, D.C.

Hager, H. A. 1998. Area-sensitivity of reptiles and amphibians: are there indicator species for habitat fragmentation. – Ecoscience 5: 139–147.

Hamer, K. C. et al. 1997. Ecological and biogeographical effects of forest disturbance on tropical butterflies of Sumba, Indonesia. – J. Biogeogr. 24: 67–75.

Hanski, I. 1994. A practical model of metapopulation dynamics. – J. Anim. Ecol. 63: 151–162.

Hanski, I. et al. 1996. The quantitative incidence function model and persistence of an endangered butterfly metapopulation. – Conserv. Biol. 10: 578–590.

Harding, P. T., Asher, J. and Yates, T. J. 1995. Butterfly monitoring 1-recording the changes. – In: Pullin, A. S. (ed.), Ecology and conservation of butterflies. Chapman and Hall, pp. 3–22.

Hecnar, S. J. and M'Closkey, R. T. 1997. Patterns of nestedness and species association in a pond-dwelling amphibian fauna. – Oikos 80: 371–381.

Heyer, R. W., Donnelly, M. and Hayek, L. C. (eds) 1994. Measuring and monitoring biological diversity: standard methods for amphibians. – Smithsonian Inst. Press, Washington, D.C.

Hill, M. O. 1979. TWINSPAN – a FORTRAN program for arranging multivariate data in an ordered two-way table by classification of individuals and attributes. – Cornell Univ., Ithaca, New York.

Holt, R. D. 1997. From metapopulation dynamics to community structure: some consequences of spatial heterogeneity. – In: Hanski, I. A. and Gilpin, M. E. (eds), Metapopulation biology: ecology, genetics, and evolution. Academic Press, pp. 149–164.

Hooson, J. R. 1995. Case study. The Morecambe Bay carboniferous limestone hills: a unique butterfly "hot spot". – Biol. J. Linn. Soc. 56 (Suppl.): 97–98.

Jonsson, B. G. and Jonsell, M. 1999. Exploring potential biodiversity indicators in boreal forests. – Biodiv. Conserv. 8: 1417–1433.

Kadmon, R. 1995. Nested species subsets and geographic isolation: a case study. – Ecology 76: 458–465.

Kremen, C. 1992. Assessing the indicator properties of species assemblages for natural areas monitoring. – Ecol. Appl. 2: 203–217.

Kremen, C. 1994. Biological inventory using target taxa: a case study of the butterflies of Madagascar. – Ecol. Appl. 4: 407–422.

Kremen, C. et al. 1993. Terrestrial arthropod assemblages: their use in conservation planning. – Conserv. Biol. 7: 796–808.

Kruys, N. and Jonsson, B. G. 1997. Insular patterns of calicioid lichens in a boreal old-growth forest-wetland mosaic. – Ecography 20: 605–613.

Kruys, N. et al. 1999. Dead spruce trees and wood-inhabiting plants and fungi in managed Swedish boreal forests. – Can. J. For. Res. 29: 178–186.

Lambeck, R. J. 1997. Focal species: a multi-species umbrella for nature conservation. – Conserv. Biol. 11: 849–856.

Landres, P. B., Verner, J. and Thomas, J. W. 1988. Ecological uses of vertebrate indicator species: a critique. – Conserv. Biol. 2: 316–328.

Link, W. A. and Sauer, J. R. 1998. Estimating population change from count data: application to the North American Breeding Bird Survey. – Ecol. Appl. 8: 258–268.

Lomolino, M. V. 1996. Investigating causality of nestedness of insular communities: selective immigrations or extinctions? – J. Biogeogr. 23: 699–703.

Longino, J. T. and Colwell, R. K. 1997. Biodiversity assessment using structured inventory: capturing the ant fauna of a tropical rain forest. – Ecol. Appl. 7: 1263–1277.

McDonald, K. A. and Brown, J. H. 1992. Using montane mammals to model extinctions due to climate change. – Conserv. Biol. 6: 409–415.

Murphy, D. D. and Weiss, S. B. 1992. Effects of climate change on biological diversity in western North America: species losses and mechanisms. – In: Peters, R. L. and Lovejoy, T. E. (eds), Global warming and biological diversity. Yale Univ. Press, pp. 355–368.

New, T. R. et al. 1995. Butterfly conservation management. – Annu. Rev. Entomol. 40: 57–83.

Niemi, G. J. et al. 1997. A critical analysis on the use of indicator species in management. – J. Wildl. Manage. 61: 1240–1252.

Norén, M. et al. 1995. Datainsamling vid inventering av nyckelbiotoper. – Skogsstyrelsen, Jönköping, in Swedish.

Noss, R. F. 1990. Indicators for monitoring biodiversity: a hierarchial approach. – Conserv. Biol. 4: 355–364.

Noss, R. F., O'Connell, M. A. and Murphy, D. D. 1997. The science of conservation planning: habitat conservation under the Endangered Species Act. – World Wildlife Fund and Island Press, Washington, D.C.

Oliver, I. and Beattie, A. J. 1996. Designing a cost-effective invertebrate survey: a test of methods for rapid assessment of biodiversity. – Ecol. Appl. 6: 594–607.

Patterson, B. D. 1987. The principle of nested subsets and its implications for biological conservation. – Conserv. Biol. 1: 323–334.

Patterson, B. D. 1990. On the temporal development of nested subset patterns of species composition. – Oikos 59: 330–342.

Patterson, B. D. and Atmar, W. 1986. Nested subsets and the structure of insular mammalian faunas and archipelagos. – Biol. J. Linn. Soc. 28: 65–82.

Pollard, E. and Yates, T. J. 1993. Monitoring butterflies for ecology and conservation. – Chapman and Hall.

Pullin, A. S. (ed.) 1995. Ecology and conservation of butterflies. – Chapman and Hall.

Reed, J. M. 1996. Using statistical probability to increase confidence of inferring species extinction. – Conserv. Biol. 10: 1283–1285.

Sahlén, G. 1999. The impact of forestry on dragonfly diversity in central Sweden. – Int. J. Odonatol. 2: 177–186.

Scott, C. T. 1998. Sampling methods for estimating change in forest resources. – Ecol. Appl. 8: 228–233.

Silvertown, J. and Wilson, J. B. 1994. Community structure in a desert perennial community. – Ecology 75: 409–417.

Simberloff, D. 1998. Flagships, umbrellas, and keystones: is single-species management passé in the landcape era? – Biol. Conserv. 83: 247–257.

Simberloff, D. and Martin, J. L. 1991. Nestedness of insular avifaunas: simple summary statistics masking complex species patterns. – Ornis Fenn. 68: 178–192.

Sjögren-Gulve, P. and Ray, C. 1996. Using logistic regression to model metapopulation dynamics: large-scale forestry extirpates the pool frog. – In: McCullough, D. R. (ed.), Metapopulations and wildlife conservation. Island Press, Washington, D.C., pp. 111–137.

Sjögren-Gulve, P. and Hanski, I. 2000. Metapopulation viability analysis using occupancy models. – Ecol. Bull. 48: 53–71.

Skarstedt, A. 1998. Are epiphytic crustose lichens randomly distributed among small spruce forest patches? – M.S. thesis, Umeå Univ., Sweden.

Stohlgren, T. J. et al. 1995. Status of biotic inventories in US National Parks. – Biol. Conserv. 71: 97–106.

Swengel, A. B. and Swengel, S. R. 1997. Co-occurrence of prairie and barrens butterflies: applications to ecosystem conservation. – J. Insect Conserv. 1: 131–144.

Temple, S. A. and Wiens, J. A. 1989. Bird populations and environmental changes: can birds be bio-indicators? – Am. Birds 43: 260–270.

Thomas, C. D. and Hanski, I. 1997. Butterfly metapopulations. – In: Hanski, I. A. and Gilpin, M. E. (eds), Metapopulation biology: ecology, genetics, and evolution. Academic Press, pp. 359–386.

UNCED 1992. The United Nations Convention on Biological Diversity. – Rio de Janeiro.

Wilson, D. E. et al. (eds) 1996. Measuring and monitoring biological diversity: standard methods for mammals. – Smithsonian Inst. Press, Washington, D.C.

Worthen, W. B. and Rohde, K. 1996. Nested subset analyses of colonization-dominated communities: metazoan ectoparasites of marine fishes. – Oikos 75: 471–478.

Wright, D. H. and Reeves, J. H. 1992. On the meaning and measurement of nestedness of species assemblages. – Oecologia 92: 416–428.

Wright, D. H., Reeves, J. H. and Berg, J. 1990. NESTCALC ver. 1.0: a BASIC program for nestedness calculations.

Wright, D. H. et al. 1998. A comparative analysis of nested subset patterns of species composition. – Oecologia 113: 1–20.

Ecological Bulletins 48: 101–110. Copenhagen 2000

Comparative precision of three spatially realistic simulation models of metapopulation dynamics

Oskar Kindvall

Kindvall, O. 2000. Comparative precision of three spatially realistic simulation models of metapopulation dynamics. – Ecol. Bull. 48: 101–110.

I used field data from a fragmented population of the bush cricket *Metrioptera bicolor* (Orthoptera: Tettigoniidae) to evaluate whether different features of metapopulation dynamics, i.e. temporal changes in local and regional occupancy, can be accurately predicted by stochastic simulation models. Three different spatially realistic metapopulation models were considered: an incidence function model, a logistic regression model that was simulated using the METAPOP III program, and a demographic model used with the RAMAS GIS simulation package. All models gave good predictions about turnover rates and temporal changes in regional occupancy of *M. bicolor*. However, the predictions were less accurate regarding the fraction of time steps that individual habitat patches were occupied.

O. Kindvall (oskar.kindvall@entom.slu.se), Dept of Entomology, The Swedish University of Agricultural Sciences, P.O. Box 7044, SE-750 07 Uppsala, Sweden.

When working on species preservation, it is of great importance to investigate what will happen to a species if the number of habitat patches decreases or if the spatial configuration of patches changes. The metapopulation concept seems to be a promising framework for risk assessment of endangered species living on discrete habitat patches in fragmented landscapes (Burgman et al. 1993, Hanski 1994a, 1998, Wahlberg et al. 1996).

Stochastic simulations of spatially realistic metapopulation models, i.e. models constructed for specific sets of habitat patches whose geographical positions are explicitly considered (see Hanski 1999 for terminology), can generate detailed predictions about the distribution of an organism in different situations (e.g. Hanski and Thomas 1994, Hanski et al. 1995). Several simulation softwares have been developed for the purposes of viability analysis of species living in patchy environments (e.g. Lindenmayer and

Possingham 1994, Akçakaya 2000, Lacy 2000, Sjögren-Gulve and Hanski 2000). However, few attempts have been made to evaluate the accuracy of the predictions made by different types of models especially predictions made by models parameterised with data from a different patch configuration.

The bush cricket *Metrioptera bicolor* Philippi (Orthoptera: Tettigoniidae) has previously been the subject of metapopulation studies (e.g. Kindvall and Ahlén 1992, Kindvall 1993, 1995a). On the Swedish Red List it is categorised as vulnerable (Ehnström et al. 1993). In this paper, I use data from the Swedish distribution area of *M. bicolor* to investigate the accuracy of three spatially realistic models and their predictions considering temporal changes in patch occupancy. The models differ in complexity and amount of information needed for parameterisation.

It is of practical importance to know how much data

and what kind of information that is required for reliable predictions of future scenarios. Two of the models applied in this paper employ regression techniques to estimate the parameters used to model local extinction and colonisation. These models only require presence/absence data of the species on a set of habitat patches and information on the spatial configuration of habitats in the landscape. The third, demographic model incorporates details about local population dynamics and interpatch dispersal. This model requires estimates of temporal changes of individual abundance on different patches and information about the dispersal behaviour of the species.

Materials and methods

Simulation approach

The basic idea was to use information sampled from one of two halves of the Swedish distribution area of *Metrioptera bicolor* to parameterise the different models and then use occupancy data from the other half to evaluate the predictive precision of the models. Thus, for each of the three simulation approaches, I parameterised one western and one eastern version of the model. Thereafter, separate simulations were made with each model, adjusted for the landscape composition and habitat configuration of the other half. I simulated the occupancy of *M. bicolor* in five successive years, starting with the known situation in 1989. The results from 100 replicates were then compared with the observed variation of occupancy between 1990 and 1994.

The Swedish occurrence of *M. bicolor*, 55°40′N, 13°35′E (Kindvall and Ahlén 1992), can be divided into two areas of habitat patches, one western and one eastern

half, that are separated from each other by a much greater distance than the average interpatch distance observed within these two areas (Fig. 1). The interpatch distance between the two nearest patches belonging to the western and eastern halves was 550 m and the median interpatch distance within each area was 30 m (n = 116).

Only once in a period of seven years has *M. bicolor* been observed to successfully colonise an empty patch situated > 500 m from the nearest occupied patch (Kindvall 1993). The greatest colonisation distance observed in the years of this study, i.e. from 1989 to 1994, was 350 m. Thus, even if some interpatch exchange between the two areas may have occurred during this study, the metapopulation dynamics observed in the two areas were not interdependent.

The observed metapopulation dynamics differed between the eastern and the western area (see Results). It is possible that this difference to some extent was explained by differences in the spatial configuration of habitat patches. Habitat patches were, on average, larger in the western area (mean ± SD = 6.2 ± 10.9 ha, n = 66) than in the eastern area (1.6 ± 2.6 ha, n = 50; t-test on log-transformed data: t = 3.06, p < 0.01). However, there was no apparent difference in the degree of isolation of habitat patches between the two areas (t-test on log-transformed data: t = 0.084, p = 0.93, NS).

The spatial configuration of habitat patches changed during the study period. A large field of tillage became suitable habitat within the western area in 1990, after the tilling ceased. This event created a contiguous patch of suitable habitat joining three previously separate patches (Kindvall and Ahlén 1992). For the same reason, a previously small and unoccupied patch became larger in 1990 in the eastern area. Another patch (eastern area) was divided in two between 1989 and 1990, then restored again by

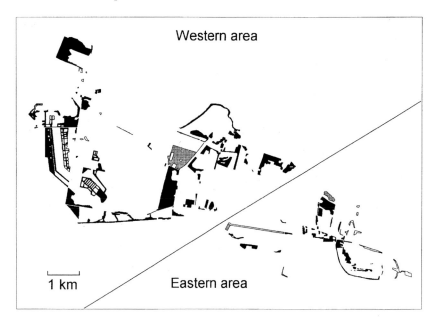

Fig. 1. A map of the Swedish distribution area of *Metrioptera bicolor* that shows the western and the eastern halves. Black patches were occupied in 1989, while white patches were vacant. The new habitats that appeared during the study period are indicated in grey.

clearance work before the onset of the 1992 season. All changes of habitat configuration that occurred between 1989 and 1994 are shown in Fig. 1.

Unfortunately, most simulation programs of metapopulation dynamics do not address changes of habitat configuration that occur during the period that is simulated (but see Lindenmayer and Possingham 1994, Akçakaya 1997). Instead, these models usually generate predictions about what will happen after habitat configuration has been altered. For simplicity, I assumed that all habitat changes occurring between 1989 and 1994 actually occurred in 1989, just prior to the beginning of the simulations.

The incidence function model

The incidence function model, pioneered by Hanski (1994b) and described in detail by Hanski (1999), is potentially the simplest spatially realistic model of metapopulation dynamics. In its simplest form, it requires only presence/absence data of a species from a set of discrete habitat patches observed in a single generation, thus observations of turnover are not necessary. However, the information about presence or absence must be sampled from a system where the fraction of patches occupied is determined by a dynamic equilibrium between local extinction and recolonisation (Hanski 1994b).

In Hanski's incidence function model, a single equation is used to calculate the individual probability of occupancy of patch i (J_i). The environmental information required for this model is the area of each available habitat patch ($Area_i$) and all interpatch distances (i.e., all D_{ij}) in the system. Both local population size and local extinction probability are assumed to be proportional to patch area. The area of the occupied patches is also assumed to affect the number or migrants in the system. Analysing how J_i relates to $Area_i$ and D_{ij} (e.g. Hanski 1999, Moilanen 1999, Sjögren-Gulve and Hanski 2000) allows calculation of patch-specific extinction and colonisation probabilities (E_i and C_i, respectively) that can be used in simulations.

Four to five model parameters are used to relate J_i to $Area_i$ and D_{ij}, depending on whether or not a rescue effect is assumed to affect local extinction risk (Hanski 1994b). Two parameters, denoted α and y affect the colonisation process, while e and x are used to model the extinction probability (Hanski 1994b, Sjögren-Gulve and Hanski 2000). For a given value of α it is possible to estimate y, e and x by fitting an equation expressing J_i to empirical data of occupancy using non-linear regression technique (Moilanen 1994, 1999). To choose the most appropriate value of α, the equation of J_i is fitted to the presence/absence data for different values of α, and then the best fitting value is selected. When including the rescue effect one also has to specify a parameter called the E to x multiplier (Moilanen 1994, 1999), which is defined as the patch size for which the extinction probability is equal to one.

An attempt to parameterise the incidence function model with a minimum amount of data, i.e. occupancy status in a single year (1989), and with the simplest version of the model, i.e. without the rescue effect, failed. It was possible to parameterise models for both the eastern and the western area. However, the resulting predictions where not realistic (Kindvall 1995a). Especially the turnover rate was severely overestimated. This experience suggests that one 'snapshot' sample of occupancy data is not enough. Data from a longer period of time is needed to provide representative parameter estimates. Also, Hanski (pers. comm.) suggested that the slightly more complicated version of the incidence function model that includes the rescue effect is probably more reliable than the simplest version, even when no apparent rescue effect has been detected, as is the case with *M. bicolor* (Kindvall 1996).

A more data-intensive model was developed, using patch occupancy information from 1989 to 1994. The fraction of time each habitat patch was observed to be occupied by *M. bicolor* was used as the dependent variable (J_i) when performing the non-linear regression analysis required to obtain model parameter values (see Hanski 1999, Moilanen 1999). The parameter estimates that best fit the occupancy status of *M. bicolor* are presented in Table 1.

The logistic regression model

The logistic regression model, included in the METAPOP III simulation package developed by Ray et al. (e.g. Sjögren-Gulve and Ray 1996), is a simple occupancy model of metapopulation dynamics. Patch-specific probabilities of local extinction and colonisation are estimated from two separate logistic regression functions: one models the local colonisation probability (C_i) in relation to correlated environmental variables, and the other models local extinction probability (E_i; see Sjögren-Gulve and Hanski 2000).

Different kinds of environmental factors that are known to affect the turnover probabilities in a given system can easily be built in to the logistic regression model. This simulation approach may therefore fit a broad spectrum of patchily distributed organisms. The input information needed is observations of both extinctions and colonisations, that can be interpreted from presence/absence data collected during a number of generations, and estimates of some environmental factors that are expected to influence turnover (Sjögren-Gulve and Ray 1996).

At least two environmental factors are known to affect the extinction risk of local populations of *M. bicolor*, i.e. patch area, $Area_i$ (Kindvall and Ahlén 1992) and habitat heterogeneity, H_i (Kindvall 1996). In contrast to some other metapopulation systems (e.g. Sjögren-Gulve 1994) there is no apparent effect of isolation on local extinction probability of *M. bicolor* (Kindvall and Ahlén 1992, Kindvall 1996).

Table 1. Results of fitting the incidence function model (including rescue effect) to the occupancy data from the eastern and western area of the Swedish distribution of *Metrioptera bicolor* observed between 1989 and 1994. The fraction of available patches that were occupied is denoted by P. Estimates of the following parameters in the incidence function model (Eq. 8 in Hanski 1994b) are presented: α, x, y, e and the E to x multiplier. The input environmental variables, patch area and interpatch distances were measured in square metres and in kilometres (closest edge-to-edge distance), respectively.

Parameters	Western area	Eastern area
Number of patches	66	50
P (1989-94)	0.82	0.71
E to x multiplier	0.05	0.001
α	2.0	6.0
x	0.876	0.514
y	7.278	2.571
e	0.072	0.029
Maximum likelihood	33.0	14.5

The model describing the annual extinction probability (E_i) was formulated according to the results from the logistic regression analyses (Table 2) as

$$E_i = \frac{\exp\left(-0.000140 \cdot \text{Area}_i - 0.0118 \cdot H_i\right)}{1 + \exp\left(-0.000140 \cdot \text{Area}_i - 0.0118 \cdot H_i\right)} \quad (1)$$

based on the western area, and as

$$E_i = \frac{\exp\left(-0.0000900 \cdot \text{Area}_i - 0.0129 \cdot H_i\right)}{1 + \exp\left(-0.0000900 \cdot \text{Area}_i - 0.0129 \cdot H_i\right)} \quad (2)$$

based on the eastern area.

The colonisation probability of *M. bicolor* is related to the distance to the nearest occupied patch, D_{ij} (Kindvall 1993a), and according to the logistic regression analyses

(Table 2) annual colonisation probability (C_i) was modelled as

$$C_i = \frac{\exp\left(-0.00587 \cdot D_{ij}\right)}{1 + \exp\left(-0.00587 \cdot D_{ij}\right)} \quad (3)$$

based on the western area, and as

$$C_i = \frac{\exp\left(-0.00889 \cdot D_{ij}\right)}{1 + \exp\left(-0.00889 \cdot D_{ij}\right)} \quad (4)$$

based on the eastern area.

The regression coefficients in eqs 1–4 (Table 2) were estimated from analyses of occupancy data collected during 1989–1994 in the two respective areas of the Swedish distribution of *M. bicolor*. Each change from occupancy

Table 2. Results from stepwise logistic regression analyses of environmental factors that influence annual probabilities of local extinction, E_i, and colonisation, C_i, in the eastern and western part of the Swedish distribution area of *Metrioptera bicolor*. The following independent variables were considered: patch area, Area_i (m²); habitat heterogeneity, H_i; and distance to the nearest occupied patch, D_{ij} (m, edge-to-edge distance). Intercepts did not enter the models at the 0.10 significance level.

Model	Variables	Estimate	SE	χ^2	p-value
Western area					
E_i (n = 268)	Area_i	−0.00014	0.000058	5.43	< 0.05
	H_i	−0.0118	0.0032	14.09	< 0.001
C_i (n = 61)	D_{ij}	−0.00587	0.0018	10.08	< 0.01
Eastern area					
E_i (n = 180)	Area_i	−0.00009	0.000045	3.74	0.053
	H_i	−0.0129	0.0027	23.14	< 0.0001
C_i (n = 74)	D_{ij}	−0.00889	0.022	16.72	< 0.0001

status 1 (= presence) to occupancy status 0 (= absence) was considered as local extinction, and each change of the occupancy status in the opposite direction is considered as local colonisation. When performing the logistic regression analyses for the extinction models, the extinction sites within the area of concern were compared with the patches where the local population survived. Similarly, patches that became colonised were compared with patches with continued vacancy, generating eqs 3–4. It should be mentioned that some (50%) of the observed turnover, given the above definition, may not be real, due to the possibility of diapausing eggs being present even when the adult population is absent (Kindvall and Ahlén 1992).

I used all annual observations of unchanged status of occupancy as controls in the logistic regression analyses. Thus, each year a patch was still occupied it was used as an observation of non-extinction (i.e. $E_i = 0$) with its annual values of $Area_i$ and H_t. The main consequence of this treatment is that most patches were used more than one time, thus all observations are not statistically independent. For example, some patches appeared both as examples of sites where local populations went extinct and sites with extant local populations. Thus, an instant relationship was assumed between the year-specific values of patch area, interpatch distances and habitat heterogeneity, and the presence or absence of turnover.

The demographic model

The RAMAS GIS simulation package, developed by Akçakaya (1997), can be used for spatially realistic simulations of demographic models of metapopulation dynamics. In demographic models, changes in numbers of individuals with time are described within patches (Akçakaya 2000). Such local population dynamics are modelled mechanistically as a demographic process with discrete (non-overlapping) generations. With the RAMAS GIS metapopulation program, interpatch exchange of individuals can be modelled by using an interpatch migration matrix. Accordingly, the change in abundance in a local population (N) from year t to t+1 was modelled as

$$N_{t+1} = R_t \cdot N_t + i_t - e_t \qquad (5)$$

where i_t and e_t are the time-specific numbers of immigrants and emigrants entering and leaving the local population, respectively, and R_t is the realised annual population growth rate. Generally, dispersal and immigration was sparse and i_t and e_t assumed low numbers.

Eggs laid by *M. bicolor* may hatch after one or two successive cold periods (Ingrisch 1986). This implies that overlapping generations may occur. When modelling local dynamics of *M. bicolor*, I made the simplifying assumption that overlapping generations do not occur, or are of minor importance. This assumption is probably justified during the years of this study because no statistically significant relationship was found between the observed annual growth rate R_t and the local population density in the previous year, N_{t-1} (Kindvall 1995a).

Because the growth rate R_t observed in the local populations of *M. bicolor* between 1989 and 1994 was negatively correlated with current population density, N_t, (Kindvall 1995a), it was appropriate to assume that local population dynamics were regulated in a density dependent way. I used the Ricker equation to model this density dependence (Akçakaya 1997), i.e.

$$R_t = R_{max} \cdot \exp(-\beta \cdot N_t) \qquad (6)$$

where R_{max} is the maximum annual growth rate, and β is a parameter estimated from linear regression of observed R_t versus N_t.

According to Burgman et al. (1993), intercepts of the R_t versus N_t regression can be used as estimates of carrying capacity, K, and maximum growth rate, R_{max}. The estimated values of K and R_{max} in the *M. bicolor* system were 59 and 7.09 males ha^{-1}, respectively (Kindvall 1995a). These calculations are based on local trajectories sampled on 45 relatively small patches (< 4.58 ha) that are mainly (76%) situated within the eastern area. In contrast to the other models, I did not specify different versions of the structured model for the western and eastern area, respectively, because the data used for parameterisation of this model was not the same kind as those used for the evaluation of model predictions.

Local population dynamics of *M. bicolor* are known to be influenced by weather fluctuations (Kindvall 1995b, 1996). In the simulation model, environmental stochasticity was imposed on the local population dynamics by letting the value of K of each patch vary randomly according to a normal distribution with a specified standard deviation, SD–K. To estimate SD–K, I used a linear regression model that relates observed levels of local population variability, measured as the coefficient of variation of local population abundance, with the two environmental factors habitat heterogeneity, H_i, and patch area, $Area_i$ (Kindvall 1996).

Estimates of interpatch dispersal rates were obtained from stochastic simulations of a spatially realistic model that describe movements of *M. bicolor* in a heterogeneous environment (Kindvall 1999). For each habitat patch (n = 116) lifetime movement pathways of at least 100 individuals were simulated. The relative number of individuals that reached other habitat patches or were lost in intervening "matrix" habitats was estimated. In contrast to most current metapopulation models (e.g., Thomas and Hanski 1997), the per capita emigration rates are not assumed to be equal among habitat patches in this study. As a consequence of the behaviour of the movement model, smaller patches have higher emigration rates than larger patches (cf. Stamps et al. 1987).

Estimated interpatch dispersal rates were so low that the average expected number of dispersers was less than one. Despite this, recolonisation and immigration was possible in the model because RAMAS GIS incorporates demographic stochasticity in the dispersal modelling, using the average dispersal rates as long-term averages, whereas at any give time step the number of dispersers between two patches can be zero, one or more according to a binomial distribution (Akçakaya 1997, 2000).

To be able to model emigration and immigration independently, and thereby taking losses of individuals during dispersal into account, I specified an imaginary habitat patch in the metapopulation program that received the proportion of emigrants from all other populations that did not reach new patches of suitable habitat. There was no migration from this "sink" patch back to the suitable patches and this imaginary "patch" was excluded from the simulation summary statistics.

Results

Regional occupancy

The fraction of available habitat patches that are occupied by a species is generally denoted by P (Hanski 1991). The mean P-values observed during 1990–94 in the western and eastern halves of the Swedish distribution area of *M. bicolor* were compared with the P-values predicted by the three types of simulation models investigated in Fig. 2. In the western area, the observed P-value increased after 1989, while the fraction of occupied patches remained almost the same in the eastern area.

All eastern models were able to predict the positive trend of the P-value in the western area (Fig. 2a). The western logistic regression model and the demographic model also made accurate predictions, i.e. observed P-values were included in the range of predicted values (Fig. 2b). However, the western incidence function model failed to predict the mean P-value of the eastern area.

Extinctions and colonisations

The eastern incidence function model and both logistic regression models predicted the number of turnover events with not more than an 18% error (difference between predicted mean and observed number) (Fig. 3). The demographic model also made good predictions about turnover rates in the western area (< 10% error). However, the number of turnover events observed in the eastern area, and especially the number of extinctions, was underestimated by the demographic model (35% error, Fig. 3d). The western incidence function model predicted on average 26% more colonisations in the eastern area than was actually observed (Fig. 3b).

Local occupancy

To evaluate how accurately the different models predicted the fraction of time each habitat patch was occupied by *M. bicolor*, I calculated the difference between the observed and the predicted occupancy. This was done by subtracting the proportion of time steps the patch was occupied in the simulations with the proportion observed in the field

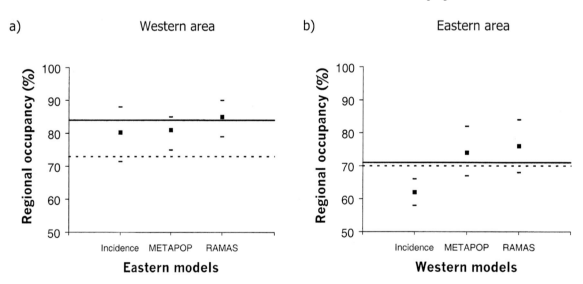

Fig. 2. Model predictions of the fraction of occupied patches in (a) the western and (b) eastern half of the Swedish distribution area of *Metrioptera bicolor*. The mean and max/min values of 100 replicates are given. The dotted line represents the starting occupancy of 1989, while the solid line represents the mean occupancy observed between 1990 and 1994. Models: the incidence function model, the logistic regression model (METAPOP III) and the demographic model (RAMAS) of metapopulation dynamics.

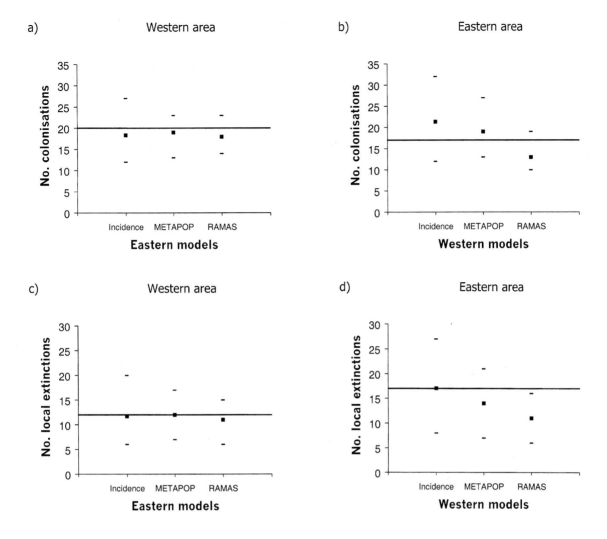

Fig. 3. Observed and predicted number (mean, max, min, n = 100) of (a, b) colonisations and (c, d) local extinctions during the five-year period 1990–94 in (a, c) the western and (b, d) eastern half of the distribution area of *Metrioptera bicolor*. Three models were compared, i.e. the incidence function model, the logistic regression model (METAPOP III) and the demographic model (RAMAS) of metapopulation dynamics.

during 1990–94. The frequency distribution of the absolute values of these differences is presented in Fig. 4. These frequency distributions can be compared with the corresponding misclassification made with the most simplistic prediction about temporal variability of local occupancy, i.e. that all patches will remain at the same occupancy status as observed in 1989.

All three models showed modest (< 0.2) deviations from the observed occupancy in > 89% of the western cases, meaning that on average the species' presence or absence at a particular patch was correctly predicted in more than 4 out of the 5 yr during 1990–94. In the eastern area, the RAMAS GIS model had a similar accuracy (88%), but the western logistic regression model had a < 0.2 deviation

in 82% of the cases and the incidence function model only in 66%.

Apparently, the western incidence function model was unable to classify local occupancy better than simply assuming no temporal change in occupancy (Fig. 4b). However, all the other models made better predictions about the fraction of time steps each habitat patch was occupied than simply assuming static occupancy. The maximum difference between observed and predicted local occupancy was never worse than 40%, i.e. 2 yr, with the demographic model (Fig. 4). However, the two occupancy models sometimes made substantial mistakes when occupancy was correctly estimated only in a single year.

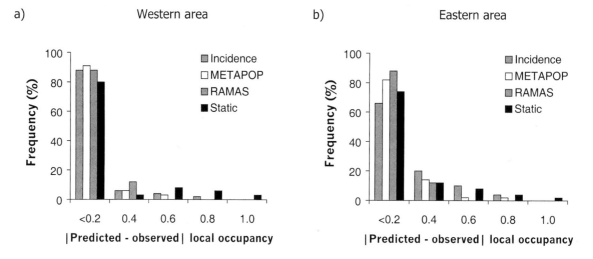

Fig. 4. Frequency distributions of the absolute values of differences between predicted and observed occupancy of individual patches in (a) the western and (b) eastern half of the distribution area of *Metrioptera bicolor*. Models: the incidence function model, the logistic regression model (METAPOP III) and the demographic model (RAMAS) of metapopulation dynamics. Black bars indicate the misclassification obtained if predicting no temporal change (static prediction) in patch occupancy from 1989 to 1994.

Discussion

This study demonstrates that spatially realistic metapopulation models can make predictions about occupancy patterns of a species in a fragmented landscape that do not deviate too much from reality. Although it could be argued that the two areas of the Swedish occurrence of *M. bicolor* are not totally independent of each other, it appears that all investigated models are able to respond realistically to differences concerning spatial configuration of patches. This result is encouraging, because these models are intended for comparative risk assessments where different landscape scenarios and management alternatives are investigated. Models of this kind must be able to make realistic predictions for other landscapes than the one used for parameter estimation.

The increase in the fraction of patches occupied by *M. bicolor* that was observed in the western area can be interpreted as a gradual response of the metapopulation dynamics to changed habitat configuration. Several new and large habitat patches were created between 1985 and 1989 (Kindvall and Ahlén 1992). This suggests that the dynamics of occupancy were probably not at metapopulation equilibrium in 1989. Thus, the critical assumption of equilibrium that underlies the incidence function model was violated. This violation may explain the fact that the eastern incidence function model made better overall predictions than the western model, since greater changes of habitat configuration occurred in the western area (Kindvall and Ahlén 1992).

The logistic regression models of patch turnover made slightly better predictions about the number of colonisations and extinctions than the demographic metapopula-

tion model (Fig. 3). This result suggests that detailed information on dispersal behaviour and local population dynamics may not be required. The logistic regression function used to describe the colonisation probability of *M. bicolor* only considered the interpatch distance to the nearest occupied patch. This measure is only expected to be a rough estimate of isolation (Hanski et al. 1994). Actually, a more realistic approach to estimate isolation is used by the simulation program of the incidence function model. In that model, all occupied patches are considered as potential sources of migrants. The number of individuals reaching vacant patches is then modified by the pairwise interpatch distances and the area of the source patches (Hanski 1994b). Obviously, it would be possible to use Hanski's connectivity index also for the logistic regression model. Furthermore, it would be possible to use the detailed information on interpatch dispersal that was used in the demographic model, when parameterising the logistic colonisation model. Nevertheless, the measure of interpatch distance used in the logistic regression model seems to be good enough.

The main problem when interpreting colonisation probabilities from simple measures of interpatch distances is that the observed frequency distribution of isolation distances is affected by the current landscape composition rather than individual dispersal behaviour (Porter and Dooley 1993). Therefore, a phenomenological model, e.g. the logistic regression model, that is parameterised with information from a set of patches situated close to each other may not be able to predict occupancy patterns in a set of patches situated far from each other. Although the accuracy of the logistic regression colonisation models was numerically better than that of the demographic model, the latter model may perhaps be superior when making predictions in future landscapes with totally different configurations of patches. It may be argued that the logistic regression models were able to predict the observed colonisation events just because the western and eastern areas are similar with respect to isolation of patches.

One of the difficulties associated with the demographic model of metapopulation dynamics is how to incorporate information about temporal variability of local population sizes. Without a detailed k-factor analysis it is not possible to discriminate between intrinsic and extrinsic factors that regulate population densities. In this study, I used information from local trajectories of *M. bicolor* both to estimate the average maximum growth rate and to model environmental stochasticity. It is likely that either the degree of environmental stochasticity or the intrinsic regulation becomes over-emphasised when using this approach. Of course, the ambiguity of how to describe local population dynamics will affect the reliability of model predictions.

When making conservation plans for endangered species, models should be able to predict extinction risks over several decades. The main problem in constructing predictive models aimed for risk assessment lies in parameterising the impact of environmental stochasticity. Short-term investigations are not likely to detect population dynamic responses to the vast array of possible environmental disturbances that may occur in the long run (Solbreck 1991). There is evidence that habitat heterogeneity is an environmental factor that can influence the relative impact of extreme weather situations (e.g., Weiss et al. 1988, Thomas 1994, Kindvall 1995b, 1996). Therefore, models of metapopulation dynamics that incorporate habitat heterogeneity are probably better for long-term viability analysis than simpler models that only deal with patch area and isolation (Beissinger 1995).

The models differed slightly in their ability to predict regional occupancy, numbers of colonisations and extinctions, and local occupancy. Summarising their performance in Figs 2–4, the logistic regression model on average made the most accurate predictions. However, since all models generated acceptable predictions I am convinced that each of them is worth considering for risk assessments of *M. bicolor* in Sweden. When trying to predict long-term extinction probabilities of *M. bicolor* using a demographic model, it is probably necessary to take the biannual egg cycles into consideration.

In conclusion it is important to make more empirical evaluations of the different simulation packages developed so far. Although several encouraging results were obtained from this study, there is certainly a need for further improvements to present models. In the aim of species preservation, I think it is especially important to develop simple models similar to the incidence function model, that require as little empirical data as possible. This study clearly demonstrated that more sophisticated models may not necessarily be better for patch-level viability analysis than relatively simple phenomenological models.

Acknowledgements – This project has been financially supported by the Swedish Council for Forestry and Agricultural Research and the World Wildlife Fund, Sweden. I am grateful for all the help I received from the persons behind the simulation programs, i.e. I. Hanski, P. Sjögren-Gulve and R. Akçakaya. I thank I. Ahlén, R. Akçakaya, A. Carlson, T. Pärt, C. Ray and P. Sjögren-Gulve for their comments on the manuscript.

References

Akçakaya, H. R. 1997. RAMAS GIS: Linking landscape data with population viability analysis (ver. 2.0). – Appl. Biomath., Setauket, New York.

Akçakaya, H. R. 2000. Population viability analyses with demographically and spatially structured models. – Ecol. Bull. 48: 23–38.

Beissinger, S. R. 1995. Modelling extinction in periodic environments: Everglades water levels and snail kite population viability. – Ecol. Appl. 5: 618–631.

Burgman, M. A., Ferson, S. and Akçakaya, H. R. 1993. Risk assessment in conservation biology. – Chapman and Hall.

Ehnström, B., Gärdenfors, U. and Lindelöw, Å. 1993. Rödlistade evertebrater i Sverige 1993. [Swedish Red List of Invertebrates 1993]. – Databanken för hotade arter, Uppsala.

Hanski, I. 1991. Single-species metapopulation dynamics: concepts, models and observations. – In: Gilpin, M. E. and Hanski, I. (eds), Metapopulation dynamics. Academic Press, pp. 17–38.

Hanski, I. 1994a. Patch-occupancy dynamics in fragmented landscapes. – Trends Ecol. Evol. 9: 131–135.

Hanski, I. 1994b. A practical model of metapopulation dynamics. – J. Anim. Ecol. 63: 151–162.

Hanski, I. 1998. Metapopulation dynamics. – Nature 396: 41–49.

Hanski, I. 1999. Metapopulation ecology. – Oxford Univ. Press.

Hanski, I. and Thomas, C. D. 1994. Metapopulation dynamics and conservation: a spatially explicit model applied to butterflies. – Biol. Conserv. 68: 167–180.

Hanski, I., Kuussaari, M. and Nieminen, M. 1994. Metapopulation structure and migration in the butterfly *Melitaea cinxia*. – Ecology 75: 747–762.

Hanski, I. et al. 1995. Metapopulation persistence of an endangered butterfly in a fragmented landscape. – Oikos 72: 21–28.

Hastings, A. 1991. Structured models of metapopulation dynamics. – In: Gilpin, M. E. and Hanski, I. (eds), Metapopulation dynamics. Academic Press, pp. 57–71.

Ingrisch, S. 1986. The plurennial life cycles of European Tettigoniidae (Insecta: Orthoptera) 2. The effect of photoperiod on the induction of an initial diapause. – Oecologia 70: 617–623.

Kindvall, O. 1993. Artbevarande i fragmenterad miljö – en generell inventeringsstrategi exemplifierad med grön hedvårtbitare [Conservation in fragmented landscapes – a general inventory strategy exemplified with the bush cricket *Metrioptera bicolor*]. – Entomologisk Tidskrift 114: 75–82.

Kindvall, O. 1995a. Ecology of the bush cricket *Metrioptera bicolor* with implications for metapopulation theory and conservation. – Ph.D. thesis, Dept of Wildlife Ecology, Swedish Univ. of Agricult. Sci., Uppsala.

Kindvall, O. 1995b. The impact of extreme weather on habitat preference and survival in a metapopulation of the bush cricket *Metrioptera bicolor* in Sweden. – Biol. Conserv. 73: 51–58.

Kindvall, O. 1996. Habitat heterogeneity and survival in a bush cricket metapopulation. – Ecology 77: 207–214.

Kindvall, O. 1999. Dispersal in a metapopulation of the bush cricket, *Metrioptera bicolor* (Orthoptera: Tettigoniidae). – J. Anim. Ecol. 68: 172–185.

Kindvall, O. and Ahlén, I. 1992. Geometrical factors and metapopulation dynamics of the bush cricket, *Metrioptera bicolor* Philippi (Orthoptera: Tettigoniidae). – Conserv. Biol. 6: 520–529.

Lacy, R. C. 2000. Considering threats to the viability of small populations using individual-based models. – Ecol. Bull. 48: 39–51.

Lindenmayer, D. B. and Possingham, H. P. 1994. The risk of extinction ranking management options for leadbeater's possum using population viability analysis. – Cent. for Resour. and Environ. Stud., The Australian National Univ., Canberra.

Moilanen, A. 1994. Incidence function model. Software manual. – Dept of Zool., Div. of Ecol., Univ. of Helsinki, Finland.

Moilanen, A. 1999. Patch occupancy models of metapopulation dynamics: efficient parameter estimation using implicit statistical inference. – Ecology 80: 1031–1043.

Porter, J. H. and Dooley, J. L. Jr 1993. Animal dispersal patterns: a reassessment of simple mathematical models. – Ecology 78: 2436–2443.

Sjögren-Gulve, P. 1994. Distribution and extinction patterns within a northern metapopulation of the pool frog, *Rana lessonae*. – Ecology 75: 1357–1367.

Sjögren-Gulve, P. and Ray, C. 1996. Using logistic regression to model metapopulation dynamics: large-scale forestry extirpates the pool frog. – In: McCullough, D. R. (ed.), Metapopulations and wildlife conservation. Island Press, Washington D.C., pp. 111-137.

Sjögren-Gulve, P. and Hanski, I. 2000. Metapopulation viability analysis using occupancy models. – Ecol. Bull. 48: 53–71.

Solbreck, C. 1991. Unusual weather and insect population dynamics: *Lygaeus equestris* during an extinction and recovery period. – Oikos 60: 343–350.

Stamps, J. A., Buechner, M. and Krishnan, V. V. 1987. The effect of edge permeability and habitat geometry on emigration from patches of habitat. – Am. Nat. 129: 533–552.

Thomas, C. D. 1994. The ecology and conservation of butterfly metapopulations in the fragmented British landscape. – In: Pullin, A. S. (ed.), Ecology and conservation of butterflies. Chapman and Hall, pp. 46–63.

Thomas, C. D. and Hanski, I. 1997. Butterfly metapopulations. – In: Hanski, I. and Gilpin, M. E. (eds), Metapopulation biology, ecology, genetics, and evolution. Academic Press, pp. 359–386.

Wahlberg, N., Moilanen, A. and Hanski, I. 1996. Predicting the occurrence of endangered species in fragmented landscapes. – Science 273: 1536–1538.

Weiss, S. J., Murphy, D. D. and White, R. R. 1988. Sun, slope and butterflies: topographic determinants of habitat quality in *Euphydryas editha*. – Ecology 69: 1486–1496.

Ecological Bulletins 48: 111–121. Copenhagen 2000

Management and population viability of the pasture plant *Gentianella campestris*: the role of interactions between habitat factors

Tommy Lennartsson

Lennartsson, T. 2000. Management and population viability of the pasture plant *Gentianella campestris*: the role of interactions between habitat factors. – Ecol. Bull. 48: 111–121.

Population dynamics of *Gentianella campestris* was studied in a management experiment, involving three levels of grazing intensity and two levels of microsite moisture. The data were evaluated by using matrix population models in order to estimate probability of extinction under various combinations of grazing and moisture. The results showed that population viability of the species was strongly affected by the interaction effects of management and microsite. On dry sites extinction risk decreased with decreasing grazing intensity, while on mesic sites medium grazing intensity showed the lowest probability of extinction. In addition, some other factors are discussed, which may also interact with grazing intensity to affect population viability. These factors are: timing of grazing, population size, environmental stochasticity, geographic location, and flowering phenology. The results are discussed in terms of how they can be used to optimise grassland management based on knowledge of population dynamics of grassland plants.

T. Lennartsson (tommy.lennartsson@nvb.slu.se), Swedish Univ. of Agricultural Sciences, Dept of Conservation Biology, Box 7002, SE-750 07 Uppsala, Sweden.

Unfertilised fodder-producing grasslands, wooded meadows and similar biotopes are often referred to as seminatural (e.g. Duffey 1974, Rackham 1986). Such biotopes have historically been the basis for agriculture in most temperate regions (Emanuelsson et al. 1985), but introduction of artificial fertilisers and other changes in agriculture have caused a rapid decline during the last 50–100 yr. Today only a fraction remains of the former extension of seminatural grasslands and the area is still decreasing. For example, in Sweden the area of mown seminatural grassland has decreased from ca 1.2 million ha in 1880 to ca 2 400 ha today (Anon. 1990, Bernes 1994). In addition to the decrease in area, the current management is often considerably different from traditional management regimes (e.g. García 1992 ch. 6, Bernes 1993, Beaufoy et al. 1995). One especially important change is the shift from traditional mowing with autumnal grazing, to grazing only (Aronsson and Matzon 1987, Ekstam et al. 1988, Beaufoy et al. 1995). The decline of seminatural grasslands is one of the major threats to European flora and fauna (Wolkinger and Plank 1981, Fuller 1987, Baldock and Long 1987, Fry 1991, Tucker 1991, Bernes 1994, Stanners and Bourdeau 1995, Beaufoy et al. 1995). In Sweden, ca 600 species of plants and animals are threatened by ceased or changed grassland management (Bernes 1994).

When the area of seminatural grasslands decreases, it becomes increasingly important to optimally manage the remaining grassland patches. The current management practice may not always be optimal for biodiversity, as indicated by the fact that many plant species have declined in grassland patches which are still managed (Eriksson 1996, Lennartsson and Svensson 1996). In order to understand and counteract such processes, we need to evaluate how population viability of grassland species is affected by, for example, type, timing and intensity of management. Such fine-scaled variation in management has so far attracted little attention compared to the study of ceased management and loss of habitats.

Management can thus be expected to be an important determinant of population viability of grassland plants and, correspondingly, population modelling to be an important tool for guiding management (e.g. Menges 1997). Management affects population viability in combination with other biotic and abiotic factors. For example, Menges and Dolan (1998) combined fire management with effects of genetic variation, geographic location, population size, and isolation in a viability analysis of *Silene regia*. Oostermeijer (1996) combined effects of management (sod cutting) with that of genetic variation for *Gentiana pneumonanthe*. Lennartsson (1997a) estimated risk of extinction of *Gentianella campestris* by combining different grazing/mowing regimes with environmental stochasticity and population size.

In Scandinavia, most of the remaining grasslands are presently grazed. Therefore, various aspects of grazing are among the most important management factors to include in population viability analyses of grassland plants. Several studies have related grazing to plant demography and fitness (e.g. Widén 1987, Oostermeijer 1992, Bastrenta and Belhassen 1992, Bastrenta et al. 1995), but very few studies have used such data for estimating, for example, extinction risk of populations (cf. Bullock et al. 1994). Especially the intensity of grazing has been much debated among biologists and managers. On one hand, an intense grazing might be needed to counteract detrimental accumulation of detritus (Knapp and Seastedt 1986, Grubb 1988), but on the other hand, such intense grazing may be difficult to combine with economically acceptable growth and productivity of the cattle (cf. Ekstam and Forshed 1996). Intense grazing may also reduce seed production of grassland plants (Oostermeijer 1992, Simán and Lennartsson 1998) and trampling effects may increase mortality (Lennartsson 1997a), thus counteracting the benefits of reduced litter accumulation.

In this paper, I estimate how population viability of the grassland plant *Gentianella campestris* (L.) Börner is affected by grazing, alone and in interaction with other factors. I present data from a 6-yr field experiment in which three levels of grazing intensity were combined with two levels of microsite moisture in simulations with a stochastic matrix model with variable frequency of summer drought. The model also includes population size as a parameter. With this study as a base, I discuss how grazing intensity may interact with other habitat factors, such as timing of grazing, geographic location, and flowering phenology, to influence population viability of *G. campestris*.

Methods

Study species

Gentianella campestris ssp. *campestris* is a European biennial herb (Pritchard and Tutin 1972, Hultén and Fries 1986). It is spring germinating, forms a rosette the first summer, and overwinters as a taproot and a bud (Lennartsson 1997b). The plant flowers the second summer, either in early July (aestival type, Wettstein 1895) or in August (autumnal type). The size at flowering is normally 5–20 cm high with 5–15 lilac flowers (Fig. 1). The life cycle is strictly biennial as all plants flower their second summer and die after flowering, and the species lack mechanisms of vegetative reproduction (Fig. 2). The adult plant has a main stem with a top meristem and 6–12 nodes, each of which with two opposite, decussate leaves and up to six flowers. Some nodes produce branches with secondary

Fig. 1. *Gentianella campestris* ssp. *campestris* (L.) Börner, field gentian. Autumnal type from the Björnvad population.

Fig. 2. Life cycle graph and basic projection matrix of *Gentianella campestris* ssp. *campestris* (L.) Börner (Gentianaceae). F_{ij} indicate fecundity transitions, a_{ij} indicate probability of transitions from stage *j* to stage *i*. Size of rosettes is defined by the number of leaves × rosette diameter index, where rosettes with an index > 15 are considered "large rosettes". Size of adult plants is defined by the number of nodes on the main stem and on the branches. Plants with > 10 nodes are assigned "large adults".

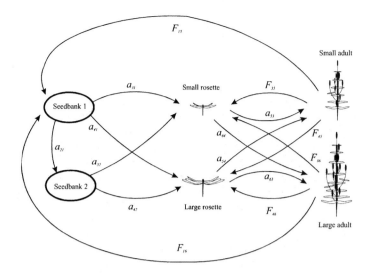

	1 Seedbank 1	2 Seedbank 2	3 Small rosette	4 Large rosette	5 Small adult	6 Large adult
1 Seedbank 1	-	-	-	-	F_{15}	F_{16}
2 Seedbank 2	a_{21}	-	-	-	-	-
3 Small rosette	a_{31}	a_{32}	-	-	F_{35}	F_{36}
4 Large rosette	a_{41}	a_{42}	-	-	F_{45}	F_{46}
5 Small adult	-	-	a_{53}	a_{54}	-	-
6 Large adult	-	-	a_{63}	a_{64}	-	-

nodes. The plants are frequently visited by bumblebees (Müller 1881, Lennartsson unpubl.), which serve as the only known pollinators, but the species is highly self-compatible and obtains a high seed set without insect pollination (Lennartsson 1997a). The seed bank is short-lived (Milberg 1994).

Gentianella campestris contains bitter glycoside substances (e.g., Inouye et al. 1967, Hostettmann-Kaldas and Jacot-Guillarmod 1978) which give some protection against herbivory, but on pasture localities 15–80% of the plants are grazed, depending on grazing agent, grazing intensity and vegetation structure. Grazing or other damage can release vigorous regrowth of new branches and flowers (Zopfi 1991, Lennartsson et al. 1998).

Gentianella campestris occurs almost exclusively in managed seminatural grasslands. Very few natural habitats are known, for example river banks and mountain scree slopes (Nilsson et al. 1999). The species has undergone a rapid decline during the last 50 yr, mainly due to ceased management of seminatural grasslands. In a studied region in central Sweden, the species has disappeared from ca 90% of its localities during the last 50 yr (Lennartsson and Svensson 1996). It is red-listed as "care demanding" in Sweden (Aronsson et al. 1995) as well as in several other European countries (e.g., Ingelög et al. 1993). *Gentianella campestris* is associated with grasslands with high biodiversity and conservation value (Anon. 1987, Lennartsson 1997a), and it may therefore serve as an appropriate indicator species for evaluating management regimes in species rich grasslands.

Experimental design and data analysis

The grazing experiment was performed in a horse pasture, Björnvad in the Province of Södermanland in central Sweden (59°20′21″N, 16°51′31″E, see Fig. 3). The site is an open, species-rich grassland pasture of ca 15 ha, dominated by *Avenula pratensis* – *Festuca rubra* meadow and herb-rich *Agrostis capillaris* meadow (Rydberg and Vik 1992, Påhlsson 1994). The topography of the grassland creates a mosaic of dry, sandy sites and mesic sites with a higher content of clay in the soil. Average population size of *Gentianella campestris* (autumnal type) is ca 5 000 adult plants, largely fluctuating between years.

The grazing intensity by horses varied within the site. In order to create three grazing intensities that were not varying between years, grazing was controlled by fencing.

Fig. 3. Part of the pasture at Björnevad in wich the study took place, just before onset of grazing in June.

Three adjacent, ca 50 × 15 m areas were selected, located in a row with the following sequence: intense grazing – weak – medium grazing. The sequence was thus nonlinear with respect to grazing intensity, which contributes to separate effects of grazing from effects of location. Grazing intensity was estimated every 10 d by measuring average vegetation height in each 2 × 2 m plot in all three treatment areas (Ekstam and Forshed 1996). Grazing was considered "intense" if average vegetation height was reduced by at least 75% during June–early September. The corresponding figures for "medium" and "weak" grazing are 40–50% reduction and 20–30% reduction, respectively. Plots which showed a varying grazing intensity during the study were omitted from the analyses. The number of omitted plots per treatment area was < 10 (< 5%). All plots with a certain grazing intensity were situated within the same treatment area.

Gentianella campestris occurred scattered over all three treatment areas. The mortality and growth of all individual adults and rosettes was followed from 1990 (start of experiment) through 1995, together with seed production of adults, herbivory and trampling. The plants were assigned to two size classes. For adult plants, these classes were defined by the number of stem and branch nodes (≥10 nodes: large adults; <10 nodes: small). For rosettes,

the size classes were defined by a size index given by the number of leaves times the rosette diameter in cm (index >15: large rosettes; <15: small, see Lennartsson 1997a for a detailed description of data sampling). Germination of last year's and older seeds was estimated from batches of seeds sown in pots in a common garden. Approximately 200 seeds from ten plants per treatment were sown each year and the germination from each batch was monitored annually throughout the study. Seedling emergence in the field was counted each year in random plots. From these data, combined with data by Milberg (1994) on seed mortality, it was possible to estimate the longevity of the seed bank (Lennartsson 1997a). Correspondingly, the recruitment of large and small rosettes from large and small adults and from the seed bank could be estimated, assuming that the origin of seeds did not affect the fate of the seedlings (cf. Lennartsson et al. 1997).

The life cycle graph for *Gentianella campestris* was translated into a projection matrix with i rows and j columns in which the matrix elements a_{ij} and F_{ij} define transitions from stage j to stage i in one-year time intervals (Fig. 2). a indicates yearly transition probabilities (ranging between 0–1), F indicates fecundity (may be >1). Transition matrices are treated in greater detail by Caswell (1989) and Akçakaya (2000). Analysis of a projection matrix at asymp-

Table 1. Finite rate of increase, λ, for *Gentianella campestris* in three grazing intensities over two types of microsites, mesic and dry, respectively.

Matrix	Intense grazing		Medium grazing		Weak grazing	
	Mesic	Dry	Mesic	Dry	Mesic	Dry
91–92	0.89	0.57	1.3	0.81	0.91	0.95
92–93	0.85	0.58	0.97	0.81	0.85	0.93
93–94	0.41	0.17	0.44	0.21	0.25	0.28
94–95	0.8	0.41	0.77	0.54	0.61	0.51
Average	0.74	0.44	0.87	0.59	0.66	0.67

totic population growth gives the population's finite growth rate, $\lambda = N_{t+1}/N_t$, where N is the population size. Analysis of a population matrix also shows how population growth rate changes due to a change in a certain matrix element a_{ij}. To compensate for differences in absolute values of a_{ij}, elasticity was used, defined as the proportional change in λ caused by a proportional change in a_{ij} (de Kroon et al. 1986). Elasticities sum to one and reflect the relative importance of matrix elements for population growth rate (Silvertown et al. 1996).

The probability of extinction was estimated by combining four autumn – autumn transition matrices per treatment in a stochastic model, where the probability of summer drought (probability of the 1993–1994 matrix, Ultuna Climate and Bioclimate Station unpubl.) was varied (cf. Grime and Curtis 1976, Hopkins 1978, Bengtsson 1993). I ran 1 000 replicates per simulation over 200 yr, using the computer programs KARISMAT, developed by Kari Lehtilä, and POPPROJ2, by Menges (1992). The proportion of simulations resulting in a population size equal to zero after a certain time period gave the extinction probability for that time period (Wilcox and Murphy 1985, Quinn and Hastings 1987, Menges 1997).

When discussing how grazing intensity may interact with other factors than microsite moisture, I refer to data from two earlier clipping experiments with *G. campestris*. One of the experiments investigated how compensatory growth following clipping damage varied with the timing of damage (Lennartsson et al. 1998). The second experiment investigated how compensatory growth varied between early flowering, intermediate and late flowering seasonal types of the species (Lennartsson et al. 1997). In both experiments plants were damaged by removing about half of the biomass. Seed production was compared with that of unclipped control plants of the same initial size as the clipped ones.

Results

Grazing intensity affected λ for the studied population of *Gentianella campestris* (Table 1), that in turn influenced the probability of extinction (Fig. 4). The most unfavour-

able grazing pressure yielded an extinction risk that was up to six times higher than the optimal one, other factors being constant. However, the relationship between probability of extinction and grazing intensity differed considerably between dry and mesic sites, thus indicating an interaction effect of grazing intensity and microsite moisture (Fig. 4). At mesic sites, an intermediate grazing pressure yielded the lowest extinction risk, whereas at dry sites extinction risk decreased with decreasing grazing intensity. Extinction risk was always lower at the mesic sites than at the dry sites, the difference being largest for the medium grazing intensity.

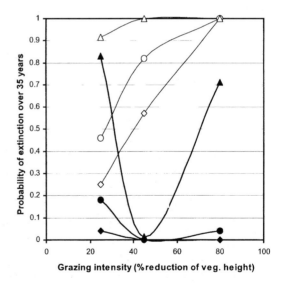

Fig. 4. Probability of extinction of *Gentianella campestris* as a function of grazing intensity. Thick lines and filled symbols indicate mesic site, thin lines and unfilled symbols indicate dry site. Lines with triangles indicate a probability of summer drought of 0.25; circles a probability of 0.1; diamonds a probability of 0.05. Initial population size, $n_0 = 1\ 000$, where the respective numbers of large and small rosettes and adults $= n_0/4$; the number of seeds in seedbank 1 $= n_0 \times 500$; and in seedbank 2 $= n_0 \times 225$.

The transition values for rosettes and adult plants growing large (transition elements a_{63-64} and F_{45-46}, respectively) were considerably higher for mesic than for dry sites (Fig. 5). The transition values $a_{31-32, 41-42}$ consequently show that a seed from the seed bank was more likely to develop a large rosette at mesic sites, and a small rosette at dry sites. These differences in size between "dry" and "mesic" plants are also reflected by the fact that production of seeds to the seed bank (F_{15-16}) was higher at mesic than at dry sites (Fig. 5).

The probability that a seed developed a large rosette (e.g. a_{41}) increased with increasing grazing intensity at mesic sites. At dry sites, growth of rosettes was on the contrary favoured by weak grazing (Fig. 5). In general, large adult plants were more often formed by large rosettes (a_{64}) than by small rosettes (a_{63}). The probability that a large rosette developed a large adult increased with decreasing grazing intensity at both mesic and dry sites. The probability that a small rosette developed a large adult plant (a_{63}) showed a peak in medium grazing intensity at mesic sites, whereas at dry sites, the probability increased with decreasing grazing pressure.

The probability of summer drought in general increased the probability of extinction in all treatments (Fig. 4). At mesic sites, an interaction effect of environmental stochasticity and grazing intensity was also shown, because summer drought increased extinction risk more in intense grazing than in weak grazing. For example, with a probability of summer drought of 0.1, weak grazing yielded a 4.5 times higher extinction risk than intense grazing. With a drought probability of 0.25, the corresponding value was 1.2 times (Fig. 4).

The risk of extinction varied with initial population size in interaction with grazing intensity and microsite moisture (Fig. 6). This interaction is reflected both by the position and the shape of the curve extinction risk as a function of population size.

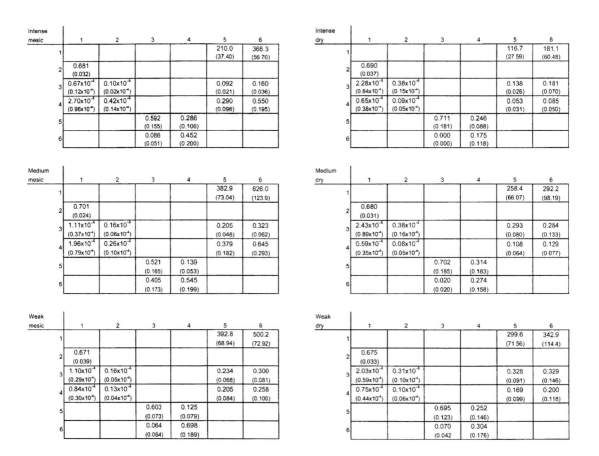

Fig. 5. Average transition matrices for each treatment. Intense, medium and weak refers to three levels of grazing intensity, mesic and dry to two levels of microsite moisture. Mean transition values and standard errors (in brackets) of four matrices.

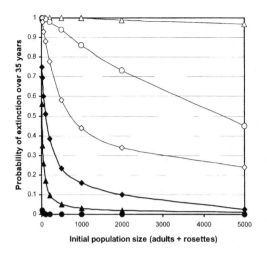

Fig. 6. Probability of extinction of *Gentianella campestris* as a function of intial population size. Thick lines and filled symbols indicate mesic site, thin lines and unfilled symbols indicate dry site. Triangles indicate intense grazing, circles medium grazing, and diamonds indicate weak grazing. Number of adults, rosettes and seeds in seedbank 1 and seedbank 2 are related to initial population size as in Fig. 4.

Discussion

This study shows that grazing is an important determinant of population viability of the grassland plant *Gentianella campestris*. It also emphasises that the effects of grazing intensity need to be evaluated together with other interacting factors. Thus, to optimise population viability by manipulating the grazing, we need some estimates on how grazing influences the population in interaction with other relevant factors. Below, I discuss some properties of environment and population, which can be assumed to interact with grazing to yield differences in population viability of *G. campestris*.

Grazing intensity in interaction with microsite and population size

Implications for management
The optimal grazing intensity for *G. campestris* differed considerably between sites with different soil moisture. This result highlights two important implications for conservation of the species: 1) if possible, dry parts of a site should be managed with a lower grazing pressure than mesic parts; 2) a viability analysis of a certain population of *G. campestris* needs to estimate the extinction risk for "dry" and "mesic" subpopulations separately, preferably by combining grazing intensity with population size (Fig. 6).

The λ values obtained by analysis of the population matrices (Table 1) were congruent with the observed

number of plants at the study site. During the first three years of the study, all subpopulations except for the intense/dry treatment were stable or increased (medium/mesic), whereas the low λ values during the last years were shown by a drastic decrease in population size in all treatments.

Even with an optimal grazing intensity, the probability of extinction was much higher for subpopulations on drier soils, and even very large populations would face a high risk of extinction, if no mesic sites are available (Fig. 6). A medium grazed, "dry" population with an initial size of 100 000 plants would still have a probability of extinction of 0.75 over 50 yr, which implies that a minimum viable population size (MVP, cf. Menges 1997) is hardly possible to find in dry habitats. In most populations of *G. campestris*, a certain proportion of the plants grow at dry microsites, in spite of reduced subpopulation viability in such habitats. Judging from the λ values, one would expect a source-sink relationship between mesic and dry patches within a population. This relationship may however be reversed during occasional periods with very weak or no grazing (cf. Fig. 4 with curves extrapolated to, for example, 10% grazing intensity). Also in very rainy years, population viability can be expected to be high at dry microsites.

It can be assumed that if grazing intensity varies between years (which is often the case), population viability of *G. campestris* may be highest in patchy grasslands, because a temporal variation in grazing seems to be buffered by a spatial variation in microsite moisture. This is supported by a field survey of 65 populations of *G. campestris* and the closely related *G. amarella* (L.) Börner in the Province of Uppland (Lennartsson and Svensson 1996). This survey showed that under managed conditions, populations were significantly larger at mesic than at dry sites, whereas unmanaged populations were less reduced at dry than at mesic sites. This may thus be explained by slightly higher λ at dry sites than at mesic, when grazing intensity is low (Table 1).

The reduced population viability in some combinations of grazing intensity and microsite is also reflected by the shape of the curves in Fig. 6, especially regarding threshold values for population size. The curve for intense grazing at mesic sites shows that a small change in population size will cause a large change in extinction risk, if population size is smaller than ca 250–500 plants (Fig. 6). The corresponding threshold size for the weak/mesic curve is 500–1 000 plants, and for the weak/dry curve 1 000–2 000 plants. For the medium/dry treatment there is a more or less linear relationship between population size and extinction probability up to 5 000 plants.

Grazing intensity in interaction with timing of grazing and environmental stochasticity

One important aspect of the life cycle of pasture plants is mechanisms for increasing the tolerance to grazing (Belsky et al. 1993, Rosentahl and Kotanen 1994, Järemo et al. 1999). In *G. campestris*, such tolerance is accomplished by high capacity for compensatory growth of branches, flowers and fruits after grazing (Lennartsson et al. 1997). Under certain conditions the plants may even overcompensate for damage, that is, damaged plants produce more seeds than undamaged ones of the same initial size. In the transition and elasticity matrices, overcompensation and compensatory growth in general is indicated by high values of the matrix element a_{63} (and to some extent a_{64}), i.e., development of large adults by small rosettes (Fig. 2). a_{63} has a high elasticity value in the treatment "medium grazing/mesic site" (Fig. 5), but lower values in all other matrices. This indicates that compensation is influenced by an interaction effect of grazing intensity and microsite moisture. Furthermore, Lennartsson et al. (1998) showed that vigorous compensatory growth could be induced by damage during a limited time period only, thus an effect of the timing of grazing. Full compensation or overcompensation was induced only if damage occurred during ca 1–25 July. Under circumstances which allow high compensation (essentially the medium/mesic treatment), this tolerance mechanism can be assumed to be an important component of the population dynamics of *G. campestris*. This implies that the same grazing intensity can be expected to yield different probabilities of extinction, depending on the timing of grazing, thus indicating an interaction effect of intensity and timing of grazing.

Implications for management

Such an interaction effect has important implications for management. In *Gentianella* populations with a significant proportion of "mesic" plants, an intermediate grazing intensity should be the objective. In sufficiently large grasslands, medium grazing intensity can be attained by using relatively few grazers during a longer time period. In small grasslands, however, even a low number of grazers would give a high grazing intensity, if grazing is extended over the whole season. In small grasslands, therefore, a shorter grazing period is needed to create an optimal grazing intensity, and the timing of that grazing period should as often as possible fall within ca 1–25 July.

Compensatory growth also varied between years, as indicated by between-year differences in the elasticity matrices (Fig. 7). During the dry summer 1994 elasticity for the matrix element a_{63} was zero in the treatment medium grazing/mesic site, which indicates that tolerance to grazing was decreased by summer drought. The matrix elements related to compensatory growth and growth of large adult plants (a_{63} and a_{64}), are in general among the most variable elements between years (Fig. 7). Between-year variation in compensation is therefore an important explanation for the interaction effect of grazing intensity and environmental stochasticity on population viability of *G. campestris*.

Grazing intensity, geographic location and flowering phenology

One prerequisite for compensatory growth is that the grazed plants have enough time for regrowth before the autumn frost. Lennartsson et al. (1998) hypothesised,

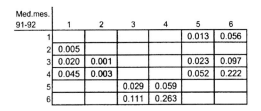

Med.mes. 91-92

	1	2	3	4	5	6
1					0.013	0.056
2	0.005					
3	0.020	0.001			0.023	0.097
4	0.045	0.003			0.052	0.222
5			0.029	0.059		
6			0.111	0.263		

Med.mes. 92-93

	1	2	3	4	5	6
1					0.010	0.071
2	0.007					
3	0.022	0.002			0.013	0.097
4	0.052	0.005			0.032	0.233
5			0.025	0.030		
6			0.110	0.292		

Med.mes 93-94

	1	2	3	4	5	6
1					0.123	0.001
2	0.021					
3	0.100	0.020			0.292	0.002
4	0.003	0.001			0.010	0.000
5			0.414	0.011		
6			0.000	0.003		

Med.mes 94-95

	1	2	3	4	5	6
1					0.019	0.063
2	0.009					
3	0.029	0.004			0.035	0.115
4	0.043	0.006			0.052	0.171
5			0.106	0.000		
6			0.077	0.271		

Fig. 7. Four elasticity matrices for the treatment medium grazing/mesic site, showing between-year variation. Note especially the variation in the matrix element a_{63} (cf. Fig. 2).

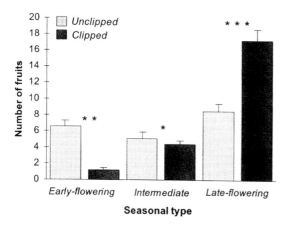

Fig. 8. Average number of mature fruits produced by clipped and unclipped *Gentianella campestris* plants in three seasonal (early-flowering, intermediate and late-flowering) subgroups from the locality Hyttan (Lennartsson et al. 1997). Clipping experiments were performed in cultivation on plants of equal initial sizes. One standard error is indicated, and t-test statistics for paired comparisons of experimental plants from the same mother plants (* p<0.05, ** p<0.01, *** p<0.001). The number of pairs in each subgroup is 21, 18 and 21, respectively.

based on a field experiment, that compensation may be weaker or more occasional at high altitudes and latitudes. This would imply that the plants' tolerance to grazing varies over the species' range of distribution, and thus an interaction effect of grazing intensity and geographic location on population viability of *G. campestris*.

As mentioned earlier, *G. campestris* exhibits a wide intraspecific variation, especially regarding flowering phenology (Lennartsson 1997b). One important difference between early and late flowering plants is that early plants show much lower tolerance to damage (Fig. 8; Lennartsson et al. 1997). This indicates an interaction effect of grazing intensity and flowering phenology, which can be assumed to decrease viability of early flowering populations under intense grazing.

Implications for management
A lower grazing pressure can be recommended for early-flowering populations of *G. campestris*. An even better alternative would be a late onset of grazing, after completed seed maturation of the plants in mid-July.

Linking population data and PVA with management and conservation

Estimation of the relative importance of single factors, and of the interactions between them, is a key task in population viability analysis and in recovery planning for threatened species (cf. Schemske et al. 1994). The viability of a

plant population is affected by a number of biotic and abiotic factors (see references in Silvertown and Franco [1993], Silvertown et al. [1996] and Menges [1997]), and *G. campestris* is no exception. In this paper I have discussed seven factors, all being important determinants of the population viability of *G. campestris*: grazing intensity, microsite, environmental stochasticity, population size, timing of grazing, geographic location, and flowering phenology. Of these factors, only intensity and timing of grazing are possible to directly influence by conservation efforts. Hence, even if other factors may be equally or more important for population viability of a grassland species, management can be regarded as the tool with which we may manipulate the effect of other factors.

If we have or intend to make a PVA for a grassland species and want to translate this knowledge to management recommendations, I suggest the following method: first, evaluate known relevant factors in terms of how they affect population viability of the target species in interaction with management. Second, design theoretical management regimes which optimise population viability for each interacting factor separately. Third, design or choose one or a set of management regime(s) which make the best compromise. In many cases, as for *G. campestris*, different types of management, varying between years, would probably be the best alternative.

Even without a PVA and population data, we usually have some knowledge about how grassland species react to management. When designing management recommendations on such a basis, I recommend that knowledge from as many different conditions as possible are used (different years, microsite, geographic locations etc.). This study of *G. campestris* clearly shows that management recommendations or recovery plans may be of little use if they are based on data from only a narrow range of biotic and abiotic conditions.

Acknowledgements – I thank the Anduri family at Björnvad for placing their gentians at our disposal, and Kari Lehtilä and Eric Menges for kindly providing their analysis programs. Eric also gave valuable comments to the manuscript, together with Thomas Nilsson. This study was financed by the Swedish Assocation for Conservation of Nature, WWF Sweden, the Foundation for Strategic Environmental Research (MISTRA, Project # 000438-97-1), and by the Swedish Council for Forestry and Agricultural Research (SJFR, Project # 34.0297/98).

References

Akçakaya, H. R. 2000. Population viability analyses with demographically and spatially structured models. – Ecol. Bull. 48: 23–38.
Anon. 1987. Inventering av ängs- och hagmarker. Handbok. – Swedish Environmental Protection Agency, Stockholm, in Swedish.
Anon. 1990. Betesmarker – historiska data. – Statistiska Centralbyrån, Statistiska Meddelanden 36: 9001, in Swedish.

Aronsson, M. and Matzon, C. 1987. Odlingslandskapet. – LT Publ., Stockholm, in Swedish.

Aronsson, M., Hallingbäck, T. and Mattsson, J. (eds) 1995. Rödlistade växter i Sverige 1995 (Swedish red-listed plants 1995). – Swedish Threatened Species Unit, Uppsala, in Swedish.

Baldock, D. and Long, A. 1987. The Mediterranean environment under pressure: the influence of the CAP on Spain and Portugal and "IMPs" in France, Greece and Italy. – Rep. to WWF, Copenhagen.

Bastrenta, B. and Belhassen, E. 1992. The effects of simulated grazing on survival and reproduction of *Anthyllis vulneraria*. – Acta Oecol. 13: 787–796.

Bastrenta, B., Lebreton, J.-D. and Thompson, J. D. 1995. Predicting demographic change in response to herbivory: a model of the effects of grazing and annual variation on the population dynamics of *Anthyllis vulneraria*. – J. Ecol. 83: 603–611.

Beaufoy, G., Baldock, D. and Clark, J. 1995. The nature of farming. Low intensity farming systems in nine European countries. – Inst. for European Environ. Policy, London.

Belsky, A. J. et al. 1993. Overcompensation by plants: herbivore optimization or red herring? – Evol. Ecol. 7: 109–121.

Bengtsson, K. 1993. *Fumana procumbens* on Öland – population dynamics of a disjunct species at the northern limit of its range. – J. Ecol. 81: 745–758.

Bernes, C. 1993. The Nordic environment – present state, trends and threats. – Nordic Council of Ministers, Stockholm, Nord 1993: 12.

Bernes, C. 1994. Biodiversity in Sweden – a land survey. – Swedish Environmental Protection Agency, Stockholm, Monitor 14.

Bullock, J. M., Hill, B. C. and Silvertown, J. 1994. Demography of *Cirsium vulgare* in a grazing experiment. – J. Ecol. 82: 101–111.

Caswell, H. 1989. Matrix population models. Construction, analysis, and interpretation. – Sinauer.

de Kroon, H. A. et al. 1986. Elasticity; the relative contribution of demographic parameters to population growth rate. – Ecology 67: 1427–1431.

Duffey, E. 1974. Grassland ecology and wildlife management. – Inst. of Terrestrial Ecology, London.

Ekstam, U. and Forshed, N. 1996. Äldre fodermarker. Betydelse av hävdregimen i det förgångna. Målstyrning, mätning och uppföljning. – Swedish Environmental Protection Agency, Stockholm, in Swedish.

Ekstam, U., Aronsson, M. and Forshed, N. 1988. Ängar. Om naturliga slåttermarker i odlingslandskapet. – LT Publ., Stockholm, in Swedish.

Emanuelsson, U. et al. 1985. Det skånska kulturlandskapet. – Signum, Lund, in Swedish.

Eriksson, O. 1996. Population ecology and conservation – some theoretical considerations with examples from the Nordic flora. – Symb. Bot. Upsal. 31: 159–167.

Fry, G. L. A. 1991. Conservation in agricultural ecosystems. – In: Spellerberg, I. F., Goldsmith, F. B. and Morris, M. G. (eds), The scientific management of temperate communities for conservation. Blackwell, pp. 415–443.

Fuller, R. M. 1987. The changing extent and conservation interest of lowland grasslands in England and Wales: a review of grassland surveys 1930–1984. – Biol. Conserv. 40: 281–300.

García, A. 1992. Conserving the species-rich meadows of Europe. – Agricult. Ecosyst. Environ. 40: 219–232.

Grime, J. P. and Curtis, A. V. 1976. The interaction of drought and mineral nutrient stress in calcareous grassland. – J. Ecol. 64: 975–988.

Grubb, P. J. 1988. The uncoupling of disturbance and recruitment, two kinds of seed bank, and persistence of plant populations at the regional and local scales. – Ann. Zool. Fenn. 25: 23–36.

Hopkins, B. 1978. The effects of the 1976 drought on chalk grassland in Sussex, England. – Biol. Conserv. 14: 1–12.

Hostettmann-Kaldas, M. and Jacot-Guillarmod, A. 1978. Xanthones et C-glucosides flavoniques du genre *Gentiana* (sous-genre *Gentianella*). – Phytochemistry 17: 2083–2086.

Hultén, E. and Fries, M. 1986. Atlas of north European vascular plants, north of the Tropic of Cancer. – Koeltz, Königstein.

Ingelög, T., Andersson, R. and Tjernberg, M. (eds) 1993. Red Data Book of the Baltic Region. Part 1. List of threatened vascular plants and vertebrates. – Swedish Threatened Species Unit, Uppsala.

Inouye, H., Ueda, S. and Nakamura, Y. 1967. Biosynthesis of the bitter glycosides of Gentianaceae, gentiopicroside, swertiamarin, and sweroside. – Tetrahedron Lett. 41: 3221–3226.

Järemo, J. et al. 1999. Plant adaptations to herbivory: mutualistic versus antagonistic coevolution. – Oikos 84: 313–320.

Knapp, A. K. and Seastedt, T. R. 1986. Detritus accumulation limits productivity of tallgrass prairie. – BioScience 36: 662–667.

Lennartsson, T. 1997a. Demography, reproductive biology and adaptive traits in *Gentianella campestris* and *G. amarella* – evaluating grassland management for conservation by using indicator plant species. – Acta Univ. Agricult. Suec., Agraria 46.

Lennartsson, T. 1997b. Seasonal differentiation – a conservative reproductive barrier in two grassland *Gentianella* (Gentianaceae) species. – Plant Syst. Evol. 208: 45–69.

Lennartsson, T. and Svensson, R. 1996. Patterns in the decline of three species of *Gentianella* in Sweden illustrating the deterioration of semi-natural grasslands. – Symb. Bot. Upsal. 31: 169–184.

Lennartsson, T., Tuomi, J. and Nilsson, P. 1997. Evidence for an evolutionary history of overcompensation in the grassland biennial *Gentianella campestris* (Gentianaceae). – Am. Nat. 149: 1147–1155.

Lennartsson, T., Nilsson, P. and Tuomi, J. 1998. Induction of overcompensation in the field gentian, *Gentianella campestris*. – Ecology 79: 1061–1072.

Menges, E. 1992. Stochastic modeling of extinction in plant populations. – In: Fiedler, P. L. and Jain, S. K. (eds), Conservation biology: the theory and practice of nature conservation, preservation, and management. Chapman and Hall, pp. 253–276.

Menges, E. 1997. Evaluating extinction risks in plant populations. – In: Fiedler, P. L. and Kareiva, P. M. (eds), Conservation biology for the coming decade. Chapman and Hall, pp. 49–65.

Menges, E. and Dolan, R. W. 1998. Demographic viability of populations of *Silene regia* in midwestern prairies: relationships with fire management, genetic variation, geographic location, population size and isolation. – J. Ecol. 86: 63–78.

Milberg, P. 1994. Germination ecology of the endangered grassland biennial *Gentianella campestris*. – Biol. Conserv. 70: 287–290.

Müller, H. 1881. Alpenblumen, ihre Befruchtung durch Insekten und ihre Anpassungen an Dieselben. – W. Engelmann, Leipzig.

Nilsson, Ö., Jonsell, B. and Lennartsson, T. 1999. *Gentianella campestris* ssp. *campestris*, fältgentiana. – In: Aronsson, M. (ed.), Artfakta kärlväxter. Swedish Threatened Species Unit, Uppsala, pp. 363–365.

Oostermeijer, J. G. B. 1992. Population biology and management of the marsh gentian (*Gentiana pneumonanthe* L.), a rare species in The Netherlands. – Bot. J. Linn. Soc. 108: 117–129.

Oostermeijer, J. G. B. 1996. Population viability of the rare *Gentiana pneumonanthe* – the relative importance of demography, genetics, and reproductive biology. – Ph.D. thesis, Univ. of Amsterdam.

Påhlsson, L. (ed.) 1994. Vegetationstyper i Norden (Nordic vegetation types). – TemaNord 1994: 665, in Swedish.

Pritchard, N. M. and Tutin, T. G. 1972. *Gentianella* Moench. – In: Tutin, T. G. et al. (eds), Flora Europaea, Vol. 3. Cambridge Univ. Press, pp. 63–67.

Quinn, J. F. and Hastings, A. 1987. Extinction in subdivided habitats. – Conserv. Biol. 1: 198–208.

Rackham, O. 1986. The history of the countryside. – Weidenfeld and Nicolson, London.

Rosentahl, J. P. and Kotanen, P. M. 1994. Terrestrial plant tolerance to herbivory. – Trends Ecol. Evol. 9: 145–148.

Rydberg, H. and Vik, P. 1992. Ängs- och hagmarker i Södermanlands län. – Länsstyrelsen Södermanlands län, Nyköping, in Swedish.

Schemske, D. W. et al. 1994. Evaluating approaches to the conservation of rare and endangered plants. – Ecology 75: 584–606.

Silvertown, J. and Franco, M. 1993. Plant demography and habitat: a comparative approach. – Plant Spec. Biol. 8: 67–73.

Silvertown, J., Franco, M. and Menges, E. 1996. Interpretation of elasticity matrices as an aid to the management of plant populations for conservation. – Conserv. Biol. 10: 591–597.

Simán, S. and Lennartsson, T. 1998. Slåtter eller bete i naturliga fodermarker? – ett skötselförsök med slåtteranpassade växter. – Svensk Bot. Tidskr. 92: 199–210, in Swedish.

Stanners, D. and Bourdeau, P. 1995. Europe's environment. The Dobris Assessment. – European Environment Agency, Copenhagen.

Tucker, G. 1991. The status of lowland dry grassland birds in Europe. – In: Goriup, P. D., Batten, L. A. and Norton, J. A. (eds), The conservation of lowland dry grassland birds in Europe. Joint Nature Conserv. Comm., Peterborough, pp. 37–48.

Wettstein, V. R. 1895. Der Saison-Dimorphismus als Ausgangspunkt für die Bildung neuer Arten im Pflanzenreiche. – Ber. Deutsch. Bot. Ges. 13: 303–313.

Widén, B. 1987. Population biology of *Senecio integrifolius* (Compositae), a rare plant in Sweden. – Nordic J. Bot. 7: 687–704.

Wilcox, B. A. and Murphy, D. D. 1985. Conservation strategy: the effects of fragmentation on extinction. – Am. Nat. 125: 879–887.

Wolkinger, F. and Plank, S. 1981. Dry grasslands of Europe. – European Comm. for the Conserv. of Nature and Natural Resources, Strasbourg.

Zopfi, H. J. 1991. Aestival and autumnal vicariads of *Gentianella* (Gentianaceae): a myth? – Plant Syst. Evol. 174: 139–158.

Ecological Bulletins 48: 123–142. Copenhagen 2000

Demography and management of relict sand lizard *Lacerta agilis* populations on the edge of extinction

Sven-Åke Berglind

Berglind, S.-Å. 2000. Demography and management of relict sand lizard *Lacerta agilis* populations on the edge of extinction. – Ecol. Bull. 48: 123–142.

The sand lizard *Lacerta agilis* has declined in most of north-western Europe during the last decades, mainly due to loss and fragmentation of its habitat. The species reaches the northern periphery of its range in central Sweden, with a few, isolated relict populations restricted to large, sandy areas dominated by pine forest. Six local populations within one of these areas were censused during 1984–1998. Two populations went extinct and the remaining four declined, each with less than ten adult females left in 1998. Efficient afforestation and fire suppression seem to be the most important factors behind the recent decline of the species, having reduced the amount of open, suitable habitat. Life table analysis of two populations implied an average 6% decline and 3% increase in population size per year, respectively (λ = 0.94 and 1.03). Simulations of stochastic future population growth for 20 yr with no management predicted a 39% and 8% risk of extinction, respectively. Projected risks of population extinction and decline were highly dependent on the population growth rate, which in turn was greatly affected by the estimates of juvenile survival. Elasticity analysis demonstrated that this latter demographic parameter contributed most to population growth rate under asymptotic conditions. Simulations of five different conservation management options ranked a programme of captive raising (increased juvenile survival in captivity during the first hibernation) or captive breeding (using a breeding stock from the two populations, respectively), in parallel with habitat management, potentially to be the most effective options to drastically reduce the risk of extinction and decline.

S.-Å. Berglind (sven-ake.berglind@ebc.uu.se), Dept of Conservation Biology and Genetics, Evolutionary Biology Centre, Uppsala Univ., Norbyvägen 18 D, SE-752 36 Uppsala, Sweden.

The conservation of species on the edge of their distribution poses particular difficulties since population densities and number of habitat types occupied tend to decrease toward the periphery of a species' geographic range (Lawton 1993, Gaston 1994). Peripheral populations can also be expected to be particularly liable to environmental and catastrophic variation, which generally have profound effects on extinction probabilities (e.g. Thomas 1990, Lande 1993, Caughley 1994), especially in species with low finite rates of increase under undisturbed conditions (Menges 1998).

In the Scandinavian forests many species are at the northern limit of their range. A large number of these are included in the national red lists, and there is an urgent need to know more about their population dynamics for long-term conservation (Berg and Tjernberg 1996, Gärdenfors 2000). The study of relict populations known to have survived since prehistoric times is of particular interest since confounding effects of recent long-distance dispersal on population survival can be ruled out. In south-central Sweden, a small number of peripheral populations of the sand lizard *Lacerta agilis* occur (Gasc et al. 1997) that are considered to be relicts from the postglacial warm period ca 7000–500 B.C. (Gislén and Kauri 1959, Gullberg et al. 1998). Most of these populations are restricted to large, sandy areas dominated by pine forest (Andrén and Nilson 1979). Berglind (1988) suggested that the survival of these populations in the past was largely dependent on recurrent forest fires and human activities which may have created suitable early successional, open habitat patches with a mosaic of exposed sand and a rich field layer of heather *Calluna vulgaris*.

Globally, the sand lizard is widely distributed from central Europe to central Asia, and occurs in a variety of open or semi-open habitats with different types of soil and ground vegetation (House and Spellerberg 1983a, Bishoff 1988). The area of occupancy in the northwestern part of its range has decreased considerably during the last decades, and there is no doubt that this is mainly due to loss of suitable habitat, principally afforestation of heathlands, overgrowth of dry meadows, and exploitation of coastal dune areas (Glandt and Bishoff 1988, Gasc et al. 1997). The life history and habitat requirements of the species have been studied in England, the Netherlands and Germany, to provide guidelines for conservation (see overview by Glandt and Bishoff 1988). In Sweden, it is classified as "vulnerable" (Gärdenfors 2000).

This paper presents results from a 14-yr study of an isolated relict population of the sand lizard in south-central Sweden that consists of a few local populations. The need for conservation action to save this geographic population was realized some years ago (Berglind 1988, 1995a), and here I perform a population viability analysis (PVA; see overview by Akçakaya and Sjögren-Gulve 2000) to evaluate five different management scenarios. The results are of particular interest since most PVAs on vertebrates so far have been applied to mammals and birds (Groom and Pascual 1998), which can be expected to be less sensitive to moderate weather variation than reptiles. Especially on the northern periphery of their range, reptiles are highly dependent on the number of sunshine hours for thermoregulation and successful egg development. Furthermore, the management of the sand lizard in central Sweden can be the foundation for conservation strategies for a habitat containing several other early successional species that have declined in the Scandinavian sandy pine forests during the last decades.

Methods

Study area and population monitoring

The field work was conducted on the 11 000 ha, glaciofluvial sand area Brattforsheden (59°40′N) in south-central Sweden (Fig. 1). The topography is characterized

Fig. 1. The study area Brattforsheden in central Sweden and the spatial distribution of local populations of the sand lizard *Lacerta agilis*. Filled circles = extant populations, open circles = extinct populations, grey = sandy soil. The lake Alstern and the Svartån stream are indicated. (Map modified from Furuholm et al. 1994.)

by flat plateaus and areas of undulating topography, the latter mainly represented by fossil sand dunes. The vegetation is dominated by Scots pine forest *Pinus sylvestris* with a ground- and field layer dominated by reindeer lichens *Cladina* spp., mosses (e.g. *Pleurozium* and *Dicranum* spp.), cow berry *Vaccinium vitis-idaea*, crow berrry *Empetrum nigrum*, and heather *Calluna vulgaris*. Presently, afforestation and effective forest-fire suppression keep most of the area wooded with managed pine forest of various ages. Brattforsheden is isolated by 50 and 160 km, respectively, from the nearest two other sandy areas where the sand lizard occurs, also in small and isolated populations (Berglind 1999). The landscape in between is unsuitable for the species at this latitude, and dominated by clayey soils or morain with agriculture or mixed spruce forests.

All occupied sand lizard patches at Brattforsheden were censused several times per season, and most potential patches at least once per season, from 1984 to 1998. Two populations separated by dense pine forest for a distance of 3 km at the sites FL (Fig. 2) and SB, were subject to demographic studies from 1988 to 1998. These sites were monitored for lizards (Fig. 3) on average 1–4 h every second day during suitable weather from mid-May to mid-September 1988–1991 (corresponding to a few weeks after adult emergence from hibernation until the beginning of next hibernation), and for ca 15 to 20 d during May–September 1992–1998. The lizards were captured by hand, measured (snout-vent length), photographed from their right (including dorsal) side, permanently marked by toe-clipping, and released on the same spot within 5–10 min. The phalanges were preserved in 4% formaldehyde, and later sectioned and stained in eosin-haematoxylin in accordance with Hemelaar (1985) for a skeletochronological analysis.

The total area of suitable habitat, i.e. open and usually south-facing sites with a mosaic of bare sand patches and a rich field layer dominated by heather, was ca 1.0 ha in 1988 at both study sites, and included sections of forest road verges, semi-open pine plantations on old burned areas, a power line corridor, and a small shore at a forest tarn (Berglind 1988). Since it was obvious that these sites had become, and were to become, smaller because of increasing shade due to canopy formation of surrounding pine stands (15 and 35 yr old respectively), the sites were subject to habitat management in the winter and spring of 1988. This included cutting of dense tree sections and scraping off patches of the humus layer to create new sand patches. This doubled the area of suitable habitat to 2.0 ha at each

Fig. 2. Sand lizard habitat patch within an area that burned some 70 yr ago and has escaped invasion by pine trees. South-oriented slope with exposed sand used for egg-laying by sand lizards, and a surrounding field layer of heather *Calluna vulgaris* used for foraging and shelter. No lizards have been observed in the closed, surrounding pine forest. Part of the study site FL in 1984. Photograph: S.-Å. Berglind.

Fig. 3. Gravid, nine year old female sand lizard, with a body length (snout-vent) of 85 mm, and a total length of 200 mm. Site FL, in May 1990. Photograph: S.-Å. Berglind.

locality. These new clearings did not decline in size over the last 10 yr, but some parts of the original habitat were reduced due to causes as above.

During the egg laying period, which normally occurred during the first two–three weeks in June or a few weeks later during cold summers, all filled burrows dug by female sand lizards on the limited number of open sand patches at the study sites were discretely marked out with small twigs in all years from 1988 to 1998. The burrows were discovered by the freshly disturbed sand, and normally accompanied by unfilled trial burrows without eggs (see Berglind 1988). In late autumn after the hatching period, which normally occurred from mid-August to late September, the clutches were excavated. Each clutch was tightly clustered ca 5–7 cm below the ground. Hence, the number of deposited eggs, their hatching success, and the minimum number of females participating in reproduction each year were monitored in detail.

Demographic modelling

Analysis of survival rates from capture-recapture data
Using the capture-recapture data gathered during 1988–1998, capture-history matrices with a time-step of one

year were constructed for animals marked at different ages (as determined by skeletochronology for those marked as subadults or adults). Reasonably large samples of hatchlings (0-yr olds) were marked only in 1988–1990. The animals were separated by locality, and those marked as adults also by sex. Multiple sightings within a year were treated as a single sighting. These matrices were used as input files for the computer software MARK 1.6 (White 2000), sub-programme "Recaptures only", designed to obtain maximum-likelihood estimates of survival and capture probability rates from the resightings of marked individuals. MARK provides parameter estimates under the Cormack-Jolly-Seber model (cf. Lebreton et al. 1993), but also under several models that appear as special cases of this model. A series of alternative models were explored.

Following Lebreton et al. (1992, 1993), model selection is often based on Akaike's Information Criterium (AIC). AIC weighs the deviance of a model and its number of parameters so as to select the most parsimonious model that adequately describes the data. Anderson et al. (1994) describe how AIC should be corrected, to account for e.g. heterogeneity in the data, by using instead QAICc, which generally leads to selecting simpler models (i.e., models with fewer parameters) than using AIC. In the present study, all models explored had rather high values of the statistic $\hat{c}$, indicating some problems with heterogeneity in the data. Due to these high $\hat{c}$ values, use of the statistical criterion QAICc would lead to models too simple to be biologically reasonable. Therefore, model selection was based on biological arguments instead.

The recapture rate of 0-yr-olds differed significantly between the populations (5 out of 77 marked at FL and 7 out of 18 marked at SB; $\chi^2 = 9.23$, DF = 1, p = 0.0024), whereas those of the other age groups did not ($\chi^2 \leq 2.43$, DF = 1, p ≥ 0.119). A biologically reasonable approach was to model age variation by separating between juveniles (0-yr olds), subadults (pre-reproductive ages; 1–2 yr olds), and adults (that have hibernated 3 times or more), respectively. Based on these data, a model accounting for age (3 age classes) and locality (except for the adult age class) was applied to the data for the animals marked as juveniles or subadults, and estimates for juveniles and subadults were taken from this model. To the data for the animals marked as adults, a model accounting only for sex differences was applied, and estimates for adult males and females were taken from this model. The survival estimates (ϕ) from these models were then used as the survival rates (p_x) in life table analysis and in Leslie matrices of the stochastic models (below).

Life table analysis

Deterministic population growth was quantified by life table analysis of the female populations at FL and SB, yielding estimates of the following demographic parameters: net reproductive rate (R_0), finite rate of increase (λ), cohort generation time (T_c; e.g. Begon et al. 1996) and population mean generation time (T; Caughley 1977).

In earlier demographic studies of lizards, mean clutch size of different age classes were estimated using a relation between body size and clutch size and assuming there is a direct relationship between body size and age (Strijbosch and Creemers 1988). However, it has recently been shown that age-specific fecundity is subject to large variation, and that age and clutch size explain only a small proportion of the variation in a female's reproductive success (i.e. number of offspring surviving to sexual maturity) in the sand lizard (Olsson et al. 1996a, Olsson and Shine 1997) and other reptiles (Madsen and Shine 1998). Therefore, instead of using age-specific fecundity, I used the overall mean clutch size, minus unhatched eggs, for all adult age classes and from each study population under field conditions.

My data on age-specific probability of reproduction were insufficient. Instead, I used data from a Dutch population studied by Strijbosch and Creemers (1988), which seemed to confirm my own observations, except for 2-yr olds, which do not seem to reproduce in my study area (and very rarely so in the Netherlands). The sex ratio among the offspring was set to 50% females, also in accordance with Strijbosch and Creemers (1988).

Elasticity analysis

A Leslie matrix parameterized with post-reproduction survival and fecundity data (below) was subjected to elasticity analysis using RAMAS GIS (Akçakaya 1998) to assess which age-specific survival or fecundity rates are the most important in contributing to λ under asymptotic population growth (for overview, see de Kroon et al. 2000).

Simulation study and management scenarios

Future population growth of the study populations with stochasticity added, and with or without management, were simulated with RAMAS GIS/Metapop (Akçakaya 1998), which can be parameterized to meet the assumptions of a single, stochastic, age-structured population model (Beissinger and Westphal 1998, Akçakaya 2000).

In all simulations, population censuses were made after reproduction. No density dependence was used, since it was not known to what extent population growth was density dependent, and if so, how it was manifested. In this situation, Ginzburg et al. (1990) concluded that estimation of extinction risks becomes more conservative when density dependence is not included. Density dependence may also be less important to consider for short-term forecasting of small, low-density populations (Boyce 1992), like in my study. Furthermore, I made several test simulations with a population ceiling set to 800 and 1000 (extrapolating from reported densities, e.g. Strijbosch and Creemers 1988), but this had only minor effects on projected population size and risk of decline.

RAMAS Metapop used a Leslie matrix parameterized with post-reproduction data. Accordingly, the age-specific fecundity (f_x) in the matrix was calculated as $f_x = p_x \times m_{x+1}$, where p_x = average yearly survival for individuals of age x, and $m_x = p_{repr} \times fem_{young}$, where p_{repr} = age-specific probability to reproduce, and fem_{young} = the mean number of hatched female offspring per reproductive female (for justification of not using age-specific fem_{young} data: see previous section).

The measure of environmental stochasticity for fecundity was calculated separately for each site using the CV of numbers of hatched eggs per clutch and year during 1988–1997 multipied by the age-specific fecundity (f_x) to give corresponding standard deviation values for RAMAS. Catastrophic events were ignored when these fecundity variances were calculated.

The environmental stochasticity for yearly survival was calculated as the standard deviation of the series of annual recapture rates, measured as captured individuals year x and number of these known to be alive in year x + 1 (i.e., recaptured in year x + 1 or later). These rates were lumped for both localities for the age classes 1–2 and 3+ year olds, respectively (1997–1998 were excluded because of less capture effort). The corresponding standard deviation for 0-yr olds under field conditions was calculated from the survival rates during 1988–1991 only, and from each population respectively (justified in Analysis of survival rates from capture-recapture data). These estimates of environmental stochasticity are crude, and also include effects of demographic stochasticity and sample error. However, this overestimated variation is probably partly compensated for by underestimated yearly variability due to my relatively short study period (less than two sand lizard generations) (cf. Brook 2000). I also tested simulations with "guesstimates" for variability in yearly survival, where the variability was set to 50, 20 and 10% of the MARK survival estimates for the 0, 1–2, and 3+ year olds, respectively. The difference in model predictions between using guessed values and those estimated from the variability in recapture rates were only minor.

Catastrophes were incorporated as 0% survival for 0-yr olds every 10th year. This estimate is based on observed regional hatching failure in 1987 and a near failure in 1998, and corroborated by captures of individuals of cohorts from 1979–1986 and 1988–1997.

Negative genetic effects were not integrated in my models since observations on egg hatching success and early juvenile viability indicated no inbreeding depression in my study populations (see Discussion). A total of 1000 replicates were run for each simulation.

Retrospective simulations

I did retrospective simulations for each population from 1988 to 1998, to compare model predictions with the observed historical population trajectories. The models were parameterized with life table data and fecundities (f_x)(Tables 2–3), and run using initial population size and the observed age structure of the female populations after reproduction in 1988. The number of 1-, 2-, and 3-yr olds were, however, estimated from half the number of hatched eggs in 1988, and the yearly survival for the corresponding age classes, using the MARK estimates, and counting backwards (the number of 1-yr olds was set to 0 based on the observed hatching failure in 1987). The numbers of individuals in these age classes from 1989 to 1998 were estimated correspondingly from observed egg numbers. These estimates of abundance probably gave less underestimated values than the low numbers caught. The number of 0-yr olds were estimated with accuracy from half the number of hatched eggs, and the number of 3+ year old females from the number of clutches and/or direct observations.

Management scenarios

I modelled future population growth for each population from 1998 and 20 yr onwards under the following five management scenarios (presented in the order from assumed low to higher economic costs).

I) No management, without taking deterministic successional factors (mainly caused by increased pine canopy formation) into effect.

II) Habitat management (cutting of dense tree stands, excavation of new sand patches, and enhancement of heather growth), creating "optimal" habitat conditions for the next 20 yr within a 8 ha large area around the present (1–2 ha large) sites. This would eliminate observed negative effects of shade from surrounding tree canopy on embryo development, i.e., the number of hatchlings is assumed to equal the number of eggs with young in late developmental stages. This increase in fecundity affects the SB-population in particular (see Results and Appendix 1). It is also reasonable to assume an increase in juvenile survival at site FL, from the low estimate of 0.24 (Table 1), after creation of this much larger habitat area. An increase in juvenile survival after habitat management has been reported from England (Corbett and Tamarind 1979), possibly via less density dependent mortality among juveniles. I chose to set the juvenile survival at FL in accordance with Strijbosch and Creemers (1988) whose estimate (0.38) was derived from a stable population over a 7-yr period, showing similar life history characteristics and occupying a similar environment as my study populations. Their estimate may be somewhat underestimated since it was not based on a statistical capture-recapture analysis, but I chose it as a conservative habitat improvement effect for FL. The "optimal" habitat management effect on the FL and SB populations, respectively, was included also in scenarios III–V below.

III) Artificial incubation for 5 or 10 yr, respectively, with eggs being dug up from the field shortly after deposition to be incubated artificially under "optimal" temperatures. Alternatively, gravid females are caught just prior to egg deposition in the field to be temporally transferred to captivity to lay their eggs. This scenario allows for earlier hatching, and larger juveniles before hibernation after release in the field (Corbett 1988). Survival of 0-yr olds for the FL-population was here set to the median value between the Strijbosch and Creemers (1988) estimate for 0-yr olds (above), and my estimate for 1-yr old survival. For the SB-population I set the already much higher 0-yr old survival estimate to the median between the estimates for 0- and 1-yr olds (Table 1 and Appendix 1).

IV) Captive raising in the form of "headstarting" (Caughley and Gunn 1996) for 5 or 10 years, with eggs being dug up from the field just prior to hatching in late summer or early autumn. Juveniles are reared and hibernated in captivity to be released the following spring. This results in substantially increased 0-yr old survival during the first hibernation. It was conservatively set to 0.80 and the standard deviation (yearly variability) to 0.08, in accordance with previous years' pilot studies of rearing juveniles in captivity.

V) Captive breeding (Caughley and Gunn 1996) for 5 or 10 yr, where 5 females are caught (the majority of the present small populations) to reproduce in captivity during winter with two clutches per female, so that twice the number of juveniles can be released in spring each year. Normally, the sand lizard produces only one clutch per year in Sweden, but under optimal conditions in the laboratory two or even three clutches can be produced (Olsson pers. comm.). The 0-yr olds reared within a captive breeding programme were set to correspond to yearly introductions of 31 female 0-yr olds (calculated from the number of potentially hatched eggs in 1998) in the simulations. Their survival during the first winter was equal to that under scenario III. I did not separate the survival of the introduced juveniles from that of the wild-reared ones during the captive breeding periods. Test simulations showed that only minor differences resulted from modelling the two juvenile groups separately.

Simulation setup
Scenarios III–V required building models with two different population growth rates: one for growth under demographic management during the initial 5 or 10 yr, respectively, and one for growth under "optimal" habitat management conditions during the rest of the 20 yr period. Technically, this was done by simulating two population sites in parallel, with differently parameterized stage and standard deviation matrices. After the initial 5 or 10 yr of demographic management at "site 1" 100% of the individuals in all stages (0- to 14-yr olds) were translocated to the vacant "site 2" (with effect of habitat management only) (Appendix 1) where population growth continued.

The simulations under scenarios I, II and V were run using an initial population size and age distribution of 1-, 2- and 3-yr olds estimated from half the number of eggs in 1997, 1996, and 1995, their hatching success, and the MARK estimates for yearly survival rates for the corresponding age classes. The observed number of clutches in 1998 was set to correspond with the number of sexually mature females (3+ yr), and half the number of hatched eggs in the field to the number of female 0-yr olds. The initial population size under scenario III and IV was larger than the others, due to a higher potential hatching success in the lab (see Discussion), resulting in a larger number of 0-yr olds (Table 4).

The period of 20 yr was chosen for three reasons: 1) it is somewhat less than the estimated average time it takes for a cleared area to become too shady for the lizards because of successional pine canopy formation (i.e., without habitat management), 2) the time period exceeds the known maximal life span of the sand lizard (14 yr), and 3) the 5 or 10 yr periods of the three demographic management scenarios III–V are within realistic time-frames to entertain a breeding programme and ensure adequate administrative continuity (cf. Snyder et al. 1996). The relatively short time period of 20 yr also render more reliable PVA projections than long-term forecasting (cf. Beissinger and Westphal 1998, Akçakaya and Sjögren-Gulve 2000).

A summary of the different assumptions and input-values used in the models is presented in Appendices 1–2.

Results

Spatial distribution and female population sizes

In total, 6 populations of the sand lizard were found in the whole area of Brattforsheden during 1984–1998. These were separated from each other by unsuitable habitat (closed pine forest or clear cuts of poor habitat quality) by a distance of 2.5–10 km (Fig. 1). The marked lizards were highly sedentary with an estimated mean home range size of 220 m^2 for adult females, measured over successive years (Berglind 1999). However, one subadult lizard moved 500 m along a sun-exposed forest road within site SB during one season. On another occasion I found an unmarked, large, adult female on a forest road almost 1 km away from site FL, with dense pine forest in between. This might have been an emigrant from FL, that dispersed as an immature and grew in transit along nearby road verges, and subsequently failed to return. Another possibility is that it was a remnant of a now extinct local population. No dispersal between the study populations was indicated. This was confirmed by annual surveys and the use of pitfall traps to collect insects between the localities. Common lizards *Lacerta vivipara* were regularly observed and found in the pitfall traps, but never any sand lizards.

Two of the populations, at the sites Källorna and St. Tjärnen, went extinct during my study period. At Källorna, only 4 large adults were observed; 2 females upon discovery of the population in 1983, 1 female in 1984, and 1 male in 1990. At St. Tjärnen, at least 5 subadult and adult lizards were observed upon discovery of the population in 1984 in a very small area of suitable habitat (0.1 ha); in 1987 and 1988, just one lizard was observed (a large female). In one of the remaining four populations, at the site Skäftdalen, 2 subadults were observed upon discovery of the population in 1990; occasional observations of egg nests and adult females were made in subsequent years, including 1998. The other three populations, including the two study populations at the sites FL and SB (discovered in 1958 and 1984, respectively) and one at the site Djäknetjärn, were larger and of approximately the same size around 1990, when the latter was discovered. The relative number of sand lizard observations did, however, decline considerably at all three sites during subsequent years, and the population at Djäknetjärn was estimated to be the smallest of them in 1998.

The estimated number of adult females at FL and SB during 1988–1998 is shown in Fig. 4. The number of females known to have been alive and the number of observed egg clutches per year were significantly correlated at site FL (Spearman rank correlation $R_S = 0.64$, DF = 9, p =

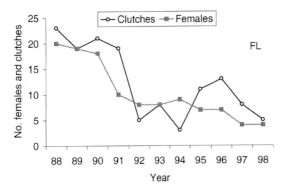

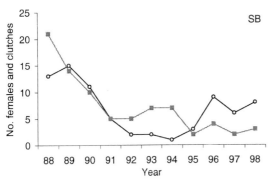

Fig. 4. The number of adult females known to have been alive and number of recorded egg clutches of the sand lizard at sites FL and SB during 1988–1998.

0.034), but not at SB (R_S = 0.37, DF = 9, p = 0.26). The lack of significant correlation at SB might be ascribed to the less intensive mark-recapture work after 1991, but the possibility that some clutches had not been found, or that some adult females had not reproduced every year, cannot be ruled out. No indication was found that individual females deposit more than one clutch per year. Adult female population size, measured as the highest value of either number of females known to have been alive or number of egg clutches as shown in Fig. 4, was significantly decreasing with time during 1988–1998 at both FL (Pearson's R = – 0.83, p = 0.0016) and SB (R = – 0.63, p = 0.037).

Clutch sizes and hatching success

Over the period 1988 to 1998, a total of 893 eggs from 133 clutches were found at FL and 420 eggs from 74 clutches at SB. There was no significant difference in egg clutch size between the sites during these years (mean clutch size = 6.50 for FL vs 6.13 for SB; Wilcoxon Matched Sign Test: p = 0.55; min-max = 3–11 eggs). However, a significantly lower hatching success was recorded at SB (mean number of hatched eggs = 5.41 for FL vs 3.71 for SB; Wilcoxon Matched Sign Test: p = 0.0117) (Fig. 5). Of the unhatched eggs, 85% contained young in late de-

velopmental stages at SB, with a corresponding figure for FL of 55%. Most of these young were dead at the time of excavation in October, but some were still alive. Because of the low temperature at that time of the year, and the winter freezing of the ground below the 5 to 7 cm depth where the eggs were deposited, there was no reason to believe the eggs would eventually have hatched. In spring 1988, after an exceptionally rainy summer in 1987, unhatched eggs with dead young in late developmental stages were also found at both localities.

Adding the number of unhatched eggs with dead or alive young in late developmental stages in autumn to the number of eggs which had hatched, the number of potentially hatched eggs per clutch was calculated. There was no significant difference between the two sites in mean number of potentially hatched eggs during 1988–1998 (mean = 6.14 for FL vs 5.54 for SB; Wilcoxon Matched Sign Test: p = 1.00) (Fig. 5).

Figure 5 shows that in the years 1987 and 1998 there was a more or less complete reproductive failure at both FL and SB. These synchronous failures coincided with cold, rainy summers with exceptionally few hours of sunshine. In 1987, females also deposited their eggs 2–3 weeks later than in normal summers. The reproductive failure in 1994

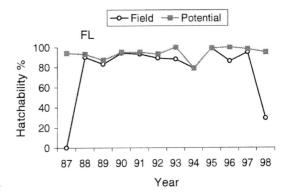

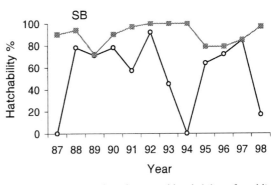

Fig. 5. Field (= natural) and potential hatchability of sand lizard clutches at sites FL and SB during 1987–1998. The potential hatchability for 1987 was calculated as the mean from 1988 to 1998. The number of clutches per year is shown in Fig. 4. Note that only one clutch was found at site SB in 1994.

solely at SB (Fig. 5) was confirmed by the fact that only a single, dried up clutch was found. A small proportion of dried up eggs in early developmental stages were found in most years, but especially after warm summers and in sand patches without any vegetation. The majority of clutches were deposited in sand sparsely colonized by the pioneer moss *Polytrichum piliferum*, which seem to offer incubation conditions of suitable humidity.

The number of observed clutches varied substantially between years at both FL and SB. As apparent from Fig. 4 there was a sudden decline at FL in 1992. This corresponds to the missing cohort of 1987 based on a calculated cohort generation time of 5.5 yr (see below). After a slight increase, the population decreased again in 1997, in accordance with the cohort generation time. At SB, there was a more gradual decline (Fig. 4). This could be a consequence of a more advanced formation of the tree canopy and subsequent shading of the oviposition sites (situated in the middle of a forest road; see Berglind 1988), with corresponding decline in hatching success and juvenile recruitment. Even though adjacent, more sun-exposed sand patches were created in 1988 at SB, there was a delay of some years until the lizards had fully begun to utilize these for oviposition.

Life table and elasticity analyses

During 1988–1998, 153 sand lizards were marked and 27 recaptured in later years at FL and 105 and 42, respectively, at SB. The estimated survival and capture probability rates of the lizards are shown in Table 1. Because of the relatively high $\hat{c}$ values of 8.5 and 5.6 for the chosen models, indicating a problematic amount of unexplained heterogeneity in the data (caused, for example, by heterogeneity in capture probabilities among individuals), these estimates must be taken cautiously.

Life tables were drawn up from these survival estimates, and the estimates on fecundity. The net reproductive rate, R_0, was 0.72 for the FL-population and 1.15 for SB (Table 2 and 3), which implies an average 28% decline in population size per generation for FL and a 15% increase for SB. The cohort generation time (T_c) was 5.55 yr for both FL and SB. The corresponding λ-values were 0.94 and 1.03 respectively, implying an average 6% decline and 3% increase in population size per year under asymptotic growth conditions. The population mean generation time (T; Caughley 1977, p. 124) was estimated to 5.99 yr for FL and 5.42 for SB.

The life expectancy of adults was high, and the two oldest individuals found were a 13 yr old female excavating an egg nest at SB in 1996, and a 14 yr old female in good condition at FL in 1998 (caught for the first time ten years before). Life expectancy at birth was 2.4 yr for the SB-population (where juvenile survival was highest).

The $l_x m_x$-values in the life tables imply that, on average, the reproduction of 3-, 4-, and 5-yr old females contributed the most to population growth (Table 2 and 3). Elasticity analysis showed that the single matrix elements contributing most to the asymptotic population growth rate (λ) were the yearly survival of juveniles ($e_{01} = 0.1849$), 1-yr olds ($e_{12} = 0.1849$), and 2-yr olds ($e_{23} = 0.1456$). These elasticity values were higher than those of adult survival (e ≤ 0.1038) and those of the fecundity elements ($e_{0x} \leq$

Table 1. Survival rates (ϕ) and capture probabilities (p) for the sand lizard *Lacerta agilis* at sites FL and SB, estimated under two models for different subsets of the data. The estimates for non-adults were derived from a model allowing survival rates and capture probabilities to vary between age classes (juveniles 0 yr old, subadults 1 or 2 yr old, adults 3 yr or older) and also between sites (except for the adult age class); for this model, the number of estimated parameters = 10, deviance = 136.15, $\hat{c}$ = 8.51, and QAICc = 54.50. The estimates for adults were derived from a model where the survival rates and capture probabilities differ only between the sexes; for this model, the number of estimated parameters = 4, deviance = 116.87, $\hat{c}$ = 5.57, and QAICc = 76.12. The standard error and confidence intervals have been adjusted to account for the high $\hat{c}$ values.

		ϕ				p			
				95% confidence interval				95% confidence interval	
Age class	Site	Estimate	SE	Lower	Upper	Estimate	SE	Lower	Upper
0	FL	0.2448	0.1793	0.0462	0.6845	0.0504	0.0607	0.0044	0.3890
1-2	FL	0.6143	0.1818	0.2614	0.8776	0.0256	0.0263	0.0033	0.1729
0	SB	0.4787	0.1953	0.1653	0.8098	0.1741	0.1272	0.0359	0.5442
1-2	SB	0.6313	0.0981	0.4285	0.7963	0.4374	0.1023	0.2561	0.6372
3+ females	FL + SB	0.6937	0.0461	0.5968	0.7761	0.3593	0.0618	0.2489	0.4869
3+ males	FL + SB	0.5384	0.0744	0.3935	0.6772	0.4953	0.1199	0.2771	0.7153

Table 2. Life table for the female sand lizard population at site FL. x = age in years; p_x = probability to survive during age interval x; l_x = proportion alive at the start of year x; p_{repr} = probability to reproduce (age 3–9 according to Strijbosch and Creemers 1988); Fem_{young} = half of overall mean hatched clutch size; $m_x = p_{repr} \times Fem_{young}$; f_x = age-specific fecundity used in RAMAS = $p_x (m_{x+1})$.

x	p_x	l_x	p_{repr}	Fem_{young}	m_x	$l_x m_x$	$x l_x m_x$	f_x
0	0.245	1.000	0.000	0.000	0.000	0.000	0.000	0.000
1	0.614	0.245	0.000	0.000	0.000	0.000	0.000	0.000
2	0.614	0.150	0.000	0.000	0.000	0.000	0.000	0.954
3	0.694	0.092	0.523	2.969	1.553	0.143	0.430	1.697
4	0.694	0.064	0.824	2.969	2.446	0.157	0.627	2.060
5	0.694	0.044	1.000	2.969	2.969	0.132	0.660	2.060
6	0.694	0.031	1.000	2.969	2.969	0.092	0.549	2.060
7	0.694	0.021	1.000	2.969	2.969	0.063	0.445	2.060
8	0.694	0.015	1.000	2.969	2.969	0.044	0.352	2.060
9	0.694	0.010	1.000	2.969	2.969	0.031	0.275	2.060
10	0.694	0.007	1.000	2.969	2.969	0.021	0.212	2.060
11	0.694	0.005	1.000	2.969	2.969	0.015	0.162	2.060
12	0.694	0.003	1.000	2.969	2.969	0.010	0.122	2.060
13	0.694	0.002	1.000	2.969	2.969	0.007	0.092	2.060
14	0.000	0.002	1.000	2.969	2.969	0.005	0.069	0.000
						R_0=0.720	T_c=5.550	
						λ=0.943		

Table 3. Life table for the female sand lizard population at site SB.

x	p_x	l_x	p_{repr}	Fem_{young}	m_x	$l_x m_x$	$x l_x m_x$	f_x
0	0.479	1.000	0.000	0.000	0.000	0.000	0.000	0.000
1	0.631	0.479	0.000	0.000	0.000	0.000	0.000	0.000
2	0.631	0.302	0.000	0.000	0.000	0.000	0.000	0.760
3	0.694	0.191	0.523	2.298	1.202	0.229	0.688	1.314
4	0.694	0.132	0.824	2.298	1.894	0.251	1.002	1.594
5	0.694	0.092	1.000	2.298	2.298	0.211	1.055	1.594
6	0.694	0.064	1.000	2.298	2.298	0.146	0.878	1.594
7	0.694	0.044	1.000	2.298	2.298	0.102	0.711	1.594
8	0.694	0.031	1.000	2.298	2.298	0.070	0.563	1.594
9	0.694	0.021	1.000	2.298	2.298	0.049	0.440	1.594
10	0.694	0.015	1.000	2.298	2.298	0.034	0.339	1.594
11	0.694	0.010	1.000	2.298	2.298	0.023	0.259	1.594
12	0.694	0.007	1.000	2.298	2.298	0.016	0.196	1.594
13	0.694	0.005	1.000	2.298	2.298	0.011	0.147	1.594
14	0.000	0.003	1.000	2.298	2.298	0.008	0.110	0.000
						R_0=1.151	T_c=5.550	
						λ=1.026		

0.0418) (all figures relating to the SB-population; the FL-population showed the same pattern).

Simulations of stochastic population growth

As demonstrated by the numbers of adult females from 1988 to 1998 (Fig. 4), both study populations were declining significantly. Simulations of stochastic population growth for this time period showed that the average predictions were optimistic in comparison with the observed population trajectories, particularly for SB (Fig. 6). The probability that population size would fall to or below the observed number in 1998 was 27% for FL (27 individuals) and 20% for SB (30 individuals).

Simulations of future stochastic population growth from 1998 to 2018 under the five management scenarios clarified the rankings of these options (Table 4). With no action taken (scenario I) at site FL (the pessimistic case), the risk of extinction was 39% and the risk of decline to 10 females or less (which I consider to be a critically low population size for recovery without demographic manage-

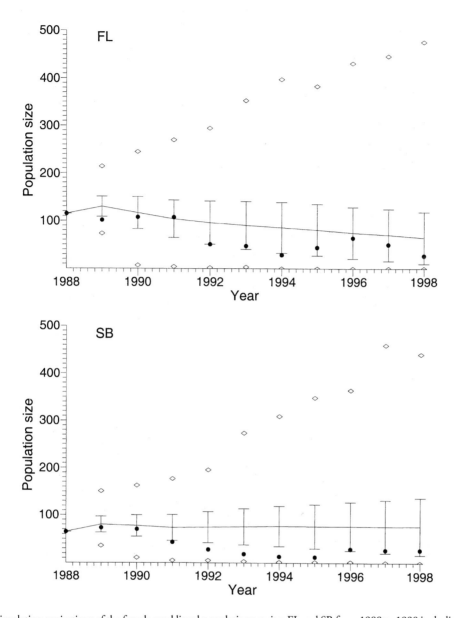

Fig. 6. Simulation projections of the female sand lizard populations at sites FL and SB from 1988 to 1998 including demographic and environmental stochasticity, and natural catastrophes (cold summers) (solid curve = average, vertical bars = ± 1 SD, diamonds = min and max values), as compared to the actual female population histories (black circles). Initial population size = 115 for FL and 65 for SB. See Methods for the calculation of number of observed individuals.

ment) sometime during this period was 84%, with a projected average population size of 9 females in yr 2018. Neither habitat management alone (II) nor artificial incubation (III) reduced the risk of decline to "acceptably" low levels, and extending the practise of scenario III from 5 to 10 yr had only a small effect. Application of captive raising (IV) or captive breeding (V) over a 10-yr period strongly reduced the risk of decline, to only 3% and 1%, respectively (Table 4).

A comparison of quasi-extinction risks for these man-

agement scenarios is illustrated in Fig. 7. It shows that the average probability for the FL-population to fall below a threshold population size of, for example, 16 individuals (right vertical bar) at least once during 1998–2018 was 93% with no management and 16% with captive breeding for 5 yr, respectively.

As expected, these rankings of management options for the FL-population are also relevant in the more "optimistic" case of SB, for which the predicted risks of extinction and decline to 10 females or less with no management was

Table 4. Effects of five different management scenarios on risk of population extinction, risk of decline, and abundance of female sand lizards during the 20 yr period 1998–2018 at sites FL and SB. Scenarios III–V were practised in parallel with scenario II, i.e. "optimal" habitat management during year 1–20. Initial abundance = total number of female lizards (including hatchlings) after reproduction at the start of the period; *under scenario III and IV, the initial abundance is higher than under scenario I and II because of the higher potential hatchability compared to the field hatchability in the unusually cold summer of 1998 (cf. Fig. 5); under scenario V, 5 adults are taken from the wild population for breeding purposes in captivity. $P_{extinction}$ = probability of population extinction, $P_{decline}$ = probability of population decline to 10 individuals or less at least once during the 20 yr period. Final abundance = predicted number of female lizards at the end of the 20 yr period.

Management scenario and site	Management period (yr)	Initial abundance*	$P_{extinction}$	$P_{decline}$	Final abundance			
					Average	± 1 SD	Min	Max
I. No management								
FL	1–20	27	0.386	0.844	9	16	0	152
SB	1–20	30	0.083	0.423	40	47	0	354
II. Habitat management								
FL	1–20	27	0.112	0.447	43	66	0	981
SB	1–20	30	0.037	0.225	90	108	0	1112
III. Artificial incubation + II								
FL	1–5	39	0.049	0.278	64	82	0	649
FL	1–10	39	0.041	0.228	86	118	0	1105
SB	1–5	50	0.023	0.153	125	159	0	1894
SB	1–10	50	0.018	0.119	143	169	0	1597
IV. Captive raising + II								
FL	1–5	39	0.013	0.102	114	134	0	1248
FL	1–10	39	0.004	0.028	234	239	0	2463
SB	1–5	50	0.001	0.050	224	253	0	2807
SB	1–10	50	0.001	0.012	341	337	0	2813
V. Captive breeding + II								
FL	1–5	22	0.020	0.093	126	147	0	1264
FL	1–10	22	0.000	0.012	248	253	1	2872
SB	1–5	25	0.002	0.032	236	231	0	2396
SB	1–10	25	0.000	0.006	394	356	1	2571

8% and 42%, respectively (Table 4). However, it appears that artificial incubation might be sufficiently "safe" to reduce the risk of further decline and rebuild this population. But again, captive raising or captive breeding for 10 yr were the most powerful actions, reducing the risk of decline to 1% for both (Table 4).

Discussion

Population subdivision and causes of decline

This northern sand lizard population exhibits several features typical for populations on the edge of extinction: 1) sharp spatial division among local populations with no dispersal between them, 2) small habitat patches, 3) small local populations, 4) low juvenile recruitment, 5) low finite rate of population increase (λ), and 6) large variation in λ (highly dependent on the effects of weather on juvenile survival) (cf. Leigh 1981, Goodman 1987, Harrison 1994, Lawton 1995, Hanski 1999). Caughley (1994) and Thomas (1994) argued that habitat dynamics rather than stochastic factors are the key to the persistence of many real populations and metapopulations, and that most extinctions are the result of deterministic population responses to a deteriorating environment. Thus, species persist in regions where they are able to track the environment, and they become extinct if they fail to keep up with the shifting habitat mosaic (Thomas 1994). This study suggests that the sand lizard population on Brattforsheden presently can be viewed as a non-equilibrium metapopulation, where the local populations have become completely isolated by deterministic environmental changes, i.e. overgrowth by pine and shading due to canopy formation in open habitat patches. The local populations are now subject to extinction without subsequent recolonization (cf. Hanski et al. 1996). Eventually the whole metapopulation will go extinct, the time limit being set by the persistence of the local populations independently. Two of the known local populations have already become extinct during my 14-yr study period, and the remaining four are so small that environmental and demographic stochasticity can lead to extinction within 20 yr.

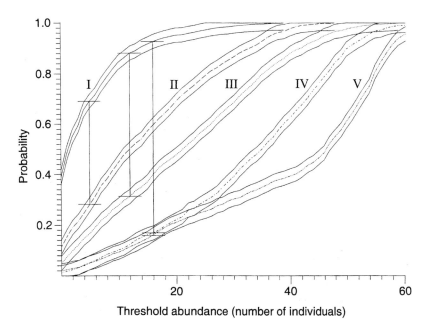

Fig. 7. Quasi-extinction risks for the female sand lizard population at site FL under the management scenarios I) no management, II) habitat management, III) artificial incubation, IV) captive raising, and V) captive breeding. The curves show, for each threshold abundance, the probability of falling below that abundance sometime during the next 20 yr from 1998. The vertical bars show the maximum difference between the indicated curves. The solid curves around the mean curves give 95% confidence intervals, based on Kolmogorov-Smirnov statistic, D. Scenarios III–V were practised during year 1–5, in parallel with II during year 1–20. See Methods for explanation.

Most probably, other local populations have gone extinct prior to my study period.

In contrast to the experiences from England (e.g. Corbett 1988), habitat management on Brattforsheden has not proven to be an efficient way of rapidly increasing local population size, at least not the small-scale management practised so far. Although the lizards have colonized cleared areas and laid eggs on newly excavated sand patches (Berglind 1999), both study populations have declined. As demonstrated, the decline is partly correlated with unfavourable weather, resulting in no juvenile recruitment in 1987, but bad weather alone cannot explain the overall population decline.

One explanation could be that the populations suffer from inbreeding depression. Mills and Smouse (1994) concluded that low growth-rate populations are extremely vulnerable to even minor inbreeding depression, and that vulnerability to extinction is affected more by survival depression than by fecundity depression. Olsson et al. (1996c) reported a 10% frequency of malformed hatchlings with zero survival in a large, natural population of the sand lizard in SW Sweden, and demonstrated that this was probably due to mating among siblings. No malformed hatchlings have been observed in the field in my populations. In early autumn 1998, 7 clutches from SB and 4 from FL were brought into the lab for hatching, which resulted in a 100 and 96% hatching success respectively. These hatchabilities are higher than the overall hatching success in the field during 1988–1997 (mean = 73% for SB vs 91% for FL) and at least as high as the potential hatchabilities in the field (mean = 86% for SB vs 96% for FL). Furthermore, all the hatched juveniles in the lab were normally developed and showed normal behaviour and growth, until hibernation in the lab two months later.

These observations suggest that the lower mean hatching success at site SB under field conditions (Fig. 5) was due to less favourable microclimatic conditions rather than inbreeding depression.

Unexpectedly, Gullberg et al. (1998) reported that the genetic variation in each population SB and FL (corresponding to "Värmland A" and "Värmland C" in Gullberg et al. 1998), measured as the number of alleles and degree of heterozygosity at six microsatellite loci, was at least as high as in populations thought to be more continuous in southern Sweden. There was a clear genetic subdivision among the Swedish populations, especially between regions, but also between local populations within the same region, where the regional population of Brattforsheden had diverged on both levels. Based on the expected heterozygosity, Gullberg et al. (1998) estimated an effective population size of 6500 individuals in the SB-population. As they conclude, it is unlikely to find such a high level of genetic variability retained in an isolated relict population after a long period at small population size. If future demographic fluctuations of these small populations are considered, rates of loss of genetic variation and accumulation of inbreeding might become a problem (cf. Lacy 2000).

As Berglind (1988) suggested, the few populations of small size we observe today on Brattforsheden are likely a relatively new phenomenon, with isolation of suitable habitats starting some 15 generations (or ca 80 yr) ago. The transformation of Brattforsheden from an "almost treeless" area by the end of the 19th century (Furuholm et al. 1994) to the managed, up to 100 yr old pine forest dominating today has accelerated the loss of suitable sand lizard habitat, with isolation and loss of local populations as a probable consequence. Extensive human activities like forest grazing by cattle, charcoal production, and tree harvesting, kept large areas of the Swedish boreal forest, including Brattforsheden, relatively open between 400 and 100 yr ago (Furuholm et al. 1994, Angelstam 1997). However, the most important factor influencing the structure of the boreal forests before the 20th century was forest fires; the time intervals between successive fires in sandy pine forests have been estimated to be some 2–3 fires 100 yr^{-1} (Zackrisson 1977). There are also many signs of former fires on Brattforsheden. Thus, the sand lizard populations face a new situation today when old human land-use and the natural disturbance regime creating open patches with early successional vegetation have been more or less eliminated, with accelerating pine succession and canopy formation as a consequence. In fact, the present small number of patches with the combination of suitable habitat and probable habitat continuity since the early 20th century is almost identical to the number of patches known to have been occupied by the sand lizard until at least 1988 (Berglind unpubl.).

The sand lizard populations presently inhabiting these small isolated patches are likely to be at risk because of factors other than merely climatic, genetic and successional ones. These patches offer pronounced habitat edges, and it is known that predators often increase the predation rate at habitat edges (Andrén 1995). Potential predators on the sand lizard include the red fox *Vulpes vulpes*, badger *Meles meles*, raven *Corvus corax*, buzzard *Buteo buteo*, grass snake *Natrix natrix* and the common toad *Bufo bufo*, all of which regularly occur at the lizard sites on Brattforsheden. Adult sand lizards may also eat their newly hatched young (Corbett and Tamarind 1979). This could probably cause major mortality among juveniles if the area of suitable habitat is small, and the surrounding matrix sharply contrasting in quality. The closed pine forest matrix is normally an inhospitable environment that severely reduces the opportunity for thermoregulation. During my 14 yr study period, I have not observed any sand lizards in closed forest, even though forest edges are much favoured. Adult predation on juveniles is reduced if appropriate management of the habitat increases the area available to the lizards (Corbett and Tamarind 1979). Such cannibalism may explain the presumed lower juvenile survival at site FL. Its structural complexity of vegetation was less than that in the part of SB where most recaptures were made. House and Spellerberg (1983b) observed that patches of high structural complexity can support higher densities of sand lizards, and Civantos et al. (1999) showed that hatchling survival in the lacertid lizard *Psammodromus algirus* was positively associated with higher vegetation cover.

Adult mortality, including that of egg-laying females, has also been observed to be caused by road traffic at my study sites. Although only a few killed animals have been found (Berglind unpubl.), this may be an unacceptable risk for these small populations.

A further reason for high risks of population decline in small patches is that per-capita emigration rate increases with decreasing patch area. Thus, populations might go extinct in small isolated habitat patches because the local growth rate plus the immigration rate is less than the emigration rate (Thomas and Hanski 1997, Hanski 1999). Although adult sand lizards are usually highly sedentary (Anon. 1983, Olsson et al. 1996b, Berglind 1999), emigration may nevertheless be important, as was stressed by Strijbosch and van Gelder (1997). They demonstrated substantial emigration in Dutch sand lizard populations, with one individual found 2 km away from an isolated population. However, within forested areas, dispersal is probably very limited outside open patches, and perhaps mostly occurring along sun-exposed forest road verges (cf. Dent and Spellerberg 1988). The longest dispersal distance observed in my study area was a subadult moving 500 m along a forest road within one site and one season. Olsson et al. (1996b) reported small-scale home range shifts between years in some adult females and males, but it is probable that most dispersal occurs among immature lizards growing in transit, and that the long life expectancy of sand lizards older than 1 yr is an important component in their dispersal biology (Strijbosch and van Gelder 1997).

Population growth

My PVA could be seen as relatively powerful concerning relevant demography and predicting future population growth since most vital rates were measured over almost two generations from the actual populations in need of immediate conservation action (cf. Beissinger and Westphal 1998, Brook et al. 2000). However, one must exercise caution in taking the deterministic growth rates and simulated projections too literally, as demonstrated by the relative discrepancy between the predicted and observed population sizes during 1988–1998, especially for site SB (Fig. 6). As PVA-models have important stochastic components, such discrepancy may be expected, but it is essential to bear in mind also the uncertainty in parameter estimation (Taylor 1995, Ludwig 1999). The growth rate of my populations were highly dependent on the survival estimates for juveniles, as indicated by the elasticity analysis and the large differences in population growth between the management scenarios (which mainly differed in their juvenile survival rates; Table 4 and Appendix 1). Hence, accurate parameterization of juvenile survival is particularly important for confident assessment of risk of population decline. My estimates of 24% and 48% yearly juvenile survival for the two study populations, respectively, had wide confidence intervals (as did the estimates for 1–2-yr olds) (Table 1), which of course entail wide confidence intervals for the probability of quasi-extinction (Ludwig 1999).

Juvenile survival also show high natural variability for many reasons, which has been clearly demonstrated in mammals (Gaillard et al. 1998). This further stresses the importance of long-term data to get reliable estimates. Strijbosch and Creemers (1988) estimated the mean juvenile survival in the sand lizard to be 38% and Olsson et al. (1996b) reported 5–15%. However, neither of these estimates were based on statistical mark-recapture analysis. Clearly, the lower of these survival figures (including my estimate of 24%) cannot reflect "true" means from viable populations, unless female reproduction starts at an earlier age, or mean fecundity or adult survival are much higher than my figures indicate. No indication of earlier reproduction was found in my study area, and the mean clutch size of other sand lizard populations in north-western Europe did not differ much from my figures (Anon. 1983: 6.1 eggs; Strijbosch 1988: 5.9; Olsson et al. 1996c: 8.5). The same applies to adult survival (Strijbosch and Creemers 1988). I suspect that juvenile survival fluctuates considerably between years on the northern limit of the species' range, and that some cohorts with high survival contribute disproportionately to population recruitment. Most probably, these fluctuations coincide with weather fluctuations, where warm summers result in early hatching and longer time for the juveniles to grow before hibernation, with less winter mortality as a consequence (cf. Civantos et al. 1999). One essential prerequisite for such a variable system of recruitment to function, especially in species with low rates of population increase and where interpopulation dispersal rates are low, is that reproductive failure in some years is buffered by high life expectancy in adults. The oldest individuals reported here (13 and 14 yr, respectively) are among the oldest reported from any natural population of this species. A similar conclusion was also reached by Strijbosch et al. (1980) for the northernmost European population of the wall lizard *Podarcis muralis* in the Netherlands.

Demographic management

The elasticity analysis showed that individual survival until the first year of reproduction was the most important factor contributing to population growth. This is fortunate from a conservation point of view since enhancement of juvenile survival is relatively easy to achieve with demographic management as outlined here.

The implications of my PVA results are straightforward. In the absence of management (scenario I), there is substantial risk that these small populations will decline further within 20 yr and even go extinct without taking deterministic pine succession into account (Table 4 and Fig. 7). Furthermore, "optimal" habitat management (II) appeared as no guarantee for population recovery. Even if larger areas of suitable habitat are created, and the natural hatchability of the eggs approaches its potential value, the risk of decline is still high. The prospects of population recovery were somewhat better if artificial incubation (III)

is practised, resulting in earlier hatching and larger juveniles before hibernation in the field. Captive raising (IV), with much higher winter survival among juveniles after being kept in captivity, had a strong positive effect on population growth. The same applies to captive breeding (V), with yearly introductions of juveniles from breeding stocks taken from the populations in need of conservation action today.

However, there are potential risks with management options III–V as well. Incubation temperatures that accelerate embryo development in the artificial incubation scenario (III) do not neccessarily maximize embryo survival and hatchling characteristics (Van Damme et al. 1992), and incubation-induced modifications to a lizard's phenotype may affect actual ability to escape from a predator (Downes and Shine 1999).

Captive raising (IV) means that the juveniles have to spend at least 6 months in captivity (including 2–4 months of controlled hibernation). This may cause adaptation to the captive environment, and perhaps lowered survival among the released animals. One potentially important condition for this method to work is to rear the juveniles under a "suboptimal" temperature regime for growth (cf. Avery 1984) in order not to let them grow too quickly. It is probably easier for juveniles compared to subadult or adult lizards to find enough food, shelter and establish home ranges in the wild after being released. In wild common lizards Massot et al. (1994) demonstrated that transplanted subadult and adult individuals survived worse than juveniles. The captively raised juveniles in scenario IV will, however, be larger at the time of release than under pure field conditions (because of a longer activity period in captivity). Based on previous years' experience of rearing juvenile sand lizards, I estimate their size at release in spring to equal that of wild 1-yr olds in autumn. Consequently they could potentially start reproducing one year earlier than normal, which might reinforce population recovery. However, for conservative reasons I did not incorporate this possibility in my simulation model since juvenile growth and survival after release from captivity may be reduced relative to that of wild-reared juveniles. If applying this management option, it is crucial to evaluate the field survival of captively raised young.

One advantage with captive breeding (V) is that time in captivity for the juveniles can be reduced to ca 2 and 0 months for the two annual broods respectively (assuming two clutches per female), depending on the timing of mating of the breeding stock. However, it is hazardous to keep the majority of a remnant adult population in captivity because of disease risks (cf. Jacobson 1994) and other unforeseen events. This management option also involves a relatively high risk of inbreeding effects due to the double annual clutches of some females. In addition, it involves higher costs than captive raising since lizards must be kept in captivity on a yearly basis.

It might seem a paradox that the mean projected population growth under captive breeding with double clutches (V) is not substantially larger than that under captive raising (IV; Table 4). This is due to the lower survival of juveniles during the first hibernation (set to 50 and 56%, respectively, vs 80% for captive raising) and the fact that the captive breeding stock is kept constant at 5 reproducing females over the 5 or 10-yr management periods. It might be possible to double the captive female breeding stock after five years, which of course would increase the average projected female population size by the end of the 20-yr period.

One plausible alternative to captive raising or breeding is to combine the two. Scenario IV starts first, in which some individuals from different clutches are selected to form the breeding colony. When they become sexually mature after 1–2 yr in captivity (Olsson pers. comm.), scenario V is initiated using these individuals simultaneously as the full adult wild populations continue to reproduce with management as in IV. In both scenarios V and IV+V, it would be necessary to cross individuals so that erosion of genetic variation in the populations is minimized (e.g. Ebenhard 1995).

Re-establishing endangered or threatened populations of reptiles has shown poor results so far (Dodd and Seigel 1991, Beck et al. 1994). One captive breeding technique that has been successful for the sand lizard in England, is to keep a breeding stock on a yearly basis in outdoor vivaria (Spellerberg and House 1982, Corbett 1988). After reproduction the eggs were located and removed to be incubated artificially, resulting in earlier hatching and larger juveniles at hibernation after release in the field (Corbett 1988, Edgar 1990). However, demographic data from the British projects are scarce, and it is crucial with follow-up studies of released animals to evaluate different management regimes in quantitative terms. This would increase the precision of future PVAs for which we need more detailed information especially on juvenile survival, age-specific probability to reproduce, and the magnitude of environmental variability on vital rates.

Conclusions and long-term conservation

The ultimate factor behind the decline of my study populations seems to be the deterministic effect of afforestation and forest fire suppression during the 20th century. This has resulted in a reduction in the amount of open, early successional habitat patches suitable for the sand lizard. The surviving populations are now so small that there is substantial risk that more or less stochastic factors, like cold summers, predation, emigration losses, traffic mortality and chance effects in birth and death rates, can lead to extinction within 20 yr.

The survival of these populations in the past seems explained principally by: 1) the regular occurrence of forest fires and extensive human activities which maintained a network of suitable habitat patches, 2) the long life expectancy of adult sand lizards that buffers against reproductive failures in cold summers, and 3) a disproportionately high recruitment from cohorts of warm summers with high juvenile survival.

In order to save these lizard populations in the long run, short-term demographic management as outlined here is of course not enough. It must be combined with long-term habitat management (e.g. Corbett and Tamarind 1979, Berglind 1999). Similar to Harrison's (1994) suggestion for succession-dependent butterflies, successful management of the sand lizard requires creating unbroken series of successional stages in a spatial mosaic fine enough to permit constant recolonization. This would also favour a number of other succession-dependent species, like the nightjar *Caprimulgus europaeus*, wood lark *Lullula arborea*, pasqueflower *Pulsatilla vernalis*, and a large number of insects, especially among the wasps, ants, flies and beetles, of which at least 30 uncommon and red-listed species have been found on Brattforsheden (Berglind 1995b, 1999).

By practising combined demographic and habitat management, there is potentially a good chance of sand lizard population recovery. This includes increasing juvenile survival during the first hibernation or release of captively bred juveniles in spring. Moreover, several hectares around the present lizard populations must be cleared from dense forest stands to create a mosaic of open sand patches and a rich and varied field layer. A practical long-term conservation strategy for the species could be: 1) to focus on recurrent habitat management at a number of particularly suitable "core" sites (with south-exposed slopes) in a network fine enough to permit interpopulation dispersal, and 2) to temporally create suitable habitat between these core sites by making clear-cuts colonizable. Clearance of trees to create broad, permanently open verges along forest roads would also be beneficial for dispersal (cf. Dent and Spellerberg 1988). Although management options IV and V seem promising for sand lizard conservation, they carry more assumptions, risks during implementation, and higher costs than scenarios I–III. Thus, more demographic data from relevant management situations should be gathered before they are fully employed. Nonetheless, without explicit and relevant management, it seems likely that we will loose these unique relict populations in the near future.

Acknowledgements – Thanks to Per Sjögren-Gulve for introducing me to PVAs, Torbjörn Ebenhard for introducing me to RAMAS GIS, and Torbjörn Nilsson for invaluable assistance with the MARK analysis. All three also gave very constructive criticism on previous drafts of this manuscript, as did Mats Olsson, Robert Paxton, Oskar Kindvall and Pekka Pamilo. Many further thanks to Jan Bengtsson and Lars Furuholm for practical and administrative assistance in the sand lizard conservation work, and the timber company Stora Enso, for leasing out the sand lizard sites for conservation purposes. Financial support was obtained from the Swedish World Wildlife Fund (WWF), the Swedish Biodiversity Centre (CBM), the County Administrative Board of Värmland, the Swedish Environmental Protection Agency, Carl Tryggers foundation, and Oscar och Lili Lamms foundation.

References

Akçakaya, H. R. 1998. RAMAS GIS: Linking landscape data with population viability analysis (ver. 3.0). – Appl. Biomath., Setauket, New York.

Akçakaya, H. R. 2000. Population viability analyses with demographically and spatially structured models. – Ecol. Bull. 48: 23–38.

Akçakaya, H. R. and Sjögren-Gulve, P. 2000. Population viability analyses in conservation planning: an overview. – Ecol. Bull. 48: 9–21.

Anderson, D. R., Burnham, K. P. and White, G. C. 1994. AIC model selection in overdispersed capture-recapture data. – Ecology 75: 1780–1793.

Andrén, C. and Nilson, G. 1979. The sand lizard (Lacerta agilis) at its northern border in Scandinavia. – Fauna och flora 74: 133–139, in Swedish with English summary.

Andrén, H. 1995. Effects of landscape composition on predation rates at habitat edges. – In: Hansson, L., Fahrig, L. and Merriam, G. (eds), Mosaic landscapes and ecological processes. Chapman and Hall, pp. 225–255.

Angelstam, P. 1997. Landscape analysis as a tool for the scientific management of biodiversity. – Ecol. Bull. 46: 140–170.

Anon. 1983. The ecology and conservation of amphibian and reptile species endangered in Britain. – NCC, London.

Avery, R. A. 1984. Physiological aspects of lizard growth: the role of thermoregulation. – Symp. Zool. Soc. Lond. 52: 407–424.

Beck, B. B. et al. 1994. Reintroduction of captive-born animals. – In: Olney, P. J. S., Mace, G. M. and Feistner, A. T. C. (eds), Creative conservation: interactive management of wild and captive animals. Chapman and Hall, pp. 266–286.

Beissinger, S. R. and Westphal, M. I. 1998. On the use of demographic models of population viability in endangered species management. – J. Wildl. Manage. 62: 821–841.

Begon, M., Harper, J. L. and Townsend, C. R. 1996. Ecology: individuals, populations and communities. 3rd ed. – Blackwell.

Berg, Å. and Tjernberg, M. 1996. Common and rare Swedish vertebrates – distribution and habitat preferences. – Biodiv. Conserv. 5: 101–128.

Berglind, S.-Å. 1988. The sand lizard, Lacerta agilis L., on Brattforsheden, south central Sweden – habitat, threats and conservation. – Fauna och flora 83: 241–255, in Swedish with English summary.

Berglind, S.-Å. 1995a. Ecology and management of relict populations of the sand lizard (Lacerta agilis) in south central Sweden. – Memoranda Soc. Fauna Flora Fennica 71: 88.

Berglind, S.-Å. 1995b. Habitat management brightens the future for Formica cinerea (Hymenoptera, Formicidae) on Brattforsheden, south central Sweden. – Entomol. Tidskr. 116: 21–25, in Swedish with English summary.

Berglind, S.-Å. 1999. Conservation of relict sand lizard (Lacerta agilis L.) populations on inland dune areas in central Sweden. – Ph.Lic. thesis, Uppsala Univ.

Bishoff, W. 1988. Zur verbreitung und systematik der Zauneidechse Lacerta agilis. – Mertensiella 1: 11–30.

Boyce, M. S. 1992. Population viability analysis. – Annu. Rev. Ecol. Syst. 23: 481–506.

Brook, B. W. 2000. Pessimistic and optimistic bias in population viability analysis. – Conserv. Biol. 14: 564–566.

Brook, B. W. et al. 2000. Predictive accuracy of population viability analysis in conservation biology. – Nature 404: 385–387.

Caughley, G. 1977. Analysis of vertebrate populations. – Wiley.

Caughley, G. 1994. Directions in conservation biology. – J. Anim. Ecol. 63: 215–244.

Caughley, G. and Gunn, A. 1996. Conservation biology in theory and practice. – Blackwell.

Civantos, E., Salvador, A. and Veiga, J. 1999. Body size and microhabitat affect winter survival of hatchling Psammodromus algirus lizards. – Copeia 1999: 1112–1117.

Corbett, K. F. 1988. Conservation strategy for the sand lizard (Lacerta agilis agilis) in Britain. – Mertensiella 1: 101–109.

Corbett, K. F. and Tamarind, D. L. 1979. Conservation of the sand lizard, Lacerta agilis, by habitat management. – Brit. J. Herp. 5: 799–823.

de Kroon, H., van Groenendael, J. and Ehrlén, J. 2000. Elasticities: a review of methods and model limitations. – Ecology 81: 607–618.

Dent, S. and Spellerberg, I. F. 1988. Use of forest ride verges in southern England for the conservation of the sand lizard Lacerta agilis L. – Biol. Conserv. 45: 267–277.

Dodd, Jr C. K. and Seigel, R. A. 1991. Relocation, repatriation, and translocation of amphibians and reptiles: are they conservation strategies that work? – Herpetologica 47: 336–350.

Downes, S. J. and Shine, R. 1999. Do incubation-induced changes in a lizard's phenotype influence its vulnerability to predators? – Oecologia 120: 9–18.

Ebenhard, T. 1995. Conservation breeding as a tool for saving animal species from extinction. – Trends Ecol. Evol. 10: 438–443.

Edgar, P. 1990. A captive breeding and release programme for sand lizards and natterjack toads at Marwell Zoological Park: an appeal for sponsorship. – Brit. Herp. Soc. Bull. 31: 3–10.

Furuholm, L., Heijkenskjöld, R. and Mellander, B. 1994. Brattforsheden – istiden i närbild. – Locus, Karlstad.

Gaillard, J.-M., Festa-Bianchet, M. and Yoccos, N. G. 1998. Population dynamics of large herbivores: variable recruitment with constant adult survival. – Trends Ecol. Evol. 13: 58–63.

Gärdenfors, U. (ed.) 2000. The 2000 Red List of Swedish species. – Threatened Species Unit, Swedish Univ. of Agricult. Sci., Uppsala.

Gasc, J.-P. et al. (eds) 1997. Atlas of amphibians and reptiles in Europe. – Societas Europaea Herpetologica and Muséum National d'Histoire Naturelle, Paris.

Gaston, K. J. 1994. Rarity. – Chapman and Hall.

Ginzburg, L. R., Ferson, S. and Akçakaya, H. R. 1990. Reconstructibility of density dependence and the conservative assessment of extinction risks. – Conserv. Biol. 4: 63–70.

Gislén, T. and Kauri, H. 1959. Zoogeography of the Swedish amphibians and reptiles. – Acta Vertebratica 1: 193–397.

Glandt, D. and Bishoff, W. (eds) 1988. Biologie und Schutz der Zauneidechse (Lacerta agilis). – Mertensiella 1: 1–257.

Goodman, D. 1987. The demography of chance extinction. – In: Soulé, M. E. (ed.), Viable populations for conservation. Cambridge Univ. Press, pp. 11–34.

Groom, M. J. and Pascual, M. A. 1998. The analysis of population persistence: an outlook on the practice of viability analysis. – In: Fiedler, P. G. and Kareiva, P. M. (eds), Conservation biology – for the coming decade. Chapman and Hall, pp. 4–27.

Gullberg, A., Olsson, M. and Tegelström, H. 1998. Colonization, genetic diversity, and evolution in the Swedish sand lizard, Lacerta agilis (Reptilia, Squamata). – Biol. J. Linn. Soc. 65: 257–277.

Hanski, I. 1999. Metapopulation ecology. – Oxford Univ. Press.

Hanski, I., Moilanen, A. and Gyllenberg, M. 1996. Minimum viable metapopulation size. – Am. Nat. 147: 527–541.

Harrison, S. 1994. Metapopulations and conservation. – In: Edwards, P. J., May, R. M. and Webb, N. R. (eds), Large-scale ecology and conservation biology. Blackwell, pp. 111–128.

Hemelaar, A. 1985. An improved method to estimate the number of year rings resorbed in phalanges of Bufo bufo (L.) and its application to populations from different latitudes and altitudes. – Amphib.- Reptilia 6: 323–341.

House, S. M. and Spellerberg, I. F. 1983a. Comparison of Lacerta agilis habitats in Britain and Europe. – Brit. J. Herp. 6: 305–308.

House, S. M. and Spellerberg, I. F. 1983b. Ecology and conservation of the sand lizard (Lacerta agilis L.) habitat in southern England. – J. Appl. Ecol. 20: 417–437.

Jacobson, E. R. 1994. Veterinary procedures for the acquisition and release of captive-bred herpetofauna. – In: Murphy, J. B., Adler, K. and Collins, J. T. (eds), Captive management and conservation of amphibians and reptiles. Society for the Study of Amphibians and Reptiles, Ithaca, NY, pp. 109–118.

Lacy, R. C. 2000. Considering threats to the viability of small populations using individual-based models. – Ecol. Bull. 48: 39–51.

Lande, R. 1993. Risks of population extinction from demographic and environmental stochasticity and random catastrophes. – Am. Nat. 142: 911–927.

Lawton, J. H. 1993. Range, population abundance and conservation. – Trends Ecol. Evol. 8: 409–413.

Lawton, J. H. 1995. Population dynamic principles. – In: Lawton, J. H. and May, R. M. (eds), Extinction rates. Oxford Univ. Press, pp. 147–163.

Lebreton, J.-D. et al. 1992. Modeling survival and testing biological hypothesis using marked animals: a unified approach with case studies. – Ecol. Monogr. 62: 67–118.

Lebreton, J.-D., Pradel, R. and Clobert, J. 1993. The statistical analysis of survival in animal populations. – Trends Ecol. Evol. 8: 91–95.

Leigh, Jr E. G. 1981. The average lifetime of a population in a varying environment. – J. Theor. Biol. 90: 213–239.

Ludwig, D. 1999. Is it meaningful to estimate a probability of extinction? – Ecology 80: 298–310.

Madsen, T. and Shine, R. 1998. Quantity or quality? Determinants of maternal reproductive success in tropical pythons (Liasis fuscus). – Proc. R. Soc. Lond. B 265: 1521–1525.

Massot, M. et al. 1994. Incumbent advantage in common lizards and their colonizing ability. – J. Anim. Ecol. 63: 431–440.

Menges, E. S. 1998. Evaluating extinction risks in plant populations. – In: Fiedler, P. G. and Kareiva, P. M. (eds), Conservation biology – for the coming decade. Chapman and Hall, pp. 49–65.

Mills, L. S. and Smouse, P. E. 1994. Demographic consequences of inbreeding in remnant populations. – Am. Nat. 144: 412–431.

Olsson, M. and Shine, R. 1997. The limits to reproductive output: offspring size versus number in the sand lizard (Lacerta agilis). – Am. Nat. 149: 179–188.

Olsson, M. et al. 1996a. Paternal genotype influences incubation period, offspring size, and offspring shape in an oviparous reptile. – Evolution 50: 1328–1333.

Olsson, M., Gullberg, A. and Tegelström, H. 1996b. Determinants of breeding dispersal in the sand lizard, Lacerta agilis (Reptilia, Squamata). – Biol. J. Linn. Soc. 60: 243–256.

Olsson, M., Gullberg, A. and Tegelström, H. 1996c. Malformed offspring, sibling matings, and selection against inbreeding in the sand lizard (Lacerta agilis). – J. Evol. Biol. 9: 229–242.

Snyder, N. F. R. et al. 1996. Limitations of captive breeding in endangered species recovery. – Conserv. Biol. 10: 338–348.

Spellerberg, I. F. and House, S. M. 1982. Relocation of the lizard Lacerta agilis: an exercise in conservation. – Brit. J. Herp. 6: 245–248.

Strijbosch, H. 1988. Reproductive biology and conservation of the sand lizard. – Mertensiella 1: 132–145.

Strijbosch, H. and Creemers, R. C. M. 1988. Comparative demography of sympatric populations of Lacerta vivipara and Lacerta agilis. – Oecologia 76: 20–26.

Strijbosch, H. and van Gelder, J. J. 1997. Population structure of lizards in fragmented landscapes and causes of their decline. – In: Böhme, W., Bishoff, W. and Ziegler, T. (eds), Herpetologia Bonnensis. Bonn, pp. 347–351.

Strijbosch, H., Bonnemayer, J. A. and Dietvorst, P. J. 1980. The northernmost population of Podarcis muralis (Lacertilia, Lacertidae). – Amphib.-Reptilia 1: 161–172.

Taylor, B. L. 1995. The reliability of using population viability analysis for risk classification of species. – Conserv. Biol. 9: 551–558.

Thomas, C. D. 1990. What do real population dynamics tell us about minimum viable population sizes? – Conserv. Biol. 4: 324–327.

Thomas, C. D. 1994. Extinction, colonization, and metapopulations: environmental tracking by rare species. – Conserv. Biol. 373–378.

Thomas, C. D. and Hanski, I. 1997. Butterfly metapopulations. – In: Hanski, I. and Gilpin, M. (eds), Metapopulation biology: ecology, genetics, and evolution. Academic Press, pp. 359–386.

Van Damme, R. et al. 1992. Incubation temperature differentially affects hatching time, egg survival, and hatchling performance in the lizard Podarcis muralis. – Herpetologica 48: 220–228.

White, G. C. 2000. Program MARK. – http://www.cnr.colostate.edu/~gwhite/mark/mark.htm.

Zackrisson, O. 1977. Influence of forest fires on the north Swedish boreal forest. – Oikos 29: 22–33.

Appendix 1. Input-values used in the simulations with RAMAS CIS/Metapop under different management scenarios for *Lacerta agilis*, sites FL and SB. p_0 = yearly survival for 0-yr old individuals (survival rates for older age classes were constant and in accordance to Table 1), SD_{p0} = standard deviation for yearly variability in survival for x-yr olds (the corresponding coefficient of variation for 0-yr olds site FL = 1.35, site SB = 0.41; 1–2-yr olds site FL+SB = 0.29; 3+ yr olds site FL+SB = 0.18), SD_{px} = age-specific fecundity = $p_x(m_{x+1})$ (see Tables 2–3), SD_{fx} = standard deviation for yearly variability in f_x, Fem_{young} = half of mean hatched clutch size (the CV for yearly variability in Fem_{young} under scenario I, site FL = 0.15, site SB = 0.24; scenario II–V, site FL = 0.18, site SB = 0.19), Intro = yearly introductions of 31 female 0-yr olds, λ = resulting finite rate of population increase under a stable age distribution.

Scenario/Site/Period (yr)	p_0	SD_{p0}	SD_{p1-2}	SD_{p3+}	f_2	f_3	f_{4-13}	SD_{f2}	SD_{f3}	SD_{f4-13}	Fem_{young}	Intro	λ
I. No management													
FL 1-20	0.2448	0.1759	0.1546	0.1094	0.9539	1.6971	2.0596	0.1435	0.2552	0.3098	2.969	0	0.94
SB 1-20	0.4787	0.1671	"	"	0.7587	1.3136	1.5941	0.1840	0.3185	0.3866	2.298	0	1.03
II. Habitat management													
FL 1-20	0.3835	0.1759	"	"	1.0001	1.7794	2.1595	0.1763	0.3137	0.3807	3.113	0	1.03
SB 1-20	0.4787	0.1671	"	"	0.9370	1.6222	1.9687	0.1742	0.3016	0.3660	2.838	0	1.07
III. Artificial incubation + II													
FL 1-5 or 1-10	0.4989	0.1759	"	"	1.0001	1.7794	2.1595	0.1763	0.3137	0.3807	3.113	0	1.08
FL 6-20 or 11-20	0.3835	"	"	"	"	"	"	"	"	"	"	0	1.03
SB 1-5 or 1-10	0.555	0.1671	"	"	0.9370	1.6222	1.9687	0.1742	0.3016	0.3660	2.838	0	1.10
SB 6-20 or 11-20	0.4787	"	"	"	"	"	"	"	"	"	"	0	1.07
IV. Captive raising + II													
FL 1-5 or 1-10	0.8	0.08	"	"	1.0001	1.7794	2.1595	0.1763	0.3137	0.3807	3.113	0	1.19
FL 6-20 or 11-20	0.3835	0.1759	"	"	"	"	"	"	"	"	"	0	1.03
SB 1-5 or 1-10	0.8	0.08	"	"	0.9370	1.6222	1.9687	0.1742	0.3016	0.3660	2.838	0	1.18
SB 6-20 or 11-20	0.4787	0.1671	"	"	"	"	"	"	"	"	"	0	1.07
V. Captive breeding + II													
FL 1-5 or 1-10	0.4989	0.1759	"	"	1.0001	1.7794	2.1595	0.1763	0.3137	0.3807	3.113	31	1.08
FL 6-20 or 11-20	0.3835	"	"	"	"	"	"	"	"	"	"	0	1.03
SB 1-5 or 1-10	0.555	0.1671	"	"	0.9370	1.6222	1.9687	0.1742	0.3016	0.3660	2.838	31	1.10
SB 6-20 or 11-20	0.4787	"	"	"	"	"	"	"	"	"	"	0	1.07

Appendix 2. Example of an input file with RAMAS GIS/Metapop ver. 3.0, applied to the FL-population of *Lacerta agilis*, under alternative I (no management), starting from 1998.

Replicates = 1000; Time = 20; Stages = 15; Environmental stochasticity (ES) distribution = Lognormal; Density dependence = Exponential; Catastrophes = probability 0.10 for 0.0 abundance of 0-yr olds (regional cold summer); Demographic stochasticity = yes; No dispersal, no correlation.

Initial abundance = 27; Starting age distribution = 6, 7, 7, 3, 0, 1, 0, 0, 2, 0, 0, 0, 0, 0, 1

Stage (Leslie) matrix

0	0	0.9539	1.6971	2.0596	2.0596	2.0596	2.0596	2.0596	2.0596	2.0596	2.0596	2.0596	2.0596	0
0.2448	0	0	0	0	0	0	0	0	0	0	0	0	0	0
0	0.6143	0	0	0	0	0	0	0	0	0	0	0	0	0
0	0	0.6143	0	0	0	0	0	0	0	0	0	0	0	0
0	0	0	0.6937	0	0	0	0	0	0	0	0	0	0	0
0	0	0	0	0.6937	0	0	0	0	0	0	0	0	0	0
0	0	0	0	0	0.6937	0	0	0	0	0	0	0	0	0
0	0	0	0	0	0	0.6937	0	0	0	0	0	0	0	0
0	0	0	0	0	0	0	0.6937	0	0	0	0	0	0	0
0	0	0	0	0	0	0	0	0.6937	0	0	0	0	0	0
0	0	0	0	0	0	0	0	0	0.6937	0	0	0	0	0
0	0	0	0	0	0	0	0	0	0	0.6937	0	0	0	0
0	0	0	0	0	0	0	0	0	0	0	0.6937	0	0	0
0	0	0	0	0	0	0	0	0	0	0	0	0.6937	0	0
0	0	0	0	0	0	0	0	0	0	0	0	0	0.6937	0

ES standard deviation matrix

0	0	0.1435	0.2552	0.3098	0.3098	0.3098	0.3098	0.3098	0.3098	0.3098	0.3098	0.3098	0.3098	0
0.1759	0	0	0	0	0	0	0	0	0	0	0	0	0	0
0	0.1546	0	0	0	0	0	0	0	0	0	0	0	0	0
0	0	0.1546	0	0	0	0	0	0	0	0	0	0	0	0
0	0	0	0.1094	0	0	0	0	0	0	0	0	0	0	0
0	0	0	0	0.1094	0	0	0	0	0	0	0	0	0	0
0	0	0	0	0	0.1094	0	0	0	0	0	0	0	0	0
0	0	0	0	0	0	0.1094	0	0	0	0	0	0	0	0
0	0	0	0	0	0	0	0.1094	0	0	0	0	0	0	0
0	0	0	0	0	0	0	0	0.1094	0	0	0	0	0	0
0	0	0	0	0	0	0	0	0	0.1094	0	0	0	0	0
0	0	0	0	0	0	0	0	0	0	0.1094	0	0	0	0
0	0	0	0	0	0	0	0	0	0	0	0.1094	0	0	0
0	0	0	0	0	0	0	0	0	0	0	0	0.1094	0	0
0	0	0	0	0	0	0	0	0	0	0	0	0	0.1094	0

Ecological Bulletins 48: 143–163. Copenhagen 2000

Population viability analyses in endangered species management: the wolf, otter and peregrine falcon in Sweden

Torbjörn Ebenhard

Ebenhard, T. 2000. Population viability analyses in endangered species management: the wolf, otter and peregrine falcon in Sweden. – Ecol. Bull. 48: 143–163.

The ability of population viability analyses to contribute to the identification of management isssues for small and isolated populations, the setting and evaluation of management goals, as well as the evaluation of management methods, is highlighted through the review of three case-studies of PVAs performed for the Swedish populations of wolf, otter and peregrine falcon using VORTEX. All three populations are under management, and both policy and management methods are contentious issues among stake-holders. Analyses for the wolf have identified the need to manage a population that is large enough to meet viability criteria, and that immigration of unrelated wolves is needed. An otter PVA showed that combatting pollutants and protecting the remaining population nuclei is not enough; the problem of critically small populations in a highly fragmented distribution must also be addressed. Analyses of the reintroduced peregrine falcon population indicated that further releases were not necessary, and that resources instead should be directed towards monitoring.

The three case-studies illustrate the application of individual-based models, which allowed a deeper analysis of the effects of inbreeding depression and loss of genetic variation, resulting in qualitatively different management recommendations. The individual-based model also offered the ability to model interactions between stochasticity, genetic processes and a detailed population structure.

The impact of PVAs on actual policy and management in Sweden is also assessed. There are problems related to the lack of perceived ownership of PVA results by government agencies, and to the communication of the probabilistic nature of PVAs. A PVA that is performed in isolation from stakeholders and relevant agencies is less likely to be taken into account in management. However, in general the PVA work positively influenced the policy adopted and action taken by agencies, and directed research efforts toward important issues.

T. Ebenhard (Torbjorn.Ebenhard@cbm.slu.se), Swedish Biodiversity Centre, Box 7007, SE-750 07 Uppsala, Sweden.

PVA in practical conservation

A population viability analysis is not a crystal ball that will tell us all about the future of a threatened population. Predicting the future is inherently extremely difficult. Given adequate observational data, a meteorologist may give a fair prediction of tomorrow's weather, but extending the time period to ten days makes the prediction highly uncertain. Predicting the fate of a population of animals or plants is similarly fraught with uncertainty. Despite this, I will argue that PVA may be applied as a valuable supporting tool in practical conservation. Unlike the weather, the fates of many threatened species are linked to deterministic processes that are not chaotic.

A number of recent papers have drawn attention to the problems of performing a meaningful PVA (e.g. Caughley 1994, Harcourt 1995, Hamilton and Moller 1995). It has been pointed out that a PVA itself cannot identify a threat, and that it focuses on modelling the consequences of having a small population instead of the causes of population decline (Akçakaya and Sjögren-Gulve 2000). Published PVA results have been shown to be sensitive to the particular modelling of demographic stochasticity (Brook et al. 1999) and density dependence (Mills et al. 1996), and may fail to include essential ecological processes. Reliable parameter estimates may be hard to obtain. Available field data sets are often inadequate for disentangling the various effects of demographic stochasticity, environmental variation and sampling variance and bias (Brook 2000, Meir and Fagan 2000).

Despite such conceptual and technical problems, a PVA can still contribute to the analysis and management of a threatened population, and produce sufficiently accurate guidance (Brook et al. 2000). Several authors have pointed out that PVA results are more meaningful when we focus on predicting the relative efficacy of different management scenarios, rather than predicting population trends or extinction probabilities (e.g. Hamilton and Moller 1995, Drechsler et al. 1998). The utility of PVA predictions is also improved by embracing uncertainties and problems with parameter estimation (Akçakaya and Sjögren-Gulve 2000).

I recognize three important facets of practical PVA application. The first involves a risk assessment of small and isolated populations, focusing on the relative sensitivities of parameters and setting limits to the demographic envelope, as was done for the New Zealand sooty shearwater *Puffinus griseus* (Hamilton and Moller 1995), the giant panda *Ailuropoda melanoleuca* in China (Zhou and Pan 1997), the samango monkey *Cercopithecus mitis* in South Africa (Swart et al. 1993) and the Cantabrian brown bear *Ursus arctos* (Wiegand et al. 1998). These analyses specifically highlight the possible consequences and risks of small population size.

The second facet of PVA application involves the setting and evaluation of conservation goals. Political conservation goals must often be translated into biological terms. For example, what does it mean to say that Sweden sets the goal to have a viable wolf population? What is population viability? In what terms can it be identified? How does one monitor population viability? The outcome of the PVA should be a clear formulation of what to achieve. An example of this kind of application is the Kenyan black rhino *Diceros bicornis* PVA (Foose et al. 1992), that helped refine the national goals for rhino conservation.

The third facet of PVA application is arguably the most important one, the evaluation and design of management methods. Such PVA results should compare the probable effects of different kinds of actions, guide field research into the causes of population decline, and facilitate the de-

sign of efficient monitoring and adaptive management, i.e. produce advice on how to achieve management goals. If the first facet of PVA is guided by the small population paradigm (Caughley 1994; see Akçakaya and Sjögren-Gulve 2000), the evaluation of management methods should be guided by the declining population paradigm. A number of recently published PVAs have demonstrated the utility of this approach, including those performed for Leadbeater's possum *Gymnobelideus leadbeateri* (Lindenmayer and Possingham 1996) and the greater bilby *Macrotis lagotis* (Southgate and Possingham 1995) in Australia, the Lower Keys marsh rabbit *Sylvilagus palustris* (Forys and Humphrey 1999), and the reintroduced warthog *Phacochoerus africanus* population in South Africa (Somers 1997).

In this paper I review the application of PVAs to the management of wolves *Canis lupus*, otters *Lutra lutra* and peregrine falcons *Falco peregrinus* in Sweden, illustrating both problems experienced and insights gained, in the three areas described above. The purpose of this review is three-fold. First, the three case-studies are shown as examples of traditional PVAs performed for large "charismatic" vertebrates, whereas other studies in this volume focus on the complexities of dealing with other types of life histories (Lennartsson 2000, Menges 2000) and species perceived as "less charismatic" (e.g., Berglind 2000). Second, all three PVAs were performed using an individual-based model (Lacy 2000a), and serve to illustrate situations where such a detailed model may be necessary, instead of choosing an occupancy model or a structured model (Akçakaya and Sjögren-Gulve 2000) with less demographic and genetic detail (Akçakaya 2000). Third, the three case-studies are also employed to illustrate problems in the implementation of PVA results in the real world. All three species have been extensively studied, and the causes of population decline are well known, including human persecution and pesticides. Although presently increasing, all three populations have gone through severe bottlenecks and still number just a few hundred or even less. All are subject to management action, and for the wolf the goals and means of this management are highly contentious political issues.

Wolf

Population history

During the early decades of the 19th century the Swedish wolf population numbered roughly 1500 individuals (Persson and Sand 1998). Towards the end of the 1830s ca 500 wolves were killed annually, and the population decreased rapidly. The number of wolves continued to decline through the early 20th century, and in 1970 the Swedish population was considered close to extinction. A few lone wolves lingered on in northernmost Sweden, and rumours of observations came from an area much further to the south, the province of Värmland and adjacent parts of Norway.

In 1983, the rumours were substantiated when a litter of at least six pups was born to a pair of wolves in northern Värmland. Reproduction continued most remaining years of the 1980s, but the population never numbered more than ten, due to illegal hunting and traffic accidents. During the early 1990s several new pairs were established, and the population started to grow rapidly. Persson et al. (1999) reported a 28% annual rate of growth for the decade. In March 1999, the Swedish-Norwegian population was estimated at minimum 62 and maximum 78 wolves, in six established packs. Even at this modest population size problems with predation on domestic reindeer, sheep and hunting dogs were encountered, and voices were raised calling for an upper limit to the number of wolves.

Assessment of a small and isolated wolf population

In 1995, the Swedish branch of the WWF commissioned a preliminary PVA for the Swedish-Norwegian wolf population, to address the issue of risk of extinction and loss of genetic variation if the population were limited to a small number, say 100 wolves. The report (Johnsson and Ebenhard 1996), contained the results from basic VOR-TEX (ver. 7; Lacy et al. 1995, Lacy 2000b) simulations (Box 1, step 1 and 2) utilizing data from field work on the Swedish population and from North American studies. Two years later, additional scenarios were constructed and analysed (Box 1, step 3), based on the present wolf situation. The VORTEX results generally indicated that a population numbering 200 wolves would meet the criteria set for avoiding extinction, including the effects of inbreeding. For preservation of genetic variation a much larger population would be needed, at least 500 wolves. The VORTEX simulations also highlighted the probable consequences to the population of having gone through a severe bottleneck, and of being isolated from other populations.

The results from such preliminary PVAs must be considered carefully. While indicating the existence of qualitative thresholds to wolf population viability, the results may be less reliable in quantitative terms. VORTEX is a generic PVA model and may lack the structural detail to capture the complex social structure of a wolf population. The way VORTEX models inbreeding effects and the integration of immigrating individuals may both be inappropriate. The ceiling algorithm that VORTEX employs to control the population size may also be inappropriate, depending on how hunting would be implemented in the real population. Furthermore, several parameters lack real data, e.g. the amount of environmental stochasticity. Finally, the time frame considered was at most 100 yr, ignoring possible effects further on. Still, the results indicated that the

wolf population size in 1995, i.e. ca 40, was not enough. Rather a population at least an order of magnitude larger would be required for viability. The PVA also highlighted the need for immigrants to the Swedish-Norwegian population.

Another problem was that the PVA was commissioned by a non-governmental organization and performed by an academic institution, without the involvement of stakeholding groups or the Swedish Environmental Protection Agency (SEPA), the government authority charged with the management of the wolf population. Although used by the WWF and other non-governmental organizations in their arguments for keeping a larger wolf population, the PVA report (Johnsson and Ebenhard 1996) did not alter the way SEPA handled the wolf issue over the following years. In 1997, a preliminary wolf management plan was produced that did not address the issues of demographic stochasticity in a socially structured population, inbreeding depression, the role of immigration, or even a minimum population size. The PVA report apparently suffered from a lack of ownership by the handling agency, as well as by stake-holder groups, such as the Same reindeer keepers and the hunter associations.

Identification of wolf management issues

The PVA report did however highlight a number of management issues that are currently much debated by scientists as well as lobby groups for and against the wolf. The first issue concerns to what degree the 1983 bottleneck has affected the present population, both regarding the risk of inbreeding depression and loss of genetic variation per se.

Ellegren et al. (1996) investigated 13 wild wolves killed during 1984–1994. Twenty-nine canine microsatellites were used to genotype all individuals, and heterozygosity was regressed on estimated year of birth. A significant relationship indicated that the population had lost 30–40% of its original heterozygosity between 1983 and 1995. This corresponds very well with the level of loss predicted by VORTEX simulations for the same time period (see Box 1). Ellegren et al. further found that the wild population was monomorphic for the investigated mtDNA D-loop. Randi et al. (2000) reported a similar monomorphism in the Italian wolf population, that has been isolated for at least 100 yr, and went through at bottleneck of 100 wolves in the 1970s. As a comparison, Bulgarian wolves showed seven different haplotypes.

Ellegren (1999a) also reported a negative relationship between microsatellite heterozygosity and degree of inbreeding in the Swedish zoo population, indicating that the level of inbreeding in the wild population is now $F = 0.2 - 0.3$. The genetic load of the wild wolf population has not been measured directly, but Laikre and Ryman (1991) reported high levels of inbreeding in the captive zoo population, with ensuing inbreeding depression affecting sever-

Box 1

Wolf PVAs

Johnsson and Ebenhard (1996) presented results from basic VORTEX (ver. 7; Lacy et al. 1995, Lacy 2000b) simulations for the Swedish-Norwegian wolf population, utilizing data from field work on the Swedish population and from North American studies.

In running VORTEX, the options to structure the population by sex, age and reproductive status were employed in an attempt to model the social structure of wolf packs. For example, only 29% of adult females participated in breeding any given year, to simulate the breeding of alpha females. Such structuring allows for a greater impact of demographic stochasticity on population fluctuations, which is essential in a wolf PVA (Vucetich et al. 1997).

There are no direct field data on the amount of environmental stochasticity in any wild wolf population; observations of between year variations in population size contain the effects of both environmental and demographic stochasticity, and sampling error. The actual level of environmental stochasticity in wolf populations has been estimated as considerable (Foley 1994) or relatively unimportant (Weaver et al. 1996). Still, the VORTEX simulations were run using "guesstimated" levels of stochasticity, instead of ignoring this fundamental source of population fluctuations.

Step 1

Johnsson and Ebenhard (1996) explored seven different demographic scenarios (A–G, Table 1) to capture the range of parameter values considered relevant, that is, the demographic envelope. The deterministic annual rate of increase employed ranged from $\lambda = 1.02$ to 1.35, based on various combinations of rates of age specific mortality, age of sexual maturation and litter size. In all simulations the population was given a ceiling (K), covering the range from 10 to 500 wolves, corresponding to upper management levels potentially imposed by government agencies, but no other density dependence. For K = 50, 100, 200 the risk of extinction was estimated using scenarios with and without inbreeding depression affecting the first year mortality. The VORTEX heterosis mode was employed to model inbreeding depression. The degree of genetic load was set to the value reported for captive wolves in Swedish zoos, 1.57 lethal equivalents per diploid genome (Laikre and Ryman 1991). This corresponds to an average 1.57 recessive lethal alleles per individual, which is below the recorded median of 3.14 for mammals (Ralls et al. 1988). The lethal equivalents parameter describes the demographic effect of inbreeding, i.e. the increment in mortality with increasing degree of inbreeding (Morton et al. 1956).

The results indicated that the risk of extinction over 100 yr is considerable in populations kept at 50 head and below (Fig. 1). Populations numbering 50 – 100 wolves may also be at risk, especially if the rate of increase is low. Inbreeding depression further increases this risk, generally by doubling it. At K = 200, VORTEX produced a negligible risk of extinction for all demographic scenarios also including inbreeding effects.

The loss of genetic variation would still be considerable at 200 wolves (Fig. 2), amounting to 11–14% over 100 yr. The immediate demographic or long-term evolutionary consequences of such a loss are not known, apart from the inbreeding depression itself. If, however, the goal is arbitrarily set at preserving 95% of heterozygosity over 100 yr, then 500 wolves are needed (Fig. 2). This result was little affected by rate of population increase. Although often applied as a goal in PVAs and captive population management, the preservation of 95% of heterozygosity over 100 yr is an arbitrary choice lacking a firm theoretical basis.

Next, Johnsson and Ebenhard (1996) addressed the effects of immigration. VORTEX was configured to let one male and one female, both 2 yr of age and unrelated to the target population, immigrate every five years. At K = 200, this modest immigration rate significantly affected the rate of loss of genetic variation (Fig. 2), but not enough to meet the 95% preservation goal. Immigration rates higher than this are hence needed if the population were restricted to 200 wolves.

Step 2

All the above simulations were performed with the initial population size set at K, addressing only the consequences of a population ceiling. The real wolf population went through a severe bottle-neck, with two or three wolves present in 1983. To incorporate this complication Johnsson and Ebenhard (1996) also built four scenarios (H–K, Table 2) with the initial population set at 2 or 3, rate of population increase close to the observed, and inbreeding effects included, to model population development during the time frame 1983–1995.

VORTEX results showed a high risk of extinction during this time period, between 38 and 59%, depending on the rate of increase (Table 2). Demographic stochasticity would have contributed heavily to this risk, especially during the early years. For surviving replicates, that is if the population were to survive until 1995, as it actually did, the amount of genetic variation preserved would be 62–68%. VORTEX hence indicated that the 1995 population had already lost 30–40% of its genetic variation.

Step 3

In 1998, Persson and Sand raised the question of future developments in the wolf population, given that it had gone through a severe bottle-neck, numbered ca 50 and still increased at a high rate. To answer this I produced another four VORTEX scenarios (L–O, Table 3), using $\lambda = 1.20$ as a basic rate of increase, and starting at 50 wolves in 1997. One possible development is that no inbreeding effects are shown (Scenario L), e.g. due to a low genetic load in the wild population or the immigration of unrelated wolves. The risk of extinction in that case is negligible as long as a population is not restricted to levels below 50 wolves (Table 4). For preservation of genetic variation the population still needs to grow to 500, or receive effective immigrants.

If the initial population of 50 wolves has a mean inbreeding coefficient $F = 0.25$, which is likely as indicated above, and the genetic load is 1.57 lethal equivalents, this translates into an increased first year mortality, from 22% to 36%. The resulting rate of increase would be $\lambda = 1.15$. With this level of inbreeding depression, but no further accumulation of inbreeding effects (Scenario M), the risk of extinction would still be small (Table 4). Similar results were obtained for Scenario N, which included further increased mortality rates due to the increase in the mean inbreeding coefficient that occurred during the simulation run. Finally, in Scenario O the genetic load was increased to 3.00, to simulate the possible consequences of having more severe inbreeding effects in the wild than in the captive population. Under this scenario the population would have to grow to at least 100 head to avoid extinction. These results hence indicated that even if subject to inbreeding depression, the wolf population would continue to grow, albeit at a lower rate, and the risk of extinction would be low.

al demographic variables. It is probable that this genetic load is also present in the wild population, since the zoo population was founded by wolves from Sweden, Finland and more recently Estonia. Thus it seems likely that the current wild population is inbred at levels that cause noticeable demographic effects in the zoo population. Despite this, no inbreeding depression has been detected in the wild population.

Natural selection is not likely to eliminate deleterious alleles from a wild population (Hedrick 1994), but the severe bottleneck may have randomly purged the population of its genetic load, through a founder effect. If that is the case, the level of inbreeding is not necessarily cause for any concern. On the other hand, inbreeding depression is inherently difficult to demonstrate in a wild population (Forbes and Boyd 1996), and the apparently sustained high rate of increase may be just apparent. Even a significant reduction in the rate of increase can be hard to detect among the stochastic variation in a small population.

The conclusion is that the bottleneck did affect the population. There is a loss in both nuclear and mitochon-

drial DNA, and the level of inbreeding is probably of the same order as that resulting from the pairing of siblings. The consequences of these genetic problems for the demography of the population are less clear. Currently, advocates for the wolf are concerned with the loss of genetic variation and the potential risk of inbreeding depression. Other stake-holders who are more concerned with the risk of having too many wolves tend to minimize these genetic problems, and point to the fact that many populations survive without any measurable genetic variation. A balanced management plan should be guided by consideration of a full range of possible demographic scenarios, including those that assume severe genetic impacts on demography.

The second management issue concerns the apparent isolation of the Swedish-Norwegian wolf population. The genetic problems described above would be alleviated by frequent immigration, whereas continued isolation would aggravate them (Johnsson and Ebenhard 1996). Disjunct wolf populations recolonizing parts of Montana have been shown to retain high levels of genetic variation (Forbes and Boyd 1996, 1997) due to frequent immigration from pop-

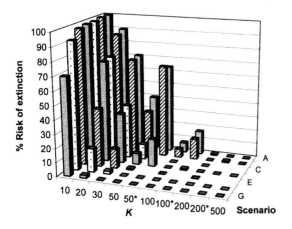

Fig. 1. Simulation results for wolf scenarios A–G. Risk of extinction (in %) over 100 yr with input data as in Table 1. For K = 50, 100 and 200 results are shown without and with* the effects of inbreeding depression (with 1.57 lethal equivalents). No immigration was simulated. Missing bars denote scenarios that were not simulated.

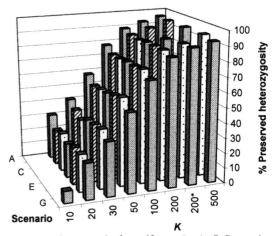

Fig. 2. Simulation results for wolf scenarios A–G. Proportion of preserved genetic variation (% of heterozygosity) over 100 yr with input data as in Table 1. For K = 200 results are shown without and with* the effects of immigration, otherwise there was no immigration. No inbreeding depression was simulated. Missing bars denote scenarios that were not simulated.

ulations to the North. Pletscher et al. (1997) and Haight et al. (1998) argue that continued immigration is necessary for the future viability of these and other recently established populations.

The present Swedish-Norwegian wolf population is isolated by ca 900 km from the main Finnish population, numbering 120–130 wolves in the south-east corner of the country. A smaller number of wolves occur in west-central Finland, just 70 km across the Baltic Sea from the Swedish population. The fact that the wolf is a very good disperser, with mean juvenile dispersal distances at ca 300 km (Persson et al. 1999), makes natural immigration likely, especially if the wolf could negotiate the Baltic Sea during winter, when it is ice-covered. The hunter associations and their affiliated press take every opportunity to point out this possibility, to establish the non-isolated nature of the wolf population.

There has however been no confirmed immigration of

any wolf to Sweden or Norway during the last two decades. No radio-marked wolf has dispersed between Finland and Sweden, and no wolf has been observed to cross the Baltic Sea. No genetic evidence suggests the presence of immigrants (Ellegren 1999b). Furthermore, a wolf dispersing over land from south-east Finland to central Sweden would have to move through the main areas of reindeer keeping. In Finland, the killing of wolves to protect reindeer herds is legal, and in both countries additional illegal killings are suspected to be frequent (10–15 wolves are known to have been illegally killed in Sweden since 1980; Anon. 1999). Again, the implication for any management plan is that a wide range of scenarios must be considered, including those that treat the population as isolated.

The third issue, concerning the hunting of wolves, is probably the most contentious. Several modelling attempts (Feingold 1996, Johnsson and Ebenhard 1996,

Table 1. Wolf scenarios A-G, with demographic input data for VORTEX-simulations (Johnsson and Ebenhard 1996). λ is the deterministic annual rate of increase. In all scenarios an average 29% of adult females participated in breeding. Populations were initially at K (10–500), with even sex ratio and stable age distribution. Time of simulation = 100 yr with 1000 replicates. Environmental stochasticity was introduced as an annual variation (SD) in the percentage of adult females participating in breeding (7 %-units), and in the age specific mortality. Maximum age = 16 yr. Monogamous matings. Dashes denote scenarios that were not simulated.

Scenario	λ	Age of maturation	Litter size	Mortality % (SD in %-units)			
				Year 1	Year 2	Year 3	Adult
A	1.02	3	4.9	40 (10)	60 (15)	30 (7)	5 (2)
B	1.05	2	4.9	40 (10)	60 (15)	–	10 (5)
C	1.07	2	6.6	20 (5)	30 (7)	–	10 (5)
D	1.10	2	6.6	40 (10)	60 (15)	–	10 (5)
E	1.17	3	4.9	20 (5)	30 (7)	20 (5)	5 (2)
F	1.20	3	6.6	20 (5)	30 (7)	20 (5)	5 (2)
G	1.35	2	6.6	20 (5)	30 (7)	–	5 (2)

Table 2. Wolf scenarios H-K, with demographic input data for VORTEX-simulations (Johnsson and Ebenhard 1996) and simulation results in terms of risk of extinction (%) and proportion of preserved genetic variation (% of heterozygosity). λ is the deterministic annual rate of increase. In all scenarios an average 29% of adult females participated in breeding, litter size was 4.94, and age of maturation 2 yr. Populations were initially 2 or 3 wolves (1 pair or 1 pair + 1 female, all adults). K = 200, and time of simulation = 13 yr (1983–1995). Environmental stochasticity was introduced as an annual variation (SD) in the percentage of adult females participating in breeding (7 %-units), and in the age specific mortality. Effects of inbreeding included, with 1.57 lethal equivalents per individual. No immigration simulated.

Scenario	λ	Mortality % (SD in %-units)			Initial population	Risk of extinction	Preserved genetic variation
		Year 1	Year 2	Adult			
H	1.16	22 (5)	40 (9)	13 (5)	3	43	67
I	1.17	22 (5)	38 (9)	12 (5)	3	38	68
J	1.17	22 (5)	38 (9)	12 (5)	2	59	62
K	1.20	22 (5)	33 (9)	11 (5)	2	55	63

Angerbjörn et al. 1999) have shown that even low levels of increased adult mortality in such a small population may have profound effects (Lacy 2000a). Haber (1996) reported effects on the stability and persistence of social units resulting in lower fecundity, less efficient hunting and territorial behaviour, as well as increased levels of natural mortality, in a study of intensive hunting (15–50% yr^{-1}) in large wolf populations.

The sensitivity of population models to increased mortality of breeding alpha wolves depends on how the modeller conceives the behavioural processes. If the loss of a breeding female causes severe social disruption, no reproductive success for at least a year, and increased mortality among the remaining pack members, then of course the prediction would be that the killing of specific adult wolves has a great impact. On the other hand, as modelled e.g. by Vucetich et al. (1997), if the loss of an alpha breeder is rapidly compensated for by the promotion of another pack member, then the number of packs is a much more impor-

tant parameter than the number of individual wolves.

Kokko et al. (1997) investigated the outcome of hunting in the Baltic grey seal *Halichoerus grypus* population, focusing on the influence of age structure and demographic stochasticity. They found that simpler models generally gave overconfident results in terms of hunting policy, i.e. allowing non-sustainable harvesting levels.

There are also genetic consequences of hunting. PVA models predict that faster population growth raises the ratio of effective to census population size and slows the loss of genetic diversity and the accumulation of inbreeding. These results suggest that hunting should be incorporated as an explicit parameter in any management model. Such models must reflect the social structure of a wolf population and allow for both demographic and environmental stochasticity. The smaller the population, the higher the demands for detail in harvesting models, which in turn calls for better field data on population size, structure and fluctuations.

Table 3. Wolf scenarios L–O, with demographic input data for VORTEX-simulations. λ is the deterministic annual rate of increase. In all scenarios an average 29% of adult females participated in breeding, litter size was 4.94, and age of maturation 2 yr. Populations were initially 50 wolves, with even sex ratio and stable age distribution. K = 25 – 200, and time of simulation 100 yr (1997–2096). Environmental stochasticity was introduced as an annual variation (SD) in the percentage of adult females participating in breeding (7 %-units), and in the age specific mortality. The effect of already accumulated inbreeding was simulated in scenarios M–O as initially increased first year mortality, based on F = 0.25. In scenarios N–O, the effect of future increases in inbreeding coefficients were also simulated. No immigration simulated.

Scenario	λ	Mortality % (SD in %-units)			Inbreeding effects
		Year 1	Year 2	Adult	
L	1.20	22 (5)	33 (9)	11 (5)	Not included
M	1.15	36 (5)	33 (9)	11 (5)	Static. First year mortality increased by 14 %-units (1.57 lethal equivalents)
N	1.15	36 (5)	33 (9)	11 (5)	Dynamic. First year mortality initially increased by 14 %-units (1.57 lethal equivalents), and further increased according to individual inbreeding coefficients
O	1.12	46 (5)	33 (9)	11 (5)	Dynamic. First year mortality initially increased by 24 %-units (3.00 lethal equivalents), and further increased according to individual inbreeding coefficients

Table 4. Simulation results for wolf scenarios L–O (Persson and Sand 1998). Risk of extinction (in %) over 100 yr with input data from Table 3.

Scenario	λ	K			
		25	50	100	200
L	1.20	5	0	0	0
M	1.15	8	0	0	0
N	1.15	27	1	0	0
O	1.12	71	7	0	0

Formulation of wolf management goals

The overall management goal for the Swedish wolf population is already stated in several binding legal acts and related documents. The Berne Convention prohibits the hunting of wolf, except under exceptional circumstances, and never in a way that threatens the viability of the population. The European Union legislation fully implement the Berne Convention in this respect. According to the Swedish Action Plan for Biodiversity (Anon. 1995), implementing the Convention on Biological Diversity, the goal is to keep a viable wolf population.

Already in 1991 the SEPA issued a policy document that stated the goal of maintaining a viable wolf population. Six years later, in the agency's preliminary wolf management plan, this was revised to say that such a wolf population can only be allowed in central and southern Sweden, outside the area of reindeer keeping (Persson and Sand 1998). Most stake-holder organizations, including WWF Sweden, the Swedish Society for Nature Conservation, the Swedish Carnivore Association, the Swedish Association for Hunting and Wildlife Management, and the Swedish Same Association have similarly stated that their policy is to contribute to a viable Swedish-Norwegian wolf population.

Everybody seems to agree that the goal is to keep a viable wolf population. It is less apparent that stakeholders agree on what a viable population is, and what it entails in terms of number of wolves. The interests of hunters, reindeer keepers and farmers must of course be considered in a management plan for the wolf, but not in deliberations of what constitutes a viable population.

The carrying capacity for wolves in Sweden is estimated at 8000–12 000, or 4700–7400 if northern Sweden is ceded to reindeer (Persson and Sand 1998). These estimates assume that the wolf can utilize all of the wild ungulate production. If the production is to be shared with the hunters, corresponding lower carrying capacities must be set. For example, allowing wolves to consume 10% of the production results in a carrying capacity at 500–700. Such a large population is however unacceptable to many stakeholders. A questionnaire survey showed that 51% of 1500 randomly sampled citizens would not accept a population > 200 wolves (Karlsson et al. 1999). Obviously, there is a need to address the issue of the size of a viable wolf population.

The government of Sweden appointed a commission of enquiry in 1998 to conduct a formal investigation into the biological, economical and political aspects of the management of all large predators, including the wolf, and to suggest a comprehensive policy. The report was presented late in 1999 (Anon. 1999; English summary in Anon. 2000), containing inter alia a special section on population viability, written by a working group of five independent scientists in ecology and genetics (Andrén et al. 1999).

The working group was asked to define what a viable wolf population is, and how many wolves it must contain. The group reviewed all genetic and demographic analyses relevant to the Swedish-Norwegian wolf population, as well as wider approaches to population viability. Because the different criteria for population viability are not easily integrated, the group arrived at five alternative definitions of a viable wolf population (definitions A–E below).

Definition A emphasised long-term preservation of genetic variation in an isolated population, and was based on Franklin's (1980) estimate of the rate of mutation needed to compensate for genetic drift, expressed as an effective population size (N_e) of 500. Other studies (e.g. Lande 1995) call for much larger effective population sizes, but Franklin's model was accepted by the group. The ratio of N_e to N was estimated as 0.10, which translates into a critical population size of 5000 wolves. Analyses using VORTEX (e.g. Johnsson and Ebenhard 1996) usually express a critical population size based on an acceptable loss of e.g. 5% of heterozygosity over 100 yr. The working group did not adopt such a definition, since the amount of acceptable loss is purely subjective.

Definition B was an extension of A, relaxing the assumption of complete isolation. Depending on the actual amount of immigration to the Swedish-Norwegian population, definition B was expressed as 30–100% of A, i.e. 1500–5000 wolves. There are currently no models, nor data, that would give a more exact relationship between the actual amount of immigration and population size required for a mutation-drift balance. For short-term (< 50 yr) avoidance of inbreeding depression, Franklin's critical population size of $N_e = 50$ was adopted for definition C, which translated into 500 wolves. This critical size was deemed to be rather insensitive to the actual rate of immigration.

Definition D was based on the risk of extinction due to demographic and environmental stochasticity in a small population. The risk of extinction accepted by the group was 5% over 100 yr. The preliminary PVA by Johnsson and Ebenhard (1996) was used as a starting point. The VORTEX simulations suggested that a population of 200 would meet the risk criterion, but given the uncertainties concerning the model structure and data quality mentioned above, the working group doubled the number to 400 wolves. Definition E was simply a combination of C and D, offering a short-term critical population size (500 wolves) based on both genetic and demographic processes.

The working group further recommended that no hunting should take place until the wolf population had reached its management level, based on genetic considerations and the 1983 bottleneck event. The group identified research on wolf migration rates, a detailed monitoring scheme and the development of tools to detect inbreeding depression in the wild population to be essential parts of a management strategy.

The final report of the commission was however not based solely on the results of the working group. All stakeholders, both for and against large numbers of predators, were given the opportunity to be heard, since a comprehensive policy was called for. The commissioner stated that he had not found enough support among the general public even for 500 wolves, the critical population size for short-term viability given by the working group. Instead a compromise of 200 wolves was suggested as a minimum number. Below that number, conservation of the wolf would be the first priority (although permitting hunting in exceptional circumstances); otherwise, management would be oriented towards interests other than conservation. No wolves would be allowed to stay permanently in northern Sweden, because of conflicts with reindeer herding. Although this restriction would decrease the probability for immigration from Finland, the report called for sufficient gene flow, and listed population connectivity as an objective. How this objective would be accomplished was not detailed.

In June 2000, while the government was still considering the commission report, the SEPA published a policy decision by issuing its own updated wolf action plan (http://www.environ.se/dokument/press/2000/juni/p000628.htm). While taking the commissioner's report into account, this action plan deviates from its suggested policy on a number of important points. The SEPA rejects 200 wolves as a minimum management goal. The SEPA aims at a minimum 500 wolves, and outlines efforts to mitigate the problems caused by wolves. Furthermore, the SEPA rejects the plan to exclude wolves from northern Sweden, due to concerns about restricting wolf habitat and migration. In contrast with the commissioner report, the SEPA plan appears to embrace those management issues highlighted by the PVAs reviewed during the government investigation.

Otter

Population history

Throughout Europe the otter has decreased in numbers due to habitat changes along lakes and streams, excessive hunting, and pollution with PCB and other agents. From 1945–1955, up to 1500 otters were harvested annually in Sweden (Erlinge 1972). By 1976, the entire Swedish population was estimated at 500–1500. The decrease continued during the 1980s and early 1990s. Figure 3 shows areas in Sweden where signs of otters were found during

surveys 1983–1993. Otters were present only in restricted areas, often widely separated from each other.

Assuming optimal otter habitat within these areas, the carrying capacity per area was estimated as ranging from 5 to 200 (Larsson and Ebenhard 1994), but the actual

Fig. 3. Areas in Sweden where signs of otters have been found during surveys 1983–1993. Provinces mentioned in the text are indicated: B Bohuslän, D Dalarna, G Gästrikland, Ha Halland, Hä Hälsingland, Sk Skåne, Sm Småland, Sö Södermanland, U Uppland, Vr Värmland, Vs Västergötland, Ö Östergötland. Supplied by Swedish Museum of Natural History.

Box 2

Otter PVAs

Larsson and Ebenhard (1994) addressed the potential problems of the small and fragmented otter population in Sweden. The general question asked was: If the deterministic threats are removed, and a positive rate of increase is restored, can the otter survive within its present distribution in Sweden?

The amount of data available was rather restricted. Larsson and Ebenhard (1994) assembled demographic data from European and North American *Lontra canadensis* otter populations, with emphasis on boreal freshwater populations not showing severe effects of pollutants, to construct a general demographic envelope. A subset of four demographic scenarios are shown here (Table 5), based on North American (Tabor and Wight 1977) and Norwegian (Heggberget 1991) mortality rates, and litter sizes from Byelorussia (Sidorovich 1991) and the European zoo population (Reuther 1991). For a number of parameters, such as amount of environmental stochasticity and genetic load, values were "guesstimated".

Step 1

VORTEX (ver. 6; Lacy et al. 1994, Lacy 2000b) was employed to simulate the effects of low carrying capacity and small initial population sizes for demographic scenarios with deterministic annual rates of increase varying between $\lambda = 1.022$ and 1.130, with K varying between 10 and 200 to match the areas of distribution indicated in Fig. 3.

Under one set of simulations initial population sizes were set equal to K. At lower rates of increase, K = 100 is needed to achieve a risk of extinction at 5% or below over 50 yr (Fig. 4). If the rate of increase is at least 4% yr^{-1} ($\lambda = 1.04$), K = 50 would meet the same risk level. If inbreeding effects are added (with 3.0 lethal equivalents per individual, close to the median value recorded for mammals; Ralls et al. 1988) and employing the heterosis mode of VORTEX, significantly larger populations are needed to meet the risk criterion under most scenarios. But predictions are actually more pessimistic than revealed by the risk of extinction; even surviving populations contain very few otters by year 50, so extinction would be imminent. Genetic predictions are also pessimistic (Fig. 5); even at K = 200 between 6 and 8% of the heterozygosity will be lost over 50 yr.

Under another set of simulations, with K = 50, 100 or 200, initial population sizes were set to much lower levels. The risk of extinction under these conditions was significantly higher (Fig. 6) than for simulations starting at K. Taking the Uppland population as an example, and assuming a 4% rate of increase, starting close to the carrying capacity entails a 5% risk of extinction, whereas an initial group of 10 otters would incur a 49% risk of going extinct within 50 yr. Again, assuming inbreeding effects, the situation would be even worse. Such small groups of otters would also have difficulties in maintaining heterozygosity (Fig. 7), since each simulation in effect starts with a population bottleneck.

Step 2

Larsson and Ebenhard (1994) also attempted an evaluation of the reintroduction methods employed in two release projects in Southern Sweden, using VORTEX. The principal questions asked were: can the release of relatively low numbers of otters significantly change the risk of extinction and the rate of loss of heterozygosity? Should all otters be released simultaneously or can releases over longer time periods be effective?

The first concern with reintroductions is to analyze whether it is at all feasible to attain a viable population in the target area. In their analyses Larsson and Ebenhard set the problems of habitat protection and elimination of pollutants aside and instead examined the effect of carrying capacity. The two target areas were estimated to carry ca 50 otters each. As shown in Figs 4 and 5, that carrying capacity is certainly at the lower limit or below for long-term viability. For reintroductions to be meaningful the longer term goal should hence be to restore habitat so that K could be increased, and contact made with adjoining local populations.

Step 3

Next, Larsson and Ebenhard used VORTEX to examine the effects of releasing 2, 4 or 10 otters per year during a time period of five years, for a total of 10, 20 or 50 animals, into a population initially numbering 10. This was done for K = 50 and 100, and with and without the effects of inbreeding. The simulation results showed that releases can indeed decrease the risk of extinction (Fig. 8), but to achieve a risk level as low as that for populations that start at their carrying capacity, large numbers of otters are needed. The 7 otters in Uppland were estimated to have made a small impact, unless the rate of increase were very high. In Södermanland the 36 released otters may have affected the risk of extinction significantly. In both cases this interpretation rests on the assumption that the rate of population growth is no longer negative.

The rate of loss of genetic variation is also decreased through releases (Fig. 9), due both to the more rapid alleviation of the bottle-neck, and to the introduction of new genetic material, but again rather large numbers of otters are needed. Further simulations, not shown here, indicated that if the initial population size is larger (e.g. 30), relatively small numbers of otters released can effectively reduce the risk of extinction to that predicted for populations that start at their carrying capacity.

Finally, a comparison was made between such releases distributed over five years and a clumped release of the same number of otters (Fig. 10). The risk of extinction was virtually identical under the two release strategies, indicating that for a relatively long-lived animal the production of animals for releases must not necessarily be concentrated to a short time period.

number of otters present in 1993 was probably somewhere between 0 and 30 per area. For example, the province of Uppland was estimated to carry 50 otters, but only 10 were believed to survive (Larsson and Ebenhard 1994). There was virtually no information on migration between the areas of distribution.

Assessment of small and isolated otter populations

In 1994, WWF Sweden published a report (Larsson and Ebenhard 1994) that addressed the potential problems of the small and fragmented otter population in Sweden (Box 2). The report was not intended to examine the known causes of decline, such as habitat change, hunting and pollution, or how to overcome them. Instead focus was on stochastic demographic and genetic processes in small populations, and hence VORTEX (ver. 6; Lacy et al. 1994, Lacy 2000b) was employed to model the otter populations.

The general conclusion from these analyses (Box 2, step 1) was that the then existing Swedish otter populations faced serious threats apart from those already identified. The elimination of pollutants and the protection of good habitat within nucleus areas would not be enough to safeguard the otter. The restoration of otter habitat within and

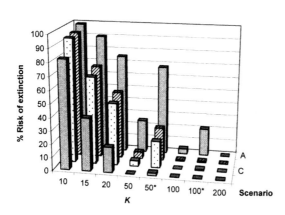

Fig. 4. Simulation results for otter scenarios A–D. Risk of extinction (in %) over 50 yr with input data from Table 5. Populations were initially at K. For K = 50 and 200 results are shown without and with* the effects of inbreeding depression (with 3.0 lethal equivalents).

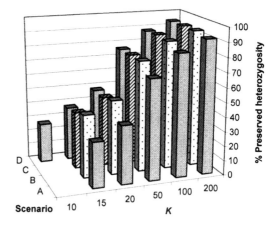

Fig. 5. Simulation results for otter scenarios A–D. Proportion of preserved genetic variation (% of heterozygosity) over 50 yr with input data from Table 5. Populations were initially at K. No inbreeding depression was simulated. Missing bars denote scenarios that were not simulated.

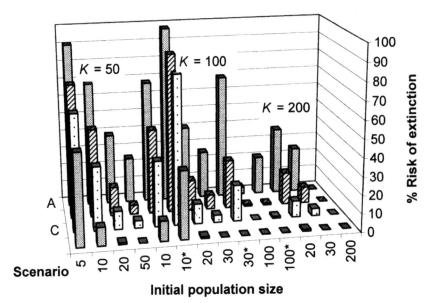

Fig. 6. Simulation results for otter scenarios A–D. Risk of extinction (in %) over 50 yr with input data from Table 5. Populations were initially below K. For K = 100 results are shown without and with* the effects of inbreeding depression (with 3.0 lethal equivalents).

between the present areas of distribution would also be essential, to increase the habitat carrying capacity of local populations and increase the probability of migration between them. Furthermore, local extinctions could be expected even in areas of optimal otter habitat.

Evaluation of otter reintroduction methods

The precarious situation of the small local populations prompted at least two reintroduction projects to be established. In Uppland, the Swedish Association for Hunting and Wildlife Management in cooperation with the Swedish Museum of Natural History released a total of seven

otters during 1988 and 1989. These otters were not equipped with radio transmittors, and their fate is not known with any certainty. Some of these otters and their offspring may be included in the ten individuals that were estimated to survive in 1993.

Between 1989 and 1992 WWF Sweden and the Swedish Association for Hunting and Wildlife Management supported another release project in the province of Södermanland. Eleven wild-caught otters from northern Norway and 25 captive-bred Swedish otters were released and monitored for a number of years. The survival rate was high, 79 and 42% the first year for Norwegian and Swedish animals, respectively (Sjöåsen 1996). The effect of this supplementation on the indigenous population

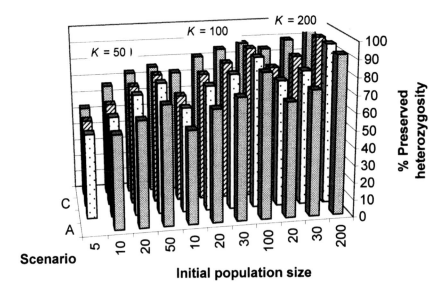

Fig. 7. Simulation results for otter scenarios A–D. Proportion of preserved genetic variation (% of heterozygosity) over 50 yr with input data from Table 5. Populations were initially below K. No inbreeding depression was simulated. The missing bar denotes a scenario that was not simulated.

Table 5. Otter scenarios A–D, with demographic input data for VORTEX-simulations (Larsson and Ebenhard 1994). λ is the deterministic annual rate of increase. In all scenarios an average 25% of adult females participated in breeding. Populations were initially at K (10–200) or below, with even sex ratio and stable age distribution. Time of simulation was 50 yr with 1000 replicates. Environmental stochasticity was introduced as an annual variation (SD) in the percentage of adult females participating in breeding (5 %-units), and in the age specific mortality. Maximum age = 12 yr. Polygamous matings. No immigration was simulated. Dashes denote scenarios that were not simulated.

Scenario	λ	Age of maturation	Litter size	Mortality % (SD in %-units)			
				Year 1	Year 2	Year 3	Adult
A	1.022	2	2.6	32 (10)	54 (5)	–	27 (5)
B	1.040	F: 3, M: 2	2.6	35 (10)	32 (5)	25 (5)	25 (5)
C	1.058	2	2.0	35 (10)	32 (5)	–	25 (5)
D	1.130	2	2.6	35 (10)	32 (5)	–	25 (5)

was however not assessed, and no extensive surveys have been carried out since.

The general PVA presented by Larsson and Ebenhard (1994) also attempted an evaluation of the reintroduction methods employed in the two release projects (Box 2, step 2 and 3). The general results were that there was a need to restore habitat, to increase the local carrying capacity, before any releases. Rather large numbers of otters would be needed for a high probability of success, but the otters may be released over a prolonged time, instead of one large release event.

VORTEX, other generic models (e.g. ULM; Legendre and Clobert 1995) and purpose-built models have been used for similar reintroduction assessments for a range of species. An initial feasibility study, to analyze whether it is possible to attain a viable population in the target area, has been done for the otter and lynx *Lynx lynx* in the Netherlands (van Ewijk et al. 1997, de Jong et al. 1997) and the wild boar *Sus scrofa* in Scotland (Howells and Edwards-Jones 1997). The composition of propagules, i.e. the number, age and sex of the group of animals to be released,

and its distribution in time, has been examined e.g. for the griffon vulture *Gyps fulvus* in France (Sarrazin and Legendre 2000), the capercaillie *Tetrao urogallus* in Scotland (Marshall and Edwards-Jones 1998), the Cape mountain zebra *Equus zebra* in South Africa (Novellie et al. 1996), and the Persian fallow deer *Dama mesopotamica* in Israel (Saltz 1996).

An increasing otter population

No extensive survey of otters has been carried out to produce updated population and distribution figures, but the general impression from several local inventories is that the population is growing. The current population is estimated at 2000 (http://www.jagareforbundet.se/forsk/viltvetande, updated 18 January 2000). Local extinctions may have taken place, but apparently, the long period of steady decline has been reversed, through persistent action to reduce the emission of pollutants and to safe-guard and restore good otter habitat. In the provinces of Gästrikland

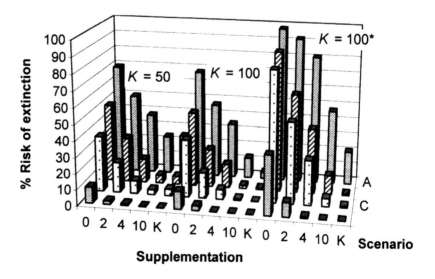

Fig. 8. Simulation results for otter scenarios A–D. Risk of extinction (in %) over 50 yr with input data from Table 5. Populations were initially at 10, but were supplemented by 0, 2, 4 or 10 otters per year during the first 5 yr. For comparison, results from simulations with initial size = K are also inserted, marked K. For K = 100 results are shown without and with* the effects of inbreeding depression (with 3.0 lethal equivalents).

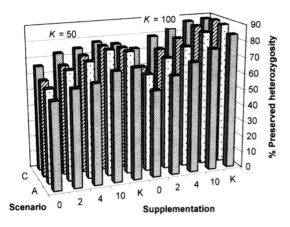

Fig. 9. Simulation results for otter scenarios A–D. Proportion of preserved genetic variation (% of heterozygosity) over 50 yr with input data from Table 5. Populations were initially at 10, but were supplemented by 0, 2, 4 or 10 otters per year during the first 5 yr. For comparison, results from simulations with initial size = K are also inserted, marked K. No inbreeding depression was simulated.

and Hälsingland (Fig. 3), the number of potential otter habitats showing signs of otters being present has doubled from 15% to 34% (Gävleborg county administrative board, http://www.xnat.se/lansstyr/verksamhet/miljovard/tillst/sjoar/utter.htm, updated 16 March 1999). In 1987–88, the otter was restricted to seven isolated areas, but now otter signs are distributed more widely, indicating that the otter has increased its area of distribution. There is however still no contact with the Uppland population to the south.

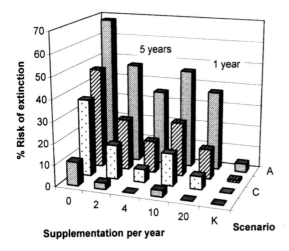

Fig. 10. Simulation results for otter scenarios A–D. Risk of extinction (in %) over 50 yr with input data from Table 5. Populations were initially at 10, but were supplemented by 0, 2, or 4 otters per year during the first 5 yr, or by 10 or 20 otters the first year. K = 100. For comparison, results from simulations with initial size = K are also inserted, marked K. No inbreeding depression was simulated.

In southern Sweden, e.g. in the province of Småland, PCB levels have also decreased, but not as much as in the north. Still, increasing numbers of otter observations are being reported to the county conservation agency (Kalmar county administrative board, http://www.f.lst.se/natv/utter.html, updated 31 May 1999). This agency has established cooperation with a number of interested parties, both NGOs, companies and other agencies, to perform surveys, habitat restoration and to identify potential migration routes between existing otter groups.

No new reintroduction projects have been initiated, but the Småland project still considers it a viable option once enough habitat is available and the PCB levels have decreased further. The above PVA has shown that reintroduction is feasible, but also that quite large efforts are necessary.

The PVA published by WWF Sweden has not resulted in any new policy or management action being taken by the SEPA, but it has focused the activities of NGOs and local conservation agencies on the problems of small populations.

Peregrine falcon

Population history

The peregrine falcon in Europe has been the victim of a sequence of threats, beginning at the early 20th century with massive persecution due to its habit of killing domestic pigeons. In the 1950s, increasing levels of heavy metals and organochlorine pesticides became a serious threat, and as the population declined, wild-caught chicks and eggs fetched higher prices, which in turn intensified the unsustainable taking of birds from the wild.

In 1900, the Swedish population was estimated at 1000 pairs, but in 1945 this had dropped to 350 pairs. At its lowest, in the early 1970s, the population was down to 9 pairs. Most of these birds were found in northernmost Sweden, but a few falcons, including one breeding pair, remained on the Swedish west coast, in the province of Bohuslän (Lindberg et al. 1988). In adjoining parts of Norway, along the Oslo Fiord, the peregrine was extinct.

Captive breeding and reintroduction

In 1972, the Swedish Society for Nature Conservation initiated its Project Peregrine Falcon. A comprehensive action plan was adopted, including research on pesticides and their effect on falcons, protection of each individual falcon during breeding, and lobbying for changes to agency policies. Two years into the project a captive breeding programme was set up, in cooperation with Gothenburg univ., WWF Sweden and the SEPA.

The explicit goal of the breeding programme was to re-establish a population numbering at least 25 breeding pairs on the Swedish west coast. A total of 42 wild-caught falcons were assembled originating from southern and northern Sweden, Finland, Norway and Scotland, but only nine of them came from the target area. Many of the birds were already in captivity at the start of the programme. Most of the breeding was conducted at a dedicated facility close to Gothenburg.

Up to 1991, a total of 214 falcons were bred and released into the wild, through hacking at the age of fledging. The majority, 186 birds, were released in the Swedish west coast area, mainly in the provinces Bohuslän, Halland and Västergötland, and adjoining parts of Norway. The remainder went to northern Sweden. Simultaneously, the wild breeding pairs were manipulated to lay double clutches and many of their chicks were hatched in captivity and later returned.

In 1992, the three sponsoring organizations decided to evaluate the breeding programme. The resulting report (Ebenhard 1992) considered the status of the captive population as well as the reintroduced one. For a sustained effort, the status of the captive population is an important aspect that is often overlooked in reintroduction programmes. The captive breeding population still numbered ca 40 birds in 1991, despite the production of ca 35 chicks per year during the recent years. Without releases, the rate of increase would have been 25% yr^{-1}, but almost all reared birds were immediately released. The average age of breeding birds was 11 for males and 10 for females. A simple deterministic projection of the life table and age specific reproduction predicted that chick production would remain stable for just three years, then decline sharply. The captive population was senescing. In this case the demographic effect of releases on the captive population was rather obvious. In other cases VORTEX simulations (Bustamante 1996, 1998, Novellie et al. 1996) or stochastic maximum sustained yield models (Saltz 1998) have been applied.

The genetic status of the captive population was also investigated (Ebenhard 1992) through gene drop analysis, a stochastic simulation of the fate of individual "alleles" from each founding bird. An examination of the studbook revealed that 12 wild-caught birds had died without leaving offspring in the current captive population, another 18 were still alive but had not produced any offspring that was still present in the captive population, and 12 were actual founders. Stochastic loss of alleles and unequal number of offspring had further reduced heterozygosity to the equivalent of 5.5 founders, a 50% loss.

Ebenhard (1992) recommended that the production potential for the coming years should be utilized, emphasizing efforts to increase the number of contributing founders. For continued releases after that, the captive population would have to be given new blood and a stable age distribution.

Evaluation of the reintroduced falcon population

The reintroduced falcon population was evaluated by Ebenhard (1992) using VORTEX (ver. 5.1, Lacy 1993, Lacy 2000b). The general results (Box 3) were that without the release programme the population would have gone extinct, but now a viable population had been restored.

As with the captive population, a genetic analysis was also performed for all 214 released birds originating from the captive population. No inbred birds were released, but the average coefficient of relatedness was 0.094, which would result in an average coefficient of inbreeding at 0.047 in the following generation. This is rather quick accumulation of inbreeding, which of course is due to the fact that only 17 founders have contributed genetically to the released birds, and among them the representation is quite unequal, resulting in just 10 founder genome equivalents, or effective founders. The captive population contained 5.5 founder genome equivalents, which means that more genetic diversity had been invested in the released population than in the breeding stock. Further releases from the same captive population would not improve founder contribution much, without more founders joining in the production of young, i.e. the captive non-breeders.

Ebenhard (1992) recommended that further releases should go on for a couple of years to utilize the production capacity remaining in the captive population, but that releases then should be phased out. Instead, available resources should be spent on intensive monitoring of the re-established population, to validate the model parameters, and to detect any deviations from the expected development. In 1994, the SEPA issued an action plan (Lindberg and Eriksson 1994) for the peregrine falcon that to large extents adopted the recommendations from the project evaluation.

Sarrazin and Barbault (1996) called for the explicit formulation of success criteria for reintroduction programmes. They suggested three such criteria, viz. a viable population size as shown by a PVA, the successful breeding of first wild born animal, and a recruitment rate faster than the adult mortality rate over three years. Saltz (1998) further made the distinction between success criteria, in the sense that a final goal for the reintroduction programme has been met, and criteria for the termination of releases. Releases may actually be terminated before the goals of the reintroduction programme have been reached. The success criteria may be formulated as a specific population size, whereas releases can be discontinued at a smaller population size. The above analysis of the peregrine falcon population demonstrated how VORTEX can be used to assess a reintroduction programme against such criteria. Green et al. (1996) reported on a similar analysis for the reintroduced white-tailed eagle *Haliaeetus albicilla* population in Scotland, using a model built specifically for this purpose.

Box 3

Peregrine falcon PVAs

Ebenhard (1992) used VORTEX (ver. 5.1, Lacy 1993, Lacy 2000b) to evaluate the reintroduced falcon population on the Swedish west coast and adjoining parts of Norway. The data input for simulations were as far as available taken from the target population, based on the monitoring programme for the released falcons.

The clutch size of fledged chicks was estimated as 2.12 in the Swedish part of the population, based on 42 clutches fledged between 1980 and 1991. The additional chicks supplemented by hacking from the captive population, as well as through double-clutching of wild birds, increased this to an average of 4.56 fledged chicks per breeding pair. Over the time period, an accumulated number of 130 adult females have been present, but only 86 have participated in the production of young, resulting in 66% of adult females breeding. Mortality rates were not available for the wild population. Instead 13% adult mortality was estimated from other studies, and juvenile and subadult mortality rates were estimated by fitting a hypothetical life table and the observed reproduction (including manipulations) to the rate of increase recorded (22% yr^{-1}). Environmental stochasticity was introduced only as a 10% coefficient of variation in the percentage of females not breeding in a particular year, based on observations that breeding failures are partly weather dependent. The genetic load of the peregrine falcon is not known; the breeding programme produced no inbred young. In the simulations 1.0 lethal equivalents per individual was applied, as a conservative estimate, employing the heterosis mode of VORTEX. The initial population size was based on number of pairs present and calculated number of juveniles assuming a stable age distribution.

A number of scenarios were constructed, five of them shown here, for the combined Swedish west coast and Norwegian population, as it has been managed as a unit and dispersal occurs over the border (Table 6).

In scenario A, the fate of the population was assessed, in absence of any releases, manipulations, or immigration, for the time period 1980–1991. The risk of extinction during that short period was predicted to be 15% (Table 7), and even if the population would have survived, it would have numbered only 11 birds. Given the same situation, but extending the time period to 2090 and adding a small immigration rate from other populations (scenario B), e.g. from the Norwegian west coast, the risk would still be high that no population was safely established for a long time. Scenario C tested the effect of the releases and manipulations that actually took place (Table 6). The rate of extinction was decreased to 2% (Table 7), and that risk would have been most pronounced during the initial years. Evidently, the action that was taken by Project Peregrine Falcon had significant effect on the remnant population.

Two scenarios examined the future of the reintroduced population over 100 yr, starting from 1991. Scenario D assumed no further releases or manipulations, starting with 68 falcons. VORTEX simulations predicted that carrying capacity would be reached within 30 yr, and the risk of extinction was nil. Further releases (scenario E) would simply speed the rise to carrying capacity. The analysis thus indicated that no further releases were needed.

Table 6. Peregrine falcon scenarios A-E, with demographic input data for VORTEX-simulations (Ebenhard 1992). λ is the deterministic annual rate of increase. In all scenarios an average 66% of adult females participated in breeding. In scenario C, the clutch size was artificially increased to match the effects of hacking of chicks. Populations were initially given an even sex ratio and stable age distribution. Environmental stochasticity was introduced as an annual variation (SD) in the percentage of adult females participating in breeding (3.4 %-units). The carrying capacity K was based on 60 pairs and number of juveniles according to a stable age distribution. Age of sexual maturation = 2 yr. Maximum age = 23 yr. Monogamous matings. All results are shown with the effects of inbreeding depression (with 1.0 lethal equivalents). All scenarios were simulated with 300 replicates.

Scenario	λ	Timespan	Initial population	K	Litter size	Mortality %		
						Year 1	Year 2	Adult
A	1.056	1980–1991	8	151	2.12	61.5	26.0	13.0
B	1.056	1980–2090	8	151	2.12	61.5	26.0	13.0
C	1.221	1980–1991	9	177	4.56	61.5	26.0	13.0
D	1.056	1991–2090	68	151	2.12	61.5	26.0	13.0
E	1.056	1991–2090	68	151	2.12	61.5	26.0	13.0

Table 7. Simulation results for peregrine falcon scenarios A–E. Risk of extinction (P, in %), time to first extinction (T, in yr), final population size (N), and proportion of preserved genetic variation (H, in % of initial heterozygosity) with input data from Table 6. For scenario B, the risk of extinction is given as the risk of not having a population present at the final year of simulation and the risk of extinction at any time during the timespan. In scenario E, 12 supplemented subadults correspond to 30 newly fledged chicks in hacking releases.

Scenario	λ	Immigration	Supplementation	P	T	N	H
A	1.056	–	–	15	7	11	80
B	1.056	1 pair every 6 yr	–	11 / 78	12	55	78
C	1.221	–	Actual	2	6	50	86
D	1.056	–	–	0	–	149	89
E	1.056	–	12 yr^{-1} for 6 yr	0	–	149	90

Present falcon management

The southern Swedish-Norwegian population has continued to grow during the 1990s. In 1991 nine pairs produced fledged chicks in the Swedish part of the area; in 1999 this had increased to 29 (Lindberg 1996, 1997, 1998, 1999), corresponding to a 16% annual rate of increase. The population has also increased its area of distribution, with the first successful breedings in the provinces of Skåne and Östergötland in 1995 and Värmland in 1997. The number of falcons being released into this population has gradually been decreased. In 1997, the last six chicks were hacked. Instead, starting in 1994, captive-bred falcons have been taken to the province Dalarna to establish a new population, in an attempt to create a contact between the southwestern and northern Swedish populations.

Close monitoring of the reintroduced population has continued, and a majority of breeding birds have been identified. In 1998, 25 breeding falcons were identified. Ten of these were earlier released as fledglings in hacking boxes, seven were hatched in captivity from eggs laid in wild nests and later returned, and the remaining eight were wild-born. No difference in breeding success or adult mortality has been recorded between these three groups of falcons.

In 1998, the captive breeding programme had produced a total of 387 fledged young, most of which were released in the Swedish west coast area. The annual production has decreased from ca 35 at the beginning of the decade to only 10, produced by three breeding pairs in 1998.

Conclusions

The three case-studies presented illustrate the wide range of applications for population viability analyses and related methods. For all three species, PVAs of small and isolated populations have examined the relative sensitivity of parameters within a demographic envelope and identified possible consequences and risks caused by low numbers, fragmentation and isolation. In this work, the assessment of absolute extinction risks have been less important than the identification of management and policy issues, and the direction of research efforts.

For the wolf, the most important issue introduced by PVA work is the existence of a minimum viable population size, i.e. that the concept of viability is actually a function of population size. It can be discussed at length exactly how many wolves are needed for viability, but the PVAs have indicated that significantly larger numbers are needed than previously accepted. The genetic effects of the severe bottle-neck, as well as the need for immigration of unrelated wolves from Finland has also been highlighted through the PVA work. Illegal hunting and the proposed legal hunting of individual "problem" wolves would restrict the current rate of population growth, and hence prolong the problems incurred in the bottle-neck. Models have also shown that maintaining a wolf population near its minimum viable size (to reduce the costs of damage to reindeer, sheep and hunting dogs) will entail higher costs for monitoring, because more precise modelling and parameterization is needed to assess the viability of minimal populations.

The otter PVA has shown that even though reducing pollutants and safe-guarding the remaining habitats are essential to otter conservation, the severe fragmentation of the population poses another threat. Restoration of good otter habitat outside the current distribution is needed. Reintroduction efforts can be effective for the maintenance of critical otter populations, but they demand relatively large resources, and cannot be seen as a long-term solution for the Swedish otter. The otter PVA has been criticized for not having predicted the current strong development of the otter population, with increases in number and distribution. Such predictions were, however, never the intention, and the demographic data needed were not available. The PVA intended to show possible consequences of the fragmented population structure, and of very low otter numbers within each fragment.

The peregrine falcon PVAs assessed the issue of continued breeding efforts and releases, guiding the use of limited resources within the project. The recommendation was to step down the releases, and concentrate on monitoring the re-established population carefully. The issue of follow-up activities was deemed most important, to verify the continued relevance of the PVA results.

The choice of VORTEX for population modelling seems appropriate, as it offered the necessary detail of an individual-based model (Lacy 2000a). For the otter, an occupancy model would have been inappropriate for the questions being addressed, and there were no data available on empty otter patches or the amount of population turnover. A structured model could have been applied for the analysis of several issues, but the individual-based approach provided further insights. The ability to model inbreeding depression and loss of genetic variation was especially important. Predictions were qualitatively affected by such considerations. Generally, much larger populations were found necessary for long-term viability. Potential problems with severe bottle-necks were also highlighted in a way that would not have resulted from a pure demographic approach.

The ability to model demographic and environmental stochasticity was also important. This can be done in a simpler structured model as well (Akçakaya 2000), but VORTEX can in addition model the interaction between stochasticity and a detailed population structure. The effect of stochasticity in a small population, in terms of fluctuations in numbers, is much affected by the social structure, mating system, sex ratio, distribution of offspring etc. in the population. A similar interaction between stochasticity and genetic processes is also modelled by VORTEX.

For the wolf, part of the reluctance among stake-holders and managing agencies to accept the results of the PVA was due to the feeling that the PVA model did not model the social structure of wolf packs well enough. To solve this, a purpose-built model for wolves would be needed. The limitations of VORTEX in this respect were however made explicit in the interpretation of the simulation results, and accounted for in recommendations made. Finally, the ability of VORTEX to model management action, such as releases and hunting, and the application of an upper population limit, was necessary for the evaluation of management options.

The impact of the PVAs on the management of the three species varies considerably. Not surprisingly, the debate on issues raised by the PVAs has been hottest for the wolf, and impact hardest to achieve. The initial PVA was performed by an academic institution on behalf of an NGO, and the results were not readily accepted by other stake-holders and government agencies. Recently, management agencies have taken more "ownership" of the PVA work. The most recent wolf action plan from the SEPA actually deals with all the important issues identified by the PVAs. The failure of the government commission of enquiry to accept the recommendations given by its own scientific working group may be seen as disappointing. It should be noted, though, that the commissioner had accepted the issue of a minimum viable population size, and specifically asked for such an analysis. Without the previous PVA work, this approach would probably not have been adopted.

With the otter PVAs, not much impact at the policy level or in national management agencies can be observed. There is, however, no real controversy about the otter, and the recommendations of the PVAs are not questioned. Instead, the recommendations are applied at the local level, in local agencies and NGOs working to improve conditions for the otter.

The peregrine falcon PVA recommendations have been fully implemented both in Project Peregrine Falcon and in the national action plan adopted by the SEPA. The PVA was performed in close cooperation with the project, and full ownership was taken. In contrast to the wolf case, the implementation of the falcon PVA has been facilitated by the fact that all relevant management is performed by one single body.

A number of obstacles to meaningful impact of PVAs can be identified. The lack of ownership of the PVA recommendations among stake-holders and management agencies is one problem (Seal and Wildt 1994). Another is the difficulty of communicating the concept of probability. The PVAs cannot deliver the truth about the future, only possible outcomes with related probabilities. This causes confusion and suspicion. The SEPA stated in their latest action plan that "there are however examples of small and isolated wolf populations that have survived in a manner that the geneticists had not expected based on their theoretical calculations". The reference was made to the Isle Royal population, which was founded by a single pair and still survives after 45 yr, as if that would not fit the predictions made by genetic models. In reality, the Isle Royal example fits very well within the predictions of both genetic and demographic stochastic modelling, as one possible outcome. That single observation cannot, however, be used to conclude that the Swedish wolf population will behave in a similar way, as there are other possible outcomes that may occur with higher probabilities.

Another problem is the failure in debates to separate the issue of viability in biological terms from the specific interests of different stake-holders. A recent article in a Swedish hunting magazine stated that "the biologists maintain that 500 wolves are needed for population viability, but that is wrong. The goal must obviously be to keep as few wolves as possible!". Of course, the actual policy adopted should certainly consider the interests of all stake-holders, and a PVA could analyze the consequences of following one or the other interest, but the analysis itself cannot consider policy or interests.

A fourth problem concerns the application of the precautionary principle. This principle has been forwarded by the Convention on Biological Diversity, and made policy in Sweden through government decisions. The scientists in the working group of the government investigation interpreted this as meaning that when a risk to biodiversity is identified, the erring should be on the safe side. For example, the VORTEX simulations gave the result that 200 wolves would be enough to avoid extinction from stochastic reasons, but still recommended that 400 wolves should be kept to avoid this risk. The reason behind this recommendation was uncertainties about the appropriateness of model structure and data input. In the process of performing PVAs, geneticists and ecologists are often pushed by policy-makers to make predictions that are as precise as possible, which means that too much confidence is put in very exact predictions, beyond what the precautionary principle would allow.

Acknowledgements – I am indebted to my coworkers in performing the PVA work, Marie Johnsson and Kjell Larsson. A large number of people have assisted in the work, among them I would especially like to thank Henrik Andrén, Anders Angerbjörn, Anders Bjärvall, Hans Ellegren, Mats Ericsson, Klas Hjelm, Erik Isakson, Ola Jennersten, Tommy Järås, Linda Laikre, Peter Lindberg, Lennart Nyman, Jens Persson, Nils Ryman, Bernt-Erik Sæther, Håkan Sand, Finn Sandegren, Thomas Sjöåsen, and Per Widén. Robert Lacy supplied the software VORTEX and support for it, as well as helpful discussions on the application of PVAs. The IUCN Conservation Breeding Specialist Group, chaired by Ulie Seal, has similarly contributed to my PVA work through seminars, workshops and discussions. I would also like to thank all participants of the PVA workshop in Stockholm, and its sponsors the Swedish Environmental Protection Agency and the Swedish Biodiversity Centre, and my co-organizer Per Sjögren-Gulve, without whose ardent work and patient support this article and report would not have been completed. I thank Lena Berg, Barry Brook and Robert Lacy for their constructive comments and helpful suggestions for improvement of the manuscript. The PVA work was financially supported by WWF Sweden, the SEPA, the Swedish Society for Nature Conservation, and the Swedish Biodiversity Centre.

References

Akçakaya, R. 2000. Population viability analyses with demographically and spatially structured models. – Ecol. Bull. 48: 23–38.

Akçakaya, R. and Sjögren-Gulve, P. 2000. Population viability analyses for conservation planning: an overview. – Ecol. Bull. 48: 9–21.

Andrén, H. et al. 1999. Rapport från Arbetsgruppen för Rovdjursutredningen. – In: Anon. (ed.), Bilagor till sammanhållen rovdjurspolitik. Statens offentliga utredningar 1999: 146, Miljödepartementet, Stockholm, pp. 65–95, in Swedish.

Angerbjörn, A., Ehrlén, J. and Isakson, E. 1999. Vargstammen i Sverige 1977–1997 – en simulering av populationens utveckling. – In: Ebenhard, T. and Höggren, M. (eds), Livskraftiga rovdjursstammar. CBM:s Rovdjursseminarium 12 oktober 1998. CBM:s Skriftserie 1, Centrum för Biologisk Mångfald, Uppsala, pp. 75–84, in Swedish.

Anon. 1995. Action plan on biological diversity. – Swedish Environ. Protection Agency Rep. 4567, Stockholm.

Anon. 1999. Sammanhållen rovdjurspolitik. – Statens offentliga utredningar 1999: 146, Miljödepartementet, Stockholm, in Swedish.

Anon. 2000. A coherent predator policy. Summary of the Carnivore Commission's final report. – Carnivore Commission, Stockholm, Sweden.

Berglind, S.-Å. 2000. Demography and management of relict sand lizard *Lacerta agilis* populations on the edge of extinction. – Ecol. Bull. 48: 123–142.

Brook, B. W. 2000. Pessimistic and optimistic bias in population viability analysis. – Conserv. Biol. 14: 564–566.

Brook, B. W. et al. 1999. Comparison of the population viability analysis packages GAPPS, INMAT, RAMAS and VORTEX for the whooping crane (*Grus americana*). – Anim. Conserv. 2: 23–31.

Brook, B. W. et al. 2000. Predictive accuracy of population viability analysis in conservation biology. – Nature 404: 385–387.

Bustamante, J. 1996. Population viability analysis of captive and released bearded vulture populations. – Conserv. Biol. 10: 822–831.

Bustamante, J. 1998. Use of simulation models to plan species reintroductions: the case of the bearded vulture in southern Spain. – Anim. Conserv. 1: 229–238.

Caughley, G. 1994. Directions in conservation biology. – J. Anim. Ecol. 63: 215–244.

de Jong, Y. A., van Olst, P. R. and de Jong, R. C. C. M. 1997. Feasibility of reintroduction of the Eurasian lynx (*Lynx lynx*) on "De Veluwe", The Netherlands, by using the stochastic simulation programme VORTEX. – Z. Säugetierk. 62 Suppl. II: 44–51.

Drechsler, M., Burgman, M. A. and Menkhorst, P. W. 1998. Uncertainty in population dynamics and ist consequences for the management of the orange-bellied parrot *Neophema chrysogaster*. – Biol. Conserv. 84: 269–281.

Ebenhard, T. 1992. Projekt pilgrimsfalk. En utvärdering. – Unpubl. report commissioned by WWF Sweden, Swedish Soc. for Nature Conserv. and Swedish Environ. Protection Agency, in Swedish.

Ellegren, H. 1999a. Inbreeding and relatedness in Scandinavian grey wolves *Canis lupus*. – Hereditas 130: 239–244.

Ellegren, H. 1999b. DNA-analyser av svenska vargar. – In: Ebenhard, T. and Höggren, M. (eds), Livskraftiga rovdjursstammar. CBM:s Rovdjursseminarium 12 oktober 1998. CBM:s Skriftserie 1, Centrum för Biologisk Mångfald, Uppsala, pp. 69–73, in Swedish.

Ellegren, H., Savolainen, P. and Rosén, B. 1996. The genetical history of an isolated population of the endangered grey wolf *Canis lupus*: a study of nuclear and mitochondrial polymorphisms. – Philos. Trans. R. Soc. Lond. B 351: 1661–1669.

Erlinge, S. 1972. The situation of the otter population in Sweden. – Viltrevy 8: 379–397.

Feingold, S. J. 1996. Monte Carlo simulation of Alaska wolf survival. – Physica A 231: 499–503.

Foley, P. 1994. Predicting extinction times from environmental stochasticity and carrying capacity. – Conserv. Biol. 8: 124–137.

Foose, T. J. et al. 1992. Kenya black rhino metapopulation workshop. Workshop report. – CBSG/SSC/IUCN and Kenya Wildl. Serv., Apple Valley, MN, USA.

Forbes, S. H. and Boyd, D. K. 1996. Genetic variation of naturally colonizing wolves in the Central Rocky Mountains. – Conserv. Biol. 10: 1082–1090.

Forbes, S. H. and Boyd, D. K. 1997. Genetic structure and migration in native and reintroduced Rocky Mountain wolf populations. – Conserv. Biol. 11: 1226–1234.

Forys, E. A. and Humphrey, S. R. 1999. Use of population viability analysis to evaluate management options for the endangered Lower Keys marsh rabbit. – J. Wildl. Manage. 63: 251–260.

Franklin, I. R. 1980. Evolutionary change in small populations. – In: Soulé, M. and Wilcox, B. (eds), Conservation biology: an evolutionary-ecological perspective. Sinauer, pp. 135–149.

Green, R. E., Pienkowski, M. W. and Love, J. A. 1996. Long-term viability of the re-introduced population of the white-tailed eagle Haliaeetus albicilla in Scotland. – J. Appl. Ecol. 33: 357–368.

Haber, G. C. 1996. Biological, conservation, and ethical implications of exploiting and controlling wolves. – Conserv. Biol. 10: 1068–1081.

Haight, R. G., Mladenoff, D. J. and Wydeven, A. P. 1998. Modeling disjunct gray wolf populations in semi-wild landscapes. – Conserv. Biol. 12: 879–888.

Hamilton, S. and Moller, H. 1995. Can PVA models using computer packages offer useful conservation advice? Sooty shearwaters Puffinus griseus in New Zealand as a case study. – Biol. Conserv. 73: 107–117.

Harcourt, A. H. 1995. Population viability estimates: theory and practice for a wild gorilla population. – Conserv. Biol. 9: 134–142.

Hedrick, P. W. 1994. Purging inbreeding depression and the probability of extinction: full-sib mating. – Heredity 73: 363–372.

Heggberget, T. M. 1991. Sex and age distribution in Eurasian otter (Lutra lutra) killed by human activity. – In: Reuther, C. and Röchert, R. (eds), Proc. V Int. Otter Colloquium, Hankensbüttel 1989, Habitat 6: 123–125.

Howells, O. and Edwards-Jones, G. 1997. A feasibility study of reintroducing wild boar Sus scrofa to Scotland: are existing woodlands large enough to support minimum viable populations? – Biol. Conserv. 81: 77–89.

Johnsson, M. and Ebenhard, T. 1996. Den skandinaviska vargpopulationen: en sårbarhetsanalys. – Världsnaturfondens rapportserie nr. 1, 1996, in Swedish.

Karlsson, J., Bjärvall, A. and Lundvall, A. 1999. Svenskarnas inställning till varg: en intervjuundersökning. – Swedish Environ. Protection Agency, Stockholm, Rep. 4933, in Swedish with English summary.

Kokko, H., Lindström, J. and Ranta, E. 1997. Risk analysis of hunting of seal populations in the Baltic. – Conserv. Biol. 11: 917–927.

Lacy, R. C. 1993. VORTEX: a computer simulation model for population viability analysis. – Wildl. Res. 20: 45–65.

Lacy, R. C. 2000a. Considering threats to the viability of small populations using individual-based models. – Ecol. Bull. 48: 39–51.

Lacy, R. C. 2000b. Structure of the VORTEX simulation model for population viability analysis. – Ecol. Bull. 48: 191–203.

Lacy, R. C., Hughes, K. A. and Kreeger, T. J. 1994. VORTEX. A stochastic simulation of the extinction process. Ver. 6. User's manual. – IUCN/SSC Conservation Breeding Specialist Group, Apple Valley, MN, USA.

Lacy, R. C., Hughes, K. A. and Miller, P. S. 1995. VORTEX: a stochastic simulation of the extinction process. Ver. 7. User's manual. – IUCN/SSC Conservation Breeding Specialist Group, Apple Valley, MN, USA.

Laikre, L. and Ryman, N. 1991. Inbreeding depression in a captive wolf (Canis lupus) population. – Conserv. Biol. 5: 33–40.

Lande, R. 1995. Mutation and conservation. – Conserv. Biol. 9: 782–791.

Larsson, K. and Ebenhard, T. 1994. Isolerade delpopulationer av utter: en sårbarhetsanalys. – Världsnaturfondens rapportserie, WWF, Stockholm, in Swedish.

Legendre, S. and Clobert, J. 1995. ULM, Unified Life Models, a software for conservation and evolutionary biologists. – J. Appl. Stat. 22: 817–834.

Lennartsson, T. 2000. Management and population viability of the pasture plant Gentianella campestris: the role of interactions between habitat factors. – Ecol. Bull. 48: 111–121.

Lindberg, P. 1996. Projekt pilgrimsfalk 1995. – Vår Fågelvärld Suppl. 25: 52–57, in Swedish.

Lindberg, P. 1997. Projekt pilgrimsfalk 1996. – Vår Fågelvärld Suppl. 27: 56–61, in Swedish.

Lindberg, P. 1998. Projekt pilgrimsfalk 1997. – Vår Fågelvärld Suppl. 30: 64–69, in Swedish.

Lindberg, P. 1999. Projekt pilgrimsfalk 1998. – Vår Fågelvärld Suppl. 32: 62–65, in Swedish.

Lindberg, P. and Eriksson, M. O. G. 1994. Åtgärdsprogram. Pilgrimsfalk (Falco peregrinus). – Swedish Environ. Protection Agency, Stockholm, in Swedish.

Lindberg, P., Schei, P. J. and Wikman, M. 1988. The peregrine falcon in Fennoscandia. – In: Cade, T. J. et al. (eds), Peregrine falcon populations. Their management and recovery. The Peregrine Fund, Boise, pp. 159–172.

Lindenmayer, D. B. and Possingham, H. P. 1996. Ranking conservation and timber management options for Leadbeater's possum in southeastern Australia using population viability analysis. – Conserv. Biol. 10: 235–251.

Marshall, K. and Edwards-Jones, G. 1998. Reintroducing capercaillie (Tetrao urogallus) into southern Scotland: identification of minimum viable populations at potential release sites. – Biodiv. Conserv. 7: 275–296.

Meir, E. and Fagan, W. F. 2000. Will observation error and biases ruin the use of simple extinction models? – Conserv. Biol. 14: 148–154.

Menges, E. 2000. Applications of population viability analyses in plant conservation. – Ecol. Bull. 48: 73–84.

Mills, L. S. et al. 1996. Factors leading to different viability predictions for a grizzly bear data set. – Conserv. Biol. 10: 863–873.

Morton, N. E., Crow, J. F. and Muller, H. J. 1956. An estimate of the mutational damage in man from data on consanguineous marriages. – Proc. Natl. Acad. Sci. USA 42: 855–863.

Novellie, P. A., Miller, P. S. and Lloyd, P. H. 1996. The use of VORTEX simulation models in a long term programme of re-introduction of an endangered large mammal, the Cape mountain zebra (Equus zebra zebra). – Acta Oecol. 17: 657–671.

Persson, J. and Sand, H. 1998. Vargen. Viltet, ekologin och människan. – Swedish Hunters Assoc., Uppsala, in Swedish.

Persson, J., Sand, H. and Wabakken, P. 1999. Biologiska karaktärer hos varg viktiga för beräkningar av livskraftig populationsstorlek. – In: Ebenhard, T. and Höggren, M. (eds), Livskraftiga rovdjursstammar. CBM:s Rovdjursseminarium 12 oktober 1998. CBM:s Skriftserie 1, Centrum för Biologisk Mångfald, Uppsala, pp. 55–68, in Swedish.

Pletscher, D. H. et al. 1997. Population dynamics of a recolonizing wolf population. – J. Wildl. Manage. 61: 459–465.

Ralls, K., Ballou, J. D. and Templeton, A. R. 1988. Estimates of lethal equivalents and the cost of inbreeding in mammals. – Conserv. Biol. 2: 185–193.

Randi, E. et al. 2000. Mitochondrial DNA variability in Italian and east European wolves: detecting the consequences of small population size and hybridization. – Conserv. Biol. 14: 464–473.

Reuther, C. 1991. Otters in captivity – A review with special reference to Lutra lutra. – In: Reuther, C. and Röchert, R. (eds), Proc. V Int. Otter Colloquium, Hankensbüttel 1989, Habitat 6: 269–307.

Saltz, D. 1996. Minimizing extinction probability due to demographic stochasticity in a reintroduced herd of Persian fallow deer Dama dama mesopotamica. – Biol. Conserv. 75: 27–33.

Saltz, D. 1998. A long-term systematic approach to planning reintroductions: the Persian fallow deer and the Arabian oryx in Israel. – Anim. Conserv. 1: 245–252.

Sarrazin, F. and Barbault, R. 1996. Reintroduction: challenges and lessons for basic ecology. – Trends Ecol. Syst. 11: 474–478.

Sarrazin, F. and Legendre, S. 2000. Demographic approach to releasing adults versus young in reintroductions. – Conserv. Biol. 14: 488–500.

Seal, U. S. and Wildt, D. E. (eds) 1994. Process design population and habitat viability assessment (PHVA) workshops. Working manual. – IUCN/SSC Conservation Breeding Specialist Group, Apple Valley, MN, USA.

Sidorovich, V. E. 1991. Structure, reproductive status and dynamics of the otter population in Byelorussia. – Acta Theriol. 36: 153–161.

Sjöåsen, T. 1996. Survivorship of captive-bred and wild-caught European otters Lutra lutra in Sweden. – Biol. Conserv. 76: 161–165.

Somers, M. J. 1997. The sustainability of harvesting a warthog population: assessment of management options using simulation modelling. – S. Afr. J. Wildl. Res. 27: 37–43.

Southgate, R. and Possingham, H. 1995. Modelling the reintroduction of the greater bilby Macrotis lagotis using the metapopulation model analysis of the likelihood of extinction (ALEX). – Biol. Conserv. 73: 151–160.

Swart, J., Lawes, M. J. and Perrin, M. R. 1993. A mathematical model to investigate the demographic viability of low-density samango monkey (Cercopithecus mitis) populations in Natal, South Africa. – Ecol. Modell. 70: 289–303.

Tabor, J. E. and Wight, H. M. 1977. Population status of river otter in western Oregon. – J. Wildl. Manage. 41: 692–699.

van Ewijk, K. Y., Knol, A. P. and de Jong, R. C. C. M. 1997. An otter PVA as a preparation of a reintroduction experiment in the Netherlands. – Z. Säugetierk. 62 Suppl. II: 238–242.

Vucetich, J. A., Peterson, R. O. and Waite, T. A. 1997. Effects of social structure and prey dynamics on extinction risk in gray wolves. – Conserv. Biol. 11: 957–965.

Weaver, J. L., Paquet, P. C. and Ruggiero, L. F. 1996. Resilience and conservation of large carnivores in the Rocky Mountains. – Conserv. Biol. 10: 964–976.

Wiegand, T. et al. 1998. Assessing the risk of extinction for the brown bear (Ursus arctos) in the Cordillera Cantabrica, Spain. – Ecol. Monogr. 68: 539–570.

Zhou, Z. and Pan, W. 1997. Analysis of the viability of a giant panda population. – J. Appl. Ecol. 34: 363–374.

Ecological Bulletins 48: 165–180. Copenhagen 2000

Incidence function modelling and conservation of the tree frog *Hyla arborea* in the Netherlands

Claire C. Vos, Cajo J. F. Ter Braak and Wim Nieuwenhuizen

Vos, C. C., Ter Braak, C. J. F. and Nieuwenhuizen, W. 2000. Incidence function modelling and conservation of the tree frog *Hyla arborea* in the Netherlands. – Ecol. Bull. 48: 165–180.

The incidence function model (IFM) is a spatially realistic metapopulation model, in which extinction and colonisation probabilities are estimated from empirical data. We tested whether an IFM can be applied to quantify spatial habitat requirements for the protection and long-term survival of the tree frog *Hyla arborea* in Zealand Flanders, using a time series of three consecutive annual distribution patterns and dispersal data. The tree frog is an endangered species in the Netherlands and its distribution pattern is highly fragmented. An extended IFM was developed in which habitat quality parameters were incorporated to estimate functions for colonisation and extinction. A new method to estimate colonisation and extinction functions was developed, using both the turnover events between years and the spatial distribution pattern of the first year of survey. A logistic regression analysis indicated that the extinction probability decreased with pond size and that colonisation probability was higher for large ponds with high connectivity. The observed dispersal distances in comparison with the distances between ponds showed that the habitat network was still connected by dispersing individuals. The extended IFM quite accurately predicted the distribution pattern within a few years. However long-term predictions of metapopulation persistence were probably optimistic. A three-year data set is too short to incorporate the total range of fluctuations of population numbers between years. In addition deterministic extinctions due to habitat destruction and the loss of suitable habitat caused by natural succession were not incorporated in the model. It is concluded that the model could be a useful tool for conservation assessment when comparing different nature restoration scenarios and to provide general guidelines for an optimal tree frog landscape.

C. C. Vos (c.c.vos@alterra.wag-ur.nl) and W. Nieuwenhuizen, Alterra, Green World Research, Dept of Landscape Ecology, P.O. Box 23, NL-6700 AA Wageningen, The Netherlands. – C. J. Ter Braak, Center for Biometry Wageningen, P.O. Box 16, NL- 6700 AA Wageningen, The Netherlands.

Species in fragmented habitats are often absent from apparently suitable locations. When these absences are correlated with landscape structure, i.e. small and isolated patches are unoccupied relatively often, the species is concluded to suffer from habitat fragmentation. Effects of habitat fragmentation have been demonstrated in empirical studies for a wide range of species and landscape combinations (for reviews, see Harrison 1991, 1994, Opdam 1991, Reich and Grimm 1996, Hanski and Gilpin 1997). Recommendations for the protection of species in fragmented habitats are often based on metapopulation concepts. Metapopulation theory implies that although small populations suffer from chance extinction due to demographic and environmental stochasticity (see Akçakaya

2000, Lacy 2000), species can survive at a regional level if local extinctions are compensated for by recolonisation (Levins 1970). For the conservation of a species it is important to know whether processes of extinction and recolonisation actually take place and, if so, at what time scale. As suitable time series of species distribution are scarce, relatively few empirical studies exist in which effects of habitat fragmentation on the processes of local extinction and recolonisation were analysed (for reviews, see Harrison 1991, 1994, Opdam 1991, Reich and Grimm 1996, Hanski and Gilpin 1997). For conservation purposes, we need models to understand the dynamics of metapopulations, in particular for long-term survival of species in fragmented habitats (Verboom et al. 1993). Many metapopulation models have been developed, based on empirical data, that link the landscape pattern to population viability (Lande 1987, Doak 1989, Verboom et al. 1991, Hanski 1994, Lindenmayer and Possingham 1995, Sjögren-Gulve and Ray 1996, Thomas and Hanski 1997, Sjögren-Gulve and Hanski 2000). The incidence function model (IFM; e.g. Hanski 1994, Ter Braak et al. 1998, Sjögren-Gulve and Hanski 2000) is an example of a patch occupancy model (Hanski and Simberloff 1997), a spatially realistic metapopulation model in which extinction and colonisation probabilities are estimated from empirical data.

The aim of this paper is to test whether an IFM can be a useful tool to quantify spatial habitat requirements for the protection and long-term survival of a species. In this case study we had access to a time series of three consecutive annual distribution patterns and dispersal data of the tree frog *Hyla arborea* in Zealand Flanders. One of the questions we wanted to answer was whether or not these data were sufficient to make reliable predictions about the future occurrence of the tree frog. This species is endangered in the Netherlands, and its abundance has declined strongly during the last decades. An earlier study (Vos and Stumpel 1996) demonstrated negative effects of habitat fragmentation on the distribution of the tree frog. In this study, we have analysed whether local extinctions and recolonisation are affected by habitat fragmentation and habitat quality factors. We also analysed over which distances dispersal took place and whether conspecific attraction played a role in the choice of immigration pond. Dispersal events between habitat patches are an indication that the present habitat network has not yet disintegrated. The occurrence of conspecific attraction can be a relevant factor in the metapopulation process, as it could influence colonisation and extinction probabilities. Next, we developed an IFM. The results of the extinction and colonisation analysis form the basis for the parameters that are incorporated into the model. To make the IFM more realistic, it was extended to include habitat quality variables that proved to be important in predicting the distribution pattern (Vos and Stumpel 1996) and extinction and colonisation probabilities (this paper).

Methods

Analysis of extinction and colonisation

The study area was 250 km^2 and is situated in the western part of Zealand Flanders (Fig. 1). The habitat of the tree frog in this mainly agricultural landscape consists of cattle drinking ponds and shrubs, bushes and vegetation of tall

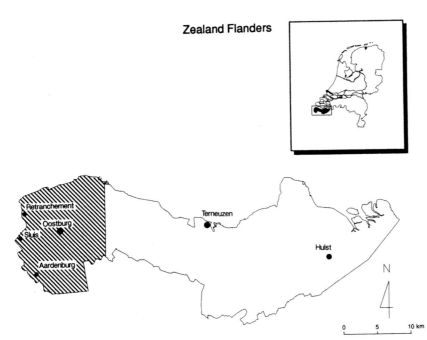

Zealand Flanders

Fig. 1. The position of the study area in The Netherlands.

herbs. The suitable habitat covers ca 1.5% of the total landscape. Census data of tree frog distribution were collected in three successive years 1981–1983 and in 1986. Tree frog presence was recorded by listening for calling males (spontaneously or in reaction to tape-recorded mating calls) and by searching for eggs, tadpoles, juveniles and adults in the pond and the pond surroundings (see Stumpel [1987] for detailed information). The ponds were visited at least three times a year during the reproduction period. In 1986, habitat quality parameters and pond area were measured in a subset of 198 ponds out of the ca 500 ponds in the study area. The subset contained all ponds where tree frogs were found. This subset was a representative sample of the variation in isolation in the study area (Vos and Stumpel 1996). In 1993 a survey of suitable terrestrial habitat, vegetation of high perennial herbs or with a developed shrub layer (Clausnitzer 1986, Stumpel 1993), was made. A control of a vegetation map of 1986 revealed that no substantial changes had taken place between 1993 and 1986, therefore the data were analysed together. GIS software (Arc-Info) was used to obtain spatial habitat parameters. The size of a suitable patch for a local population was determined by taking into account both the aquatic habitat part (the pond) and the terrestrial habitat part (the area of suitable terrestrial habitat in a radius of 250 m surrounding the pond). The suitable terrestrial habitat within a distance of 250–300 m from the reproduction site is regarded as the main terrestrial habitat of a local population (Blab 1986, Stumpel 1987, Tester 1990). For convenience, a patch, which is a combination of a pond and suitable terrestrial habitat, will be referred to subsequently as a "pond".

Connectivity of pond i was estimated by a weighted sum of distances to all occupied ponds in the surrounding landscape (Verboom et al. 1991, Hanski 1994, Ter Braak et al. 1998), taking the pattern of occupancy of the preceding year (Verboom et al. 1991). Three different connectivity measures were tested: S_i, S_i-area and S_i-area-barrier (Table 1). In the most simple connectivity measure, S_i is a weighted sum of pairwise distances between all ponds j and i, namely

$$S_i = \sum_{j \neq i} y_j \exp(-\alpha d_{ij}) \tag{1}$$

where y_j equals 1 for occupied and 0 for unoccupied (vacant) ponds and α is a constant setting the contribution of pond$_j$ given the distance d_{ij}, the distance between pond i and j. The contribution of ponds to the connectivity of pond i declines exponentially with distance (Hanski 1994). The rate of decline depends on α (Fig. 2). For the tree frog, an α of 2 was chosen, which corresponds to a maximum yearly dispersal distance of 2 km. The choice of 2 km was a practical one as the available digitised information allowed no larger distances. However a yearly dispersal distance of 2 km seems reasonable as it incorporates 80% of the observed dispersal distances (Fig. 4b).

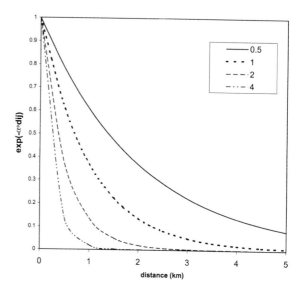

Fig. 2. The contribution of source ponds to the connectivity of a target pond given the distance (d_{ij}) between source and target pond (j and i, respectively) for different values of α (see eqs 1–3).

In the second connectivity measure, the size of the potential source pond j that may provide immigrants was taken into account:

$$S_i\text{-area} = \sum_{j \neq i} y_j \text{Area}_j^{0.5} \exp(-\alpha d_{ij}) \tag{2}$$

where the size of each source pond was incorporated by $\text{Area}_j^{0.5}$, the square root of the pond area and the amount of suitable terrestrial habitat in a radius of 250 m surrounding the pond. When the pond area was unknown, the median value (250 m²) was taken.

A negative effect of roads on amphibian survival and distribution has been demonstrated (Van Gelder 1973, Heine 1987, Fahrig et al. 1995, Vos and Chardon 1998, Hels 1999). To test the influence of roads as barriers for crossing individuals, a barrier effect was considered in the third connectivity variable:

$$S_i\text{-area-barrier} = \sum_{j \neq i} y_j \text{Area}_j^{0.5} B_{ij} \exp(-\alpha d_{ij}) \tag{3}$$

where B_{ij} equals 1 in cases with no roads in between ponds i and j, and 0.5 when a road was present. Thus the contribution to the connectivity of pond i by an occupied pond with a road in between is only 50% compared to if no road were present.

Presence-absence data of 1981–1983 were used for analysis of extinction and colonisation. An extinction of a local population was recorded when a pond was found occupied in a year t and unoccupied in year t+1. The reverse will be referred to as a colonisation. Extinction and coloni-

Table 1. Overview of the pond quality variables and habitat fragmentation variables. The mean value and the standard deviation (SD) are given for ponds that remained occupied (year t = 1; year t + 1 =1), went extinct (year t = 1; year t + 1 = 0), were colonised (year t = 0; year t + 1 = 1), or remained unoccupied (year t = 0; year t + 1 = 0) during two successive years in the survey period. Numbers are based on 197 transitions between 1981–1982 and 196 transitions between 1982–1983.

Variables	Description	Occupied ponds[•] (n=31) t=1/t+1=1 Mean±SD	Extinction ponds (n=10) t=1/t+1=0 Mean±SD	Colonised ponds (n=17) t=0/t+1=1 Mean±SD	Unoccupied ponds (n=335) t=0/t+1=0 Mean±SD
Cover_i	Coverage (%) of the pond surface by aquatic vegetation	48.2 ± 37.6 (48.3 ± 36.6)	49.1 ± 39.8	57.47 ± 41.3	21.20 ± 34.2
Water-Cond_i	Electrical conductivity of water (µS/cm) in pond i	59.7 ± 26.9 (56.0 ± 25.8)	80.9 ± 38.7	74.5 ± 41.7	98.1 ± 40.4
Area_i[†]	Pond area and area of suitable terrestrial habitat in a radius of 250 m surrounding pond i.	1.97 ± 1.80 (2.21 ± 1.92)	1.23 ± 1.13	1.48 ± 1.49	0.66 ± 1.3
Connectivity measures (S_i)	$S_i = \sum_j y_j A_j^b B_{ij} \exp(-2d_{ij})^{\circ}$ y_j = occupancy 0/1 of pond i; A_j^b = area pond j; B_{ij} = road barriers between pond i and j; d_{ij} = distance between pond i and j				
S_i[†]	Connectivity based on distance to all occupied ponds: $S_i = \sum_j y_j \exp(-2d_{ij})$	−0.37 ± 2.00 (0.49 ± 0.97)	0.27 ± 0.72	−0.77 ± 1.44	−2.53 ± 1.66
S_i-area[†]	Connectivity including size of sources: $S_i = \sum_j y_j A_j^{1/2} \exp(-2d_{ij})$	−0.39 ± 2.40 (0.65 ± 1.19)	0.46 ± 1.06	−0.92 ± 1.78	−2.95 ± 2.00
S_i-area-barrier[†]	Connectivity including size of sources and 50% road barriers: $S_i = \sum_j y_j A_j^{1/2} B_{ij} \exp(-2d_{ij})$	−0.73 ± 2.51 (0.33 ± 1.35)	0.11 ± 1.20	−1.13 ± 1.76	−3.27 ± 2.10

[•]In brackets values of 3 ponds were excluded (see text for explanation); [†]Because of skewed distribution log transformation was used; [°]cf. Verboom et al. 1991, Hanski 1994, Ter Braak et al. 1998.

sation were analysed with logistic regression (Jongman et al. 1995, Sjögren-Gulve and Ray 1996). Transitions between year 1981–1982 and 1982–1983 were analysed simultaneously (Verboom et al. 1991, Ter Braak et al. 1998). Possible differences between years in colonisation and extinction probability were tested by adding a factor "year of survey" to the regression model. Transitions were assumed to be independent under the condition that the two covariates "year of survey" and connectivity, which depends on the occupancy pattern of year 1981 or year 1982 for the first and the second transition respectively, were included in the logistic regression model. Two habitat quality variables were tested: the vegetation cover of the pond surface ($Cover_i$, Table 1) and electrical conductivity of the pond water (Water-$Cond_i$, Table 1). These variables had proved to be important in predicting the distribution pattern of the tree frog in an earlier study (Vos and Stumpel 1996). In the regression analysis, first the variable "year of survey" and the pond quality variables were entered in the model in a stepwise fashion. Subsequently the model was extended with fragmentation variables to test whether these variables contribute significantly to the regression model in addition to the quality variables (Van Apeldoorn et al. 1992a, Fahrig et al. 1995, Vos and Chardon 1998).

Analysis of dispersal data

Individuals were captured in the ponds and their direct surroundings from 1981 to 1989. During capture activities all sighted frogs were captured and marked individually (Stumpel 1987). Time between captures, the transition period, was recorded and the sex (male, female or juvenile) was registered. In the period 1981–1986 all ponds were visited at least twice, but especially ponds with large populations were visited more than ten times (Stumpel 1987). In the period 1987–1989, searches were incidental. A disperser was defined as an animal that was recaptured in a pond (immigration pond) other than the pond where it was first captured (emigration pond). An exception was made for animals that were recaptured in a different pond but had returned to their original pond at the next observation. This was interpreted to be exploratory behaviour rather than dispersal. Individuals that were captured more than once and had not moved to another pond were defined as non-dispersers or residents. To determine if conspecific attraction played a role in the choice of pond, immigration ponds were categorised depending on the probability that a pond was actually occupied when a disperser arrived. Immigration ponds were regarded as "occupied" when during the first observation of the dispersed animal also other individuals were present and more than five animals were recorded during the transition period. If that period was one day, only the number of animals during the first observation was taken into account. Ponds were regarded as "unoccupied" if the disperser was the only

individual present during the first observation and less than five observations were made during the transition period. All other ponds were categorised as "probably occupied". These ponds could have been occupied during first arrival but, if so, only by a small population. The possible role of conspecific attraction was analysed with an ordinal regression analysis, using the proportional-odds model (McCullagh and Nelder 1989). The category of occupancy of the immigration pond was the response variable and dispersal distance, transition period and sex were possible explanatory variables.

Possible difference in dispersal probability between males, females and juveniles was analysed with logistic regression. To correct for differences in (re)capture probability (calling males tend to have a higher capture probability), only animals that were captured more than once were taken into account.

The extended IFM

The IFM with rescue effect (Hanski 1994) was used to model the metapopulation dynamics. In this model, the occupancy of each patch (here called pond) is modelled by a discrete Markov chain with two states, occupied or vacant. The matrix of transition probabilities of this chain is determined by extinction and colonisation probability functions and varies in space (among ponds) and in time (among years). In the original IFM (Hanski 1994) the extinction probability of patch i (E_i) is a function of pond area ($Area_i$) only, namely

$$E_i = \min(\, e\, Area_i^{-x},\, 1) \qquad (4)$$

while the colonisation probability (C_i) depends only on the connectivity (S_i) of the pond by the function

$$C_i = 1\, /\, (1 + y\, /\, S_i^{\,z}) \qquad (5)$$

where e, x, y and z are unknown parameters. Our logistic regression of the extinction and colonisation events suggested however that the extinction and colonisation also depended on two habitat quality variables (Vos and Stumpel 1996; see results in this paper). Extinction was related to water conductivity (Water-$Cond_i$, H_1), and colonisation to percentage cover of the water vegetation of the pond ($Cover_i$, H_2). Therefore the model was extended to include these habitat variables by assuming that they modify the area and the connectivity that are effective in determining E_i and C_i respectively. Let $Area_{eff} = Area_i H_1^{p_1}$ and $S_{eff} = S_i H_2^{p_2}$, where p_1 and p_2 are additional unknown parameters. On inserting $Area_{eff}$ and S_{eff} for $Area_i$ and S_i in (4) and (5) and reparameterising to $q_1 = -xp_1$ and $q_2 = zp_2$ we obtain the extended incidence model

$$E_i = \min\, (e\, Area_i^{-x}\, H_1^{q_1},\, 1) \qquad (6)$$

and

$$C_i = 1 / (1 + y / S_i^z H_2^{q_2})$$ (7)

where e, x, y, z, q_1 and q_2 are unknown parameters that need to be estimated from the data.

With the rescue effect, immigrants increase the local population size, which is assumed to decrease the probability of extinction from E_i to $E_i (1-C_i)$. The resulting incidence probability is

$$J_i = \frac{C_i}{C_i + E_i(1 - C_i)}$$ (8)

(Hanski 1994). The unknown parameters of the extended IFM were estimated from the 1981–1983 data using a new method that extracts information from both the spatial occupancy pattern as well as the turnover events (colonisations and extinctions) between years. Our new method is based on the observation in Appendix B of Ter Braak et al. (1998) that the full likelihood of the data can be decomposed into three parts, namely 1) a spatial part, consisting of the occupancy pattern in the first year (1981), 2) an extinction part, consisting of the potential and realised extinctions in subsequent years (1981–1982 and 1982–1983) and 3) a colonisation part, consisting of the potential and realised colonisations in subsequent years (1981–1982 and 1982–1983).

Previous approaches used either only the spatial part (Hanski 1994) or the extinction and colonisation parts (Verboom et al. 1991, Sjögren-Gulve and Ray 1996). Our method extends Hanski's (1994) pseudo-likelihood, which is based on the spatial part only, with the likelihoods of the extinction and colonisation parts, thus combining the previous approaches. In accordance with Hanski (1994), a binary non-linear regression is used. The data set for the binary non-linear regression method thus consists of the following three parts. The spatial data set of presence/absence of the ponds in the first year (1981) is modelled by J_i (eq. 8). The extinction data set with potential and realised extinctions in subsequent years (1981–1982 and 1982–1883) is modelled by $E_i (1-C_i)$, as described by eq. (6). Finally, the colonisation data set with potential and realised colonisations in subsequent years is modelled by C_i (eq. 7). For the spatial data set the connectivity S_i of a pond is calculated from the occupancies of the remaining ponds in the same year (as in Hanski's method), whereas for the extinction and colonisation data sets the occupancies of the remaining ponds in the preceding year are used (as in Verboom et al. 1991, Van Apeldoorn et al. 1992b). The connectivity measure that we used in the incidence model calculations is S_i-area-barrier, as described earlier (eq. 3), because this measure takes the most elements into account that are considered to be of ecological relevance for the dispersal process (see discussion). The binary non-linear regression for the extended IFM was programmed using the procedure "Fitnonlinear" of Genstat 5 rel. 3.1 (Genstat 5

Committee 1993).

To test the predictive value of the IFM, the actual distribution pattern from 1986 and the distribution predicted by the model for year 1986 by eq. (8) were compared. To predict the viability of the metapopulation, the fraction of occupancy during the next 100 yr was simulated. For this purpose, all 500 ponds in the study area were taken into account. However, the habitat quality (Water-Cond$_i$ and Cover$_i$) of 60% of the ponds was unknown. Therefore, the missing values were assigned to ponds by the following method. If the values of both variables were missing, the values were randomly drawn from a bivariate lognormal distribution that was estimated from ponds without missing data (Rubin 1987, Bradley 1994). If only one of the values was missing, the missing value was drawn from an univariate lognormal distribution that was estimated from the non-missing data by regression on the other variable (Rubin 1987, Bradley 1994). Each simulation of 100 yr started with the occupancy pattern from 1986. In total, 1000 simulation runs were carried out.

Results

Extinction probability

During 1981–1983, ten local populations went extinct, while 31 ponds remained occupied during two successive years (Table 1, Fig. 3). Three ponds, indicated by arrows in Fig. 3, remained occupied during the survey period and had a high leverage value (0.26) in the regression analysis of extinction probability. In regression models including these three ponds, the extinction probability increased with high connectivity (S_i, S_i-area, S_i-area-barrier, Table 2a). The three ponds were already occupied in 1977, long before the survey period (Burny 1976). However, as there are no available sources in the present landscape within 2 km, the connectivity of these ponds is zero. This could indicate a time lag effect in which the occupancy is no longer correlated with the present landscape but to a landscape in the past (Tilman et al. 1994). As inclusion of an increasing extinction probability with high connectivity would cause very unstable and unrealistic simulation behaviour in the IFM, we decided to analyse extinction probability without these three ponds (Table 2b). In the first step of the logistic regression analysis, "year of survey" and the habitat quality variables were entered to the model. Only water conductivity added significantly to the model predicting a higher extinction probability with increasing conductivity (Water-Cond$_i$, Table 2b). In the second step, fragmentation variables were added to the model. Pond size was a significant explanatory variable, predicting a higher extinction probability for small ponds (Area$_i$, Table 2b). At this point, the positive effect of the connectivity variables on extinction probability had disappeared and the connectivity variables no longer contributed significantly to the model (S_i, S_i-area, S_i-area-barrier, Table 2b).

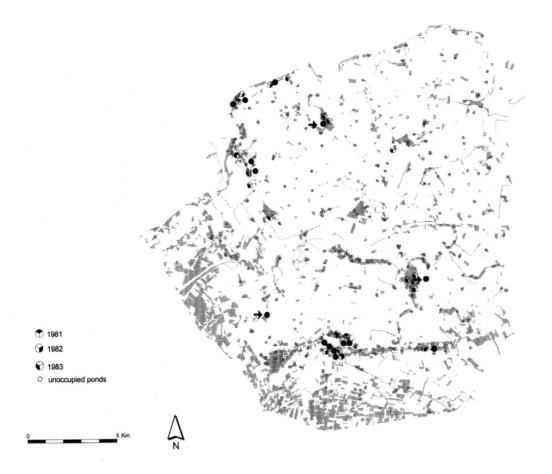

Fig. 3. Presence and absence of tree frogs in Zealand Flanders during 1981–1983. An extinction of a local population was recorded when a pond was occupied in year t and unoccupied in year t+1. A colonisation was recorded when a pond was unoccupied in year t and occupied in year t+1. The arrows indicate three ponds that were removed from the extinction analysis.

The legend in the figure reads:
- 1981
- 1982
- 1983
- ○ unoccupied ponds

0 — 5 Km

N

Colonisation probability

There were 17 ponds colonised, while 335 ponds remained unoccupied during two successive years (Table 1, Fig. 3). In the first step of the regression analysis, "year of survey" and percentage of water vegetation cover contributed significantly to the model (Year, $Cover_i$, Table 2c). The colonisation probability was higher from year 1982 to 1983 than from 1981 to 1982. A higher vegetation cover in the pond also increased the colonisation probability. In the second step, the fragmentation variables were added to the model. Pond size added significantly to the model, predicting higher colonisation probability for larger ponds ($Area_i$, Table 2c). All connectivity variables contributed significantly to the model, predicting a higher colonisation probability with increasing connectivity (S_i, S_i-area, S_i-area-barrier, Table 2c). Judged from their improvement χ^2 statistics, there was not much difference in explanatory power between them (χ^2 values were 7.56, 6.21 and 6.52, respectively).

Dispersal

In total, 89 dispersal events were registered involving 79 different tree frogs. Mean dispersal distance between ponds was 1469 m with a maximum distance of 12 570 m. Figure 4a gives an overview of the observed dispersal events. 2878 residents were registered involving 858 different tree frogs that were recaptured in the same pond. Thus 9% from the recaptured individuals dispersed successfully to another pond. Logistic regression analysis showed that males had a significantly higher probability to disperse (p < 0.001), whereas females had the lowest probability and juveniles were intermediate.

The analysis of the possible role of conspecific attraction on the choice of immigration ponds showed that 49 of the registered movements were directed towards occupied ponds, 34 towards probably occupied ponds and six movements towards probably unoccupied ponds (Fig. 4b). The ordinal regression analysis showed that dispersal distances towards occupied ponds were significantly larger

Table 2. Logistic regression models of extinction and colonisation probability. Sign, significance level of the last variable added to each model and proportion of variance explained (adjusted R^2) of the total model are listed. In the extinction analysis "minus three ponds" three ponds were removed from the data as explained in the text. Significance levels of the individual variables are from improvement χ^2 statistics * $p < 0.05$; ** $p < 0.01$; *** $p < 0.001$, ns not significant (for abbreviations see Table 1).

Variables of Multiple Logistic Regression	Sign and significance of the last variable	Adjusted R^2 of the total model (%)
2a Extinction analysis		
Year	ns	–
$Cover_i$	ns	–
$Water\text{-}Cond_i$	ns	–
$Area_i$	– *	10
$Area_i + S_i$	+ *	19
$Area_i + S_i\text{-area}$	+ *	20
$Area_i + S_i\text{-area-barrier}$	+ *	20
2b Extinction analysis minus three ponds		
Year	ns	–
$Cover_i$	ns	–
$Water\text{-}Cond_i$	+ *	11
$Water\text{-}Cond_i + Area_i$	– *	27
$Water\text{-}Cond_i + Area_i + S_i$	ns	–
$Water\text{-}Cond_i + Area_i + S_i\text{-area}$	ns	–
$Water\text{-}Cond_i + Area_i + S_i\text{-area-barrier}$	ns	–
2c Colonisation analysis		
Year	+ *	4
$Year + Water\text{-}Cond_i$	ns	–
$Year + Cover_i$	+ ***	15
$Year + Cover_i + Area_i$	+ **	21
$Year + Cover_i + Area_i + S_i$	+ **	27
$Year + Cover_i + Area_i + S_i\text{-area}$	+ *	26
$Year + Cover_i + Area_i + S_i\text{-area-barrier}$	+ *	26

than those towards vacant ponds, while those towards probably occupied ponds were intermediate ($p < 0.01$). As is illustrated in Fig. 4c the predicted fraction of dispersers that immigrate into an occupied pond increases with dispersal distance. Although females and juveniles never dispersed to vacant ponds, this was not significant. The transition period was positively correlated with dispersal distance ($r = 0.333$, $p < 0.01$) and added no further explanation to the model. Note that search intensity was higher in the large populations, which belonged to the category "occupied ponds". As higher search intensity will increase the probability of finding dispersed individuals, the delimitation between fractions (49, 34 and 6 dispersal events to occupied, probably occupied and unoccupied ponds, respectively) is probably biased towards the occupied category. However, this does not influence the result that the ratio between the three pond categories changed significantly with distance (Fig. 4c).

The extended IFM

Table 3 shows the parameter values of the extended IFM estimated from the 1981–1983 data by a non-linear binary regression. In Fig. 5 the relative effects of the habitat quality and fragmentation parameters on the colonisation and extinction probabilities are shown. Figure 5a illustrates how colonisation probability increases with increasing connectivity. High vegetation cover of the pond affected colonisation considerably. Although the logistic regression results indicated that colonisation probability also depended on size of the target pond, this was not included in the extended IFM. The reason is that in the IFM context the effect of pond area on colonisation cannot be distinguished from the effect on extinction (see eq. 8) (Ter Braak et al. 1998). The decrease in extinction probability with increasing pond size is illustrated in Fig. 5b. High water conductivity increased extinction probability greatly. Figure 5c shows the strength of the rescue effect included in the model. The extinction probability decreased only slightly with increasing pond connectivity.

The predictive power of the model was tested by com-a

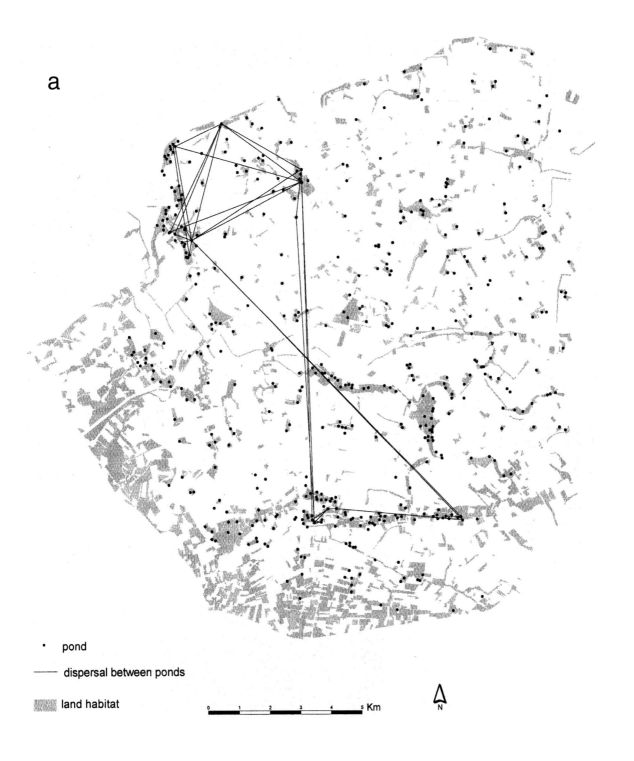

• pond

——— dispersal between ponds

▨ land habitat

0   1   2   3   4   5 Km

N

Fig. 4.

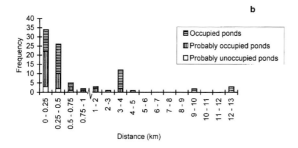

b

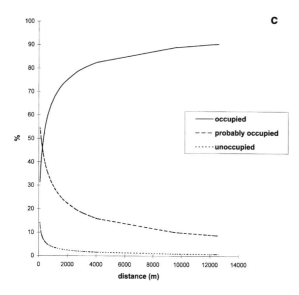

c

Table 3. Parameter estimates and standard error (SE) of the extended incidence function model estimated from the 1981–1983 data by binary non-linear regression (Hanski 1994, Ter Braak et al. 1998). See eqs 6 and 7 for explanation of the model parameters.

Model parameters	Estimate	SE
e	0.0086	0.0154
z	0.338	0.118
x	0.429	0.227
q_1	0.895	0.412
q_2	0.398	0.157

Fig. 4. a) Overview of the dispersal events registered by capture-recapture techniques in Zealand Flanders during the period 1981–1989. b) Frequency distribution of dispersal distances. The shading indicates the immigration pond type: occupied pond, probably occupied pond and probably unoccupied pond. c) Predicted fractions of pond types (already occupied, probably occupied or unoccupied when the immigrant arrived) that received an immigrant in relation to dispersal distance between ponds.

paring the occupancy probability predicted by the model with the actual distribution pattern in 1986, at both regional and individual pond levels. At the regional level the model was tested by comparing the predicted and observed occupancy in pond classes. On the horisontal axis in Fig. 6, the ponds are grouped into classes, based on their predicted occupancy probability. Note that the numbers of observations per class differ, with 34% of the ponds in the class with the lowest predicted occupancy. There was a significant correlation between the predicted and observed occupancy in 1986 (linear regression, $R^2 = 0.59$, $p < 0.05$). Thus the incidence model had a good fit at the regional level. At the individual pond level we carried out a misclassification analysis according to Sjögren-Gulve and Ray (1996). Figure 7 shows the proportion of simulations that each individual pond was occupied in 1986 (predicted oc-

cupancy) as compared to its observed status in that year. Twenty-one of the 167 ponds were misclassified (13%): 15 ponds were occupied in 1986 although the predicted occupancy was < 0.5, while six ponds were vacant although the predicted occupancy was > 0.5. Thus, the predictive power of the model was quite good also at the individual pond level. The good fit of predictions was not just due to that the distribution pattern of 1981 and 1986 largely remained the same. The number of occupied ponds increased from 22 in 1981 to 29 in 1986. Sixteen out of 22 of the ponds that were occupied in 1981 were still occupied in 1986, six ponds experienced extinction, while 13 ponds were colonised.

To test the long-term survival of the metapopulation, its dynamics during the next 100 yr were simulated in 1000 runs. The mean number of occupied ponds was predicted to increase in the first few years from 29 (the number of occupied ponds in year 1986) to 36, and stayed stable between 38 and 39 ponds until year 100. The metapopulation did not go extinct during the 100 yr.

Discussion

The results seem to justify a landscape or regional approach in conservation planning for the tree frog in Zealand Flanders and the application of metapopulation concepts for optimal spatial habitat configuration. Extinctions take place regularly and are influenced by spatial features of the landscape: extinction probability decreases with pond size. Vacant ponds are recolonised and this process is influenced by the configuration of habitat, with higher colonisation probability for large ponds with high connectivity. The observed dispersal distances in comparison with the distances between ponds and the observed recolonisations show that the habitat network is still connected by dispersing individuals. The extended IFM, parameterised with three years of occupancy data (1981–1983), quite accurately predicted the distribution pattern within a few years (1986). Although the model predicted that the metapopulation will still be viable within the next 100 yr, this prediction is uncertain as will be discussed below.

Are local extinction and colonisation related to landscape structure ?

Pond size, encompassing both pond area and suitable terrestrial habitat within a radius of 250 m around the pond, was negatively correlated with the extinction probability of a local population. This is in accordance with the metapopulation theory, if one assumes that small ponds can sustain only small populations that may go extinct by

chance processes of demographic stochasticity (Levins 1970, Richter-Dyn and Goel 1972). Generally the size of the reproduction site is thought to be a limiting factor in amphibian populations (Wilbur 1987, John-Alder and Morin 1990). However, in an earlier study (Vos and Stumpel 1996), patch size described solely by pond area only did not explain the tree frog's occupancy pattern. The extension of pond area with terrestrial habitat is likely to be a better approximation of patch size in this species, which has a pronounced terrestrial post-breeding period and for which suitable terrestrial habitat is scarce.

In many studies connectivity is found to be inversely related to extinction probability (Stacey et al. 1997); this is often attributed to the rescue effect (Brown and Kodric-Brown 1977). We found no significant effect of connectivity on the extinction probability. On the other hand, the dispersal data indicated a preference for dispersal to occupied ponds. We do not believe that the absence of a rescue effect in the extinction analysis can be attributed to the choice of radius of 2 km for the calculation of the connectivity measures. In a tree frog population in southern Sweden, a rescue effect was found using a connectivity measure with a much smaller range. The extinction probability decreased with the number of calling males within a radius of 500 m around the pond (Edenhamn 1996). Although longer dispersal distances were observed, the large majority of our recorded dispersal distances (80%) was within 2 km (Fig. 4b). Even if occupied ponds at longer distances would have been included, their contribution to the total connectivity would be small (Fig. 2). An important difference between the Swedish and Dutch situation might be that the metapopulation in Sweden is less fragmented and

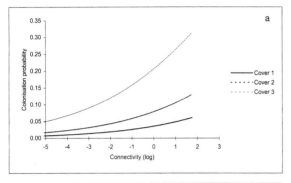

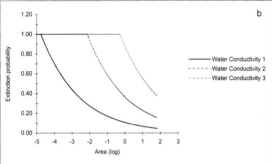

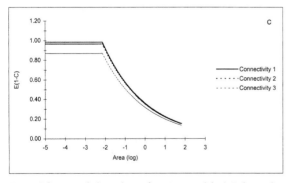

Fig. 5. The extended incidence function model. a) Relation between colonisation probability and connectivity with minimum (1), mean (2) and maximum (3) levels of vegetation cover of the pond. b) Relation between extinction probability and pond area with minimum (1), mean (2) and maximum (3) values of the water conductivity of the pond. c) Relation between extinction probability with rescue effect, $E(1-C)$, and pond area for ponds with minimum (1), mean (2) and maximum (3) connectivity. For the habitat quality variables Water conductivity$_i$ and Cover$_i$ the mean value was taken.

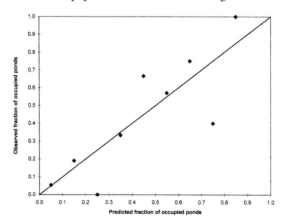

Fig. 6. Comparison of the actual occupation of the tree frog in Zealand Flanders in 1986 and the occupancy probability predicted by the incidence function model. The predicted occupancy probability was divided into classes. For each class the fraction of actually occupied ponds based on the distribution pattern of 1986 is given. The line indicates where perfect agreement between predicted and observed values occurs ($y = x$). Predicted and observed fractions of occupied ponds were significantly correlated ($R^2 = 0.59$, $p < 0.05$); the model explained 59% of the regional variation in pond occupancy.

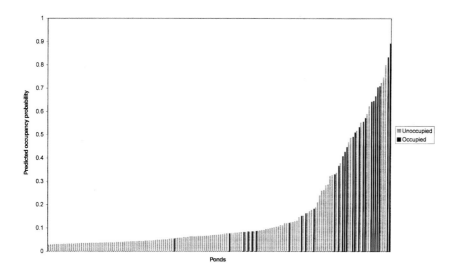

Fig. 7. Misclassification analysis at the individual pond level. The occupancy probability for each pond predicted by the incidence function model is indicated. Black bars denote occupied ponds and white bars denote unoccupied ponds according to the observations made in 1986.

much larger with 30% of 1500 ponds being occupied as compared to 10% of 500 ponds in Zealand Flanders (see Vos and Chardon [1998] for further comparison). In Zealand Flanders the number of immigrants might be too small to offset extinction of critically small populations.

The colonisation probability of a pond increased with connectivity. This is in accordance with the metapopulation theory, in which a higher number of potential sources of dispersers increases colonisation probability (Verboom et al. 1991, Hanski 1994). There was no significant difference between the three connectivity measures based on their model improvement statistics. Theoretically it is expected that a connectivity measure, which takes an area component of the potential dispersal source into account as well as a reduction of this source when roads have to be crossed, would give the most accurate estimation of connectivity. A negative effect of roads on amphibian density and occupancy of ponds has been demonstrated by Fahrig et al. (1995), Vos and Chardon (1998) and Lehtinen et al. (1999). We suspect that a measure that corrects for the complete matrix resistance between ponds would further improve the description of connectivity. For instance, corridors have been claimed to facilitate (dispersal) movements for several species, including amphibians (Forman and Godron 1986, Merriam 1991, Saunders and Hobbs 1991, Wiens 1997, Vos and Chardon 1997, Bennett 1999).

A positive effect of pond area and vegetation cover of the pond on colonisation probability was found. We interpret this as an indication that the larger the habitat patch, the higher the probability that the patch is detected by a dispersing animal. As the area of suitable terrestrial habitat within a radius of 250 m around the pond is an important component of the total size, it is plausible that dispersal to the pond is facilitated by these landscape elements as well (see also Vos and Stumpel 1996). But this does not explain why also the vegetation cover of the pond, a habitat quality

variable, increased the colonisation probability. If one assumes that better habitat quality will improve reproductive success, it might increase the probability of colonisation success of immigrants. However, it is unlikely that such differences already occur in the first season after colonisation. Increasing colonisation probability with pond size and vegetation cover could imply that dispersers choose to stay only when circumstances seem favourable and otherwise move on. This is supported by the recapture data where some individuals crossed large distances, even though many ponds were available at much shorter distances (Fig. 4a).

In the analyses every pond was regarded as a separate potential site for a local population. As was pointed out in Vos and Stumpel (1996) this simplification seems justified by the fact that even in 2/3 of the clusters of ponds, which were < 100 m apart, only one pond per cluster was occupied during the survey period. The delimitation of local populations, and hence local extinction and recolonisation, is to some extent arbitrary (Harrison and Taylor 1997). Not every presence of frogs in a pond that was unoccupied in the previous year will be a real colonisation, and not every absence at a pond will be a real extinction. There will be a continuum between distance and independent behaviour of local populations. In an extreme case of habitat fragmentation where several fragments are used by one individual, the so-called patchy population (Harrison 1994, Harrison and Taylor 1997), presence followed by absences may be the result of foraging behaviour or responses to conspecifics. However, the fact that 91% of the recaptured individuals remained resident at the site where they were first sighted does imply a reasonable independence between ponds. Also it is known that in some cases a tree frog population was found to occupy a single pond over several decades (Stumpel 1987). It is possible that pseudoturnover is scored among amphibian populations and other species that have a juvenile phase that exceeds the yearly census interval. One-year absence of calling individuals would erroneously be regarded as an extinction followed by a recolonisation if juveniles remain undetected in the pond surroundings. However, these erroneous classifications will not be correlated with landscape structure and will only weaken the effects of habitat fragmentation in explaining extinction and recolonisation events. Potential misclassifications could be avoided if one would regard only two-year absences as true extinctions. Obviously this would cause a reduction of the observed turnover events and might make an analysis based on a three-year data set statistically unfeasible.

Is the habitat network connected by dispersers?

The observed dispersal distances in relation to distances between ponds indicate that the populations are still interconnected by dispersal. The distances observed were large in comparison with the maximum distances of tree frogs mentioned by other authors: 3750 m (Clausnitzer and Clausnitzer 1984) and 5000 m (Fog 1993). The capture-recapture data showed that males and juveniles have a greater tendency than adult females to disperse. The distance covered was significantly larger for animals that had moved to occupied ponds as compared to those that moved to vacant ponds and probably occupied ponds, although many ponds were available at shorter ranges (Fig. 4a). This result implies that conspecific attraction plays a role in the choice of immigration pond. The probability to move on is higher when the immigration pond is unoccupied or probably occupied, resulting in larger distances covered before a disperser finally reaches an occupied pond. Sjögren-Gulve (1998) also found target-oriented dispersal, directed towards occupied ponds, in the pool frog *Rana lessonae*.

Conspecific attraction in metapopulations

In amphibians, male calls play an important role to advertise their presence to reproductive females over a long range (Forrest 1994, Cocroft and Ryan 1995). However, as most dispersed animals were males or juveniles (unspecified) it seems that also males are attracted to calls of conspecifics. Could this behaviour be interpreted in relation to the metapopulation structure of the species? Several benefits for conspecific attraction have been mentioned: increased mating success, predator protection, defence against intruders and information about habitat quality (Kiester 1979, Stamps 1988, Prins 1996). For anurans, increased mating success has been suggested as an explanation for the evolution of reproductive grouping. Aggregation of calling males would be advantageous because mating calls of a large chorus travel over a greater distance than

those of a small chorus (Otte 1974, Alexander 1975). There are some field observations showing that a large chorus attract more females per male (Ryan 1985, Wagner and Sullivan 1992), but in other studies no difference was found (Tejedo 1993, Schwartz 1994). The loud calls of the tree frog can be heard over a distance of 1000 m (by humans) and have an orientation function for dispersing individuals, as has been suggested for the chorus frog *Pseudacris triseriata* and the natterjack toad *Bufo calamita* (Sinsch 1990).

A second benefit of conspecific attraction that could play a role for tree frogs is that conspecifics are good indicators of habitat quality (Slatkin 1974, Alatalo et al. 1982, Stamps 1987,1988, Kiester and Shields et al. 1988). Because of natural succession, reproduction waters of amphibians will be suitable for only a limited period (Joly and Grolet 1996). In addition, the quality and location of the tree frog habitat are unpredictable because of environmental fluctuations. Species in such unpredictable habitat must have strong colonisation ability and the mechanism to find conspecifics must be well developed. The loud mating call of the tree frog could be such an adaptation to changing locations of suitable reproduction sites between years.

What might be the impact of conspecific attraction on the survival in fragmented habitats? Conspecific attraction will decrease the extinction probability of a local population, due to the rescue effect (Brown and Kodric-Brown 1977). As a result, empty ponds will receive fewer colonisers and will have a lower colonisation probability (Smith and Peacock 1990, Ray et al. 1991). Model simulations (Ray et al. 1991) predict a lowered level of occupancy within the metapopulation even with a small attraction value. Ray et al. (1991) pointed out that conspecific attraction may be a evolutionarily stable strategy, but it may also prove unstable in increasingly fragmented habitats.

The extended IFM

Two new components were introduced in the extended IFM. Habitat quality factors were incorporated to estimate colonisation and extinction functions. This will improve the quality of the of estimated functions for extinction and colonisation. As was shown in many empirical studies the distribution pattern of species can rarely be explained by habitat fragmentation only, but also depends on habitat quality factors (for reviews, see Harrison 1991, 1994, Opdam 1991, Reich and Grimm 1996, Hanski and Gilpin 1997). The second innovation was a new method to estimate extinction and recolonisation functions from consecutive annual distribution patterns. Previous approaches used either only the spatial part (Hanski 1994) or the extinction and colonisation parts (Verboom et al. 1991, Sjögren-Gulve and Ray 1996). We used both the turnover events between years as well as the spatial distribution pattern of the first year of survey. The advantage of this method is that the available information in the occupancy data is optimally applied. A comparison of parameter estimates based on either methods separately, reveals that our integrated method improves the quality of estimated functions by reducing standard errors (Ter Braak et al. unpubl.).

The fit between the predicted and the observed distribution pattern of 1986 was good, as is illustrated in Figs 6 and 7. This suggests that the metapopulation model describes relevant processes in the spatial population dynamics of the tree frog in Zealand Flanders, at least over a short time period. However, our data cannot substantiate how reliable the long-term predictions of metapopulation survival are. Some causes for local extinction were not taken into account, which may make model predictions optimistic. The extinction and colonisation probabilities are based on a three-year data set. These years are not necessarily representative years and definitely do not incorporate the total range of possible fluctuation of population numbers between years. Because strong fluctuation of population size will increase extinction probability (Gilpin and Soulé 1986), predictions based on relatively "good years" may be overoptimistic. A second complication is that local extinctions are not only due to stochastic processes in small populations. Extinction may also be deterministic, due to habitat destruction and the loss of suitable habitat caused by natural succession. The turnover of ponds is especially high in agricultural landscapes. These aspects form an extra extinction threat, which was not incorporated in the model. Longer time series (10–20 yr) would certainly improve the estimation of extinction and colonisation probabilities. For species that occur in dynamic habitat networks, the incorporation of habitat succession (Lindenmayer and Possingham 1995) and changes in habitat distribution over time would improve the predictive value of incidence models.

How can the presented IFM contribute to the protection of the species in the study area? The model could be a useful tool for conservation assessment when comparing different nature restoration scenarios. Although the exact long-term predictions have a high level of uncertainty, the qualitative results may be robust. Consequently the best alternative as predicted by the model is likely to be the best in reality, provided that the model captured the essential behaviour of species and landscape under study. As a second potential for application, the model can yield general guidelines for "an optimal tree frog landscape". Guidelines may be derived form the predicted impact of pond area, connectivity and habitat quality on extinction and colonisation rates (see Fig. 5). If the number of occupied ponds within the region increases, it is likely to enhance tree frog metapopulation stability and survival.

Acknowledgements – We thank Ton Stumpel for putting (unpubl.) tree frog data at our disposal. We thank Pim Arntzen, Rob Bugter, Paul Goedhardt, Paul Opdam, Herbert Prins, Ton Stumpel, Jana Verboom, Per Sjögren-Gulve and Torbjörn Ebenhard for critical comments on the manuscript.

References

Akçakaya, H. R. 2000. Population viability analyses with demographically and spatially structured models. – Ecol. Bull. 48: 23–38.

Alatalo, R. V., Lundberg, A. and Björklund, M. 1982. Can the song of male birds attract other males? An experiment with the pied flycatcher *Ficedula hypoleuca*. – Bird Behav. 4: 42–45.

Alexander, R. D. 1975. Natural selection and specialized chorusing behaviour in acoustical insects. – In: Pimentel, D. (ed.), Insects, science and society. Academic Press, pp. 35–77.

Bennett, A. F. 1999. Linkages in the landscape. – The IUCN Forest Conservation Programme. IUCN, Gland, Switzerland, Cambridge, U.K.

Blab, J. 1986. Biologie, Oekologie und Schutz von Amphibien. – Schriftenreihe für Landschaftspflege und Naturschutz 18, Kilda, Bonn.

Bradley, E. 1994. Missing data, imputation, and the bootstrap. – J. Am. Stat. Assoc. 89: 463–475.

Brown, J. H. and Kodric-Brown, A. 1977. Turnover rates in insular biogeography: effect of migration on extinction. – Ecology 58: 445–449.

Burny, J. M. 1976. The tree frog: a rare species of Zealand Flanders. – Zeeuws Nieuws over Natuur-Landschap-Milieu 2: 1–3, in Dutch.

Clausnitzer, H. J. 1986. Zur Oekologie und Ernährung des Laubfrosches *Hyla a. arborea* (L) im Sommerlebensraum. – Salamandra 22: 162-172.

Clausnitzer, C. and Clausnitzer, H. J. 1984. First results of a repopulation of the tree frog *Hyla arborea* in the district of Celle, Lower Saxony, West Germany. – Salamandra 20: 50–55.

Cocroft, R. B. and Ryan, M. J. 1995. Patterns of advertisement call evolution in toads and chorus frogs. – Anim. Behav. 49: 283–303.

Doak, D. 1989. Spotted owls and old growth logging in the pacific northwest. – Conserv. Biol. 3: 389–396.

Edenhamn, P. 1996. Spatial dynamics of the European tree frog in a heterogeneous landscape. – Ph.D. thesis, Swedish Univ. of Agricult. Sci., Uppsala.

Fahrig, L. et al. 1995. Effect of road traffic on amphibian density. – Biol. Conserv. 73: 177–182.

Fog, K. 1993. Migration in the tree frog *Hyla arborea*. – In: Stumpel, A. H. P. and Tester, U. (eds), Ecology and conservation of the European tree frog. Proc. first international workshop on *Hyla arborea*, 13–14 Feb. 1992, Potsdam, Germany. DLO Inst. for For. and Nat. Res., Wageningen, Schweizerische Bund für Naturschutz, Basel, pp. 55–64.

Forman, R. T. T. and Godron, M. 1986. Landscape ecology. – Wiley.

Forrest, T. G. 1994. From sender to receiver: propagation and environmental effects on acoustic signals. – Am. Zool. 34: 644–654.

Gilpin, M. E. and Soulé, M. E.1986. Minimum viable populations: processes of species extinction. – In: Soulé, M. E. (ed.), Conservation biology; the science of scarcity and diversity. Sinauer, pp. 19–34.

Hanski, I. 1994. A practical model of metapopulation dynamics. – J. Anim. Ecol. 63: 151–162.

Hanski, I. and Gilpin, M. E. (eds) 1997. Metapopulation biology: ecology, genetics, and evolution. – Academic Press.

Hanski, I. and Simberloff, D. 1997. The metapopulation approach, its history, conceptual domain, and application to conservation. – In: Hanski, I. and Gilpin, M. E. (eds), Metapopulation biology: ecology, genetics, and evolution. Academic Press, pp. 5–26.

Harrison, S. 1991. Local extinction in a metapopulation context – an empirical evaluation. – Biol. J. Linn. Soc. 42: 73–88.

Harrison, S. 1994. Metapopulations and conservation. – In: Edwards, P. J., Webb, N. R. and May, R. M. (eds), Large scale ecology and conservation biology. Blackwell, pp. 111–128.

Harrison, S. and Taylor, A. D. 1997. Empirical evidence for metapopulation dynamics. – In: Hanski, I. and Gilpin, M. E. (eds), Metapopulation biology; ecology, genetics, and evolution. Academic Press, pp. 27–43.

Heine, G. 1987. Einfache Mess- und Rechenmethode zur Ermittlung der Überlebenschance wandernder Amphibien beim Überqueren von Strassen. – In: Hoelzinger, J. and Schmid, G. (eds), Die Amphibien und Reptilien Baden-Württembergs. Naturschutz und Landschaftspflege Baden-Württembergs 41. Institut für Ökologie und Naturschutz, Karlsruhe, pp. 473–480.

Hels, T. 1999. Effects of roads on amphibian populations. – Ph.D. thesis, Univ. of Copenhagen, Min. of Environ. and Energy, Nat. Environ. Res. Inst.

John-Alder, H. B. and Morin, P. 1990. Effects of larval density on jumping ability and stamina in newly metamorphosed *Bufo woodhousii fowleri*. – Copeia 1990: 856–860.

Joly, P. and Grolet, O. 1996. Colonization dynamics of new ponds, age structure of colonizing Alpine newts, *Triturus alpestris*. – Acta Oecol. 17: 599–608.

Jongman, R. H. G., Ter Braak, C. J. F. and Van Tongeren, O. F. R. 1995. Data analysis in community and landscape ecology. – Cambridge Univ. Press.

Kiester, A. R. 1979. Conspecifics as cues: a mechanism for habitat selection in a Panamanian grass anole (*Anolis auratus*). – Behav. Ecol. Sociobiol. 5: 323–331.

Kiester, A. R. and Slatkin, M. 1974. A strategy of movement and resource utilization. – Theor. Popul. Biol. 6: 1–20.

Lacy, R. C. 2000. Considering threats to the viability of small populations using individual-based models. – Ecol. Bull. 48: 39–51.

Lande, R. 1987. Extinction thresholds in demographic models of territorial populations. – Am. Nat. 130: 624–635.

Lehtinen, R. M., Galatowitsch S. M. and Tester, J. R. 1999. Consequences of habitat loss and fragmentation for wetland amphibian assemblages. – Wetlands 19: 1–12.

Levins, R. 1970. Extinction. – In: Gerstenhaber, M. (ed.), Some mathematical questions in biology. Lectures on mathematics in life sciences. Vol. 2. Am. Math. Soc., Providence RI, pp. 77–107.

Lindenmayer, D. B. and Possingham, H. P. 1995. Modelling the viability of metapopulations of the endangered Leadbeaters's possum in south-eastern Australia. – Biodiv. Conserv. 4: 984–1018

McCullagh, P. and Nelder, J. A. 1989. Generalized linear models. – Chapman and Hall.

Merriam, G. 1991. Corridors and connectivity: animal populations in heterogeneous environments. – In: Saunders, D. A. and Hobbs, R. J. (eds), Nature conservation II: the role of corridors. Surrey Beatty, Chipping Norton: 133–142.

Opdam, P. 1991. Metapopulation theory and habitat fragmentation: a review of holarctic breeding bird studies. – Landscape Ecol. 5: 93–106.

Otte, D. 1974. Effects and functions in the evolution of signaling systems. – Annu. Rev. Ecol. Syst. 5: 385–417.

Prins, H. H. T. 1996. Ecology and behaviour of the African buffalo; social inequality and decision making. – Chapman and Hall.

Ray, C., Gilpin, M. and Smith, A. T. 1991. The effect of conspecific attraction on metapopulation dynamics. – Biol. J. Linn. Soc. 42: 123–134.

Reich, M. and Grimm, V. 1996. Das Metapopulationskonzept in Ökologie und Naturschutz: eine kritische Bestandsaufnahme. – Z. für Ökologie und Naturschutz 5: 123–139.

Richter-Dyn, N. and Goel, N. S. 1972. On the extinction of a colonizing species. – Theor. Popul. Biol. 3: 406–433.

Rubin, D. B. 1987. Multiple imputation for nonresponse in surveys. – Wiley.

Ryan, M. J. 1985. The Tungara frog. – Univ. of Chicago Press.

Saunders, D. A. and Hobbs, R. J. (eds) 1991. Nature conservation II: the role of corridors. – Surrey Beatty, Chipping Norton.

Schwartz, J. J. 1994. Male advertisement and female choice in frogs: recent findings and new approaches to the study of communication. – Am. Zool. 34: 616–624.

Shields, W. M. et al. 1988. Ideal free coloniality in swallows. – In: Slobodchikoff, C. N. (ed.), Ecology of social behavior. Academic Press, pp. 189–228.

Sinsch, U. 1990. Migration and orientation in anuran amphibians. – Ethol. Ecol. Evol. 2: 65–79.

Sjögren-Gulve, P. 1998. Spatial movement patterns in frogs: target-oriented dispersal in the pool frog, *Rana lessonae*. – Ecoscience 5: 31–38.

Sjögren-Gulve, P. and Ray, C. 1996. Using logistic regression to model metapopulation dynamics: large-scale forestry extirpates the pool frog. – In: McCullough, D. R. (ed.), Metapopulations and wildlife conservation. Island Press, Washington DC, pp. 111–138.

Sjögren-Gulve, P. and Hanski, I. 2000. Metapopulation viability analysis using occupancy models. – Ecol. Bull. 48: 53–71.

Smith, A. T. and Peacock, M. M. 1990. Conspecific attraction and the determination of metapopulation colonization rates. – Conserv. Biol. 4: 320–323.

Stacey, P. B., Johnson, V. A. and Taper, M. L. 1997. Migration within metapopulations: the impact upon local population dynamics. – In: Hanski, I. and Gilpin, M. E. (eds), Metapopulation biology; ecology, genetics, and evolution. Academic Press, pp. 267–292.

Stamps, J. A. 1987. Conspecifics as cues to territory quality: a preference of juvenile lizards (*Anolis aeneus*) for previously used territories. – Am. Nat. 129: 629–642.

Stamps, J. A. 1988. Conspecific attraction and aggregation in territorial species. – Am. Nat. 131: 329–347.

Stumpel, A. H. P. 1987. Distribution and present numbers of the tree frog *Hyla arborea* in Zealand Flanders, the Netherlands (Amphibia, Hylidae). – Bijd. Dierk. 57: 151-163.

Stumpel, A. H. P. 1993. The terrestrial habitat of *Hyla arborea*. – In: Stumpel, A. H. P. and Tester, U. (eds), Ecology and conservation of the European tree frog. Proc. first international workshop on *Hyla arborea*, 13–14 Feb. 1992, Potsdam, Germany. DLO Inst. for For. and Nat. Res., Wageningen, Schweizerische Bund für Naturschutz, Basel, pp. 47–54.

Tejedo, M. 1993. Do male natterjack toads join larger breeding choruses to increase mating success? – Copeia 1993: 75–80.

Ter Braak, C. J. F., Hanski, I. and Verboom, J. 1998. The incidence function approach to modelling of metapopulation dynamics. – In: Bascompte, J. and Solé, R. V. (eds), Modeling spatio-temporal dynamics in ecology. Springer, pp. 167–188.

Tester, U. 1990. Artenschutzerisch relevante Aspekte zur Ökologie des Laubfrosches (*Hyla arborea* L.). – Ph.D. thesis, Univ. of Basel.

Thomas, C. D. and Hanski, I. 1997. Butterfly metapopulations. – In: Hanski, I. and Gilpin, M. E. (eds), Metapopulation biology; ecology, genetics, and evolution. Academic Press, pp. 359–386.

Tilman, D. et al. 1994. Habitat destruction and the extinction debt. – Nature 371: 65–66.

Van Apeldoorn, R. C. et al. 1992a. Effects of habitat fragmentation on the bank vole, *Clethrionomys glareolus*, in an agricultural landscape. – Oikos 65: 265-274.

Van Apeldoorn, R. C., Celada, C. and Nieuwenhuizen, W. 1992b. Distribution and dynamics of the red squirrel (*Sciurus vulgaris* L.) in a landscape with fragmented habitat. – Landscape Ecol. 9: 227–235.

Van Gelder, J. J. 1973. A quantitative approach to the mortality resulting from traffic in a population of *Bufo bufo* L. – Oecologia 13: 93–95.

Verboom, J. et al. 1991. European nuthatch metapopulations in a fragmented agricultural landscape. – Oikos 61: 149-156.

Verboom, J., Metz, J. A. J. and Meelis, E. 1993. Metapopulation models for impact assessment of fragmentation. – In: Vos, C. C. and Opdam, P. (eds), Landscape ecology of a stressed environment. Chapman and Hall, pp. 172–191.

Vos, C. C. and Stumpel, A. H. P. 1996. Comparison of habitat-isolation parameters in relation to fragmented distribution patterns in the tree frog (*Hyla arborea*). – Landscape Ecol. 11: 203–214.

Vos, C. C. and Chardon, J. P. 1997. Landscape resistance and dispersal in fragmented populations: a case study of the tree frog (*Hyla arborea*) in an agricultural landscape. – In: Cooper, A. and Power, J. (eds), Species dispersal and land use processes. Proc. sixth annual IALE (U.K.) conference, Univ. of Ulster, Colleraine, 9–11 Sept. 1997, pp. 19–26.

Vos, C. C. and Chardon, J. P. 1998. Effects of habitat fragmentation and road density on the distribution pattern of the moor frog *Rana arvalis*. – J. Appl. Ecol. 35: 44–56.

Wagner, W. E. and Sullivan, B. K. 1992. Chorus organization in the Gulf Coast toad (*Bufo valliceps*): male and female behavior and the opportunity for sexual selection. – Copeia 1992: 647–658.

Wiens, J. A. 1997. Metapopulation dynamics and landscape ecology. – In: Hanski, I. and Gilpin, M. E. (eds), Metapopulation biology; ecology, genetics, and evolution. Academic Press, pp. 43–68.

Wilbur, H. M. 1987. Regulation of structure in complex systems: experimental temporary pond communities. – Ecology 68: 1437-1452.

Ecological Bulletins 48: 181–190. Copenhagen 2000

Population viability analysis in the classification of threatened species: problems and potentials

Ulf Gärdenfors

Gärdenfors, U. 2000. Population viability analysis in the classification of threatened species: problems and potentials. – Ecol. Bull. 48: 181–190.

The new Red List system from the World Conservation Union (IUCN) constitutes five sets of quantitative criteria (denoted A–E) for identifying threatened species. All criteria are based on quantitative thresholds. A–D focus on a few commonly known risk factors while only E requires a full analysis of extinction risk. To date, nearly all Red List assessments have been based on criteria A–D. Criterion E is seldomly used, partly because of assessors' unfamiliarity with existing population viability analysis (PVA) methods, but also because data are usually inadequate for such analyses. However, to more rigorously examine the causes of the limited application, I compared estimated risk levels using PVAs to levels estimated using other criteria. I assessed risk levels for Swedish animal and plant populations and also properties of these analysed species that may influence the listing process. With very few exceptions, the threat category suggested according to the PVAs (criterion E) indicated a lower risk level than the category met by the criteria A–D. I also found that PVAs were performed on a biased set of species, i.e., most often for vertebrates with small populations that were threatened by factors affecting individuals in a way that can be described by a pure stochastic model. In contrast, PVAs were rarely performed for plants or invertebrates, species with large but decreasing populations, or species affected by deterministically negative changes in habitat or other resources. Additional reasons for the dominance of criteria A–D in the Red Lists may be a lack of congruence between the extinction risks expressed by the different criteria A–E. For instance, as a consequence of criterion E operating with fixed time windows, the threat category met may represent a too conservative view in cases where the PVA extinction risk trajectory has a sigmoid shape over time.

U. Gärdenfors (ulf.gardenfors@dha.slu.se), Swedish Threatened Species Unit, Swedish Univ. of Agricultural Sciences, Box 7007, SE-750 07 Uppsala, Sweden.

In Red List context threatened species have over approximately the past 30 yr been identified and classified according to qualitative definitions, like "Taxa in danger of extinction and whose survival is unlikely if the causal factors continue operating" (Endangered). Such definitions worked as long as a limited number of people made the assessments of putative threatened species. However, as the number of assessors grew, as more and more countries compiled Red Lists and Red Data Books for an increasing variety of organisms, the interpretation of the qualitative definitions diverged (Mace and Collar 1994).

Recognising the unreliable comparability between different Red Lists, the World Conservation Union (IUCN) initiated a review process of the Red List Categories. In a first proposal (Mace and Lande 1991) category definitions were explicit, quantitatively defined, extinction risks. After several workshops and consultations, and a series of evolving criteria, a final set of quantitative definitions was adopted in November 1994 (IUCN 1994; also available in the world-wide web: http://iucn.org/themes/ssc/redlists/ssc-rl-c.htm). Threatened species could be classified into three categories – Critically Endangered (CR), Endangered (EN) and Vulnerable (VU) – if they qualify for any one of five sets of criteria, denoted A–E. Only criterion E is defined by a probability of extinction within a specified timeframe. For CR criterion E reads: "Quantitative analysis showing the probability of extinction in the wild is at least 50% within 10 yr or 3 generations, whichever is the longer". The corresponding threshold for EN is 20% within 20 yr or 5 generations and for VU 10% within 100 yr. In addition to these categories classifying threatened species, the new system encompasses two categories for extinct species (Extinct – EX and Extinct in the wild – EW), three for species facing lower risk (LR: Conservation Dependent, Near Threatened and Least Concern [the criteria have been the subject of review and the complex category Lower Risk will be replaced from 2001 by the two categories Near Threatened (NT) and Least Concern (LC)(IUCN 2000)]), one category called Data Deficient (DD), and a final category called Not Evaluated (NE). The category Near Threatened (LR:nt) is defined as: "Taxa which do not qualify for Conservation Dependent, but which are close to qualifying for Vulnerable". Species classified in the categories EX, EW, CR, EN, VU, LR:cd and LR:nt are collectively denoted red-listed.

Quantitative analysis according to the criterion E usually means a population viability analysis (PVA), but may in effect be any kind of analysis quantifying the extinction risk. For instance, if an analysis of meteorological data shows that there is a 20% risk that a small forest, harbouring the only remaining population of a species, will be destroyed by a hurricane within the next 100 yr, the species could be classified as VU according to criterion E. In this paper, however, I always refer to a PVA when talking about criterion E.

Criteria A–D also have quantitative thresholds, but do not require a quantitative analysis. Conceptually, they can be viewed as rules of thumb that were set to roughly correspond to extinction risks as defined in criterion E, based on a limited set of information. Criterion A is based on rate of population decrease (observed or estimated, inferred or suspected). Criterion B is based on a small distribution area in conjunction with fragmented occurrence, population decline and/or fluctuation. Criterion C is based on a small declining population, and finally D is based solely on a very small population. Empirically, as well as theoretically, all these cases are prone to extinction. None of the criteria A–E have predominance over any of the others: the category with the highest risk level should be adopted. For example, if available information meets criterion A at the VU level, criterion C at the EN level and criterion E at the VU level, the species should be classified as Endangered.

In practice, the vast majority of the Red List assessments apply criteria A–D, while very few seem to utilise

the criterion E option (cf. Baillie and Groombridge 1996, Oldfield et al. 1998, Gärdenfors 2000). Does the general lack of application of criterion E in the Red List assessment work result from unfamiliarity of the assessors with existing quantitative risk analysis methods, from a real or perceived deficiency in the data, or are there other reasons as well?

I examined some possibilities and problems with applying quantitative analyses in the Red List classification procedure. I compared estimated threat categories according to PVAs of potentially threatened species that recently were conducted in Sweden to estimated threat categories using criteria A–D. I also examined whether PVAs were conducted on a biased set of species, either because PVAs are performed primarily on certain taxa or because there is a bias towards taxa being the target of certain kinds of threat. For the latter, I compared the most common factors threatening the persistence of "PVA-species" to the incidence of threat factors of all red-listed species in Sweden.

Another, quite different, problem may arise when restricting the risk analysis to a subpopulation of a country or a region: if a geo-political border divides a population, a risk analysis of merely the part of the population occurring within the country may quite obviously lead to an erroneous result (in most cases an overestimation of the risk). This issue was not further explored in this study, but has been discussed by Gärdenfors (1995, 1996), Gärdenfors and Kindvall (1999), and Gärdenfors et al. (1999).

Methods

Comparison of outcome of criteria A–D and criterion E

I gathered reports from PVAs of Swedish (Scandinavian) populations of potentially threatened species conducted during the last 5 yr. The objective of some of the PVAs was to conduct a sensitivity analysis, so as to evaluate the effectivity of different assumptions or conservation options rather than to assess the risk of extinction within a specified timeframe. In some analyses the extinction risk was not given at the timeframes used by IUCN (10 yr/3 generations, 20 yr/5 generations and 100 yr). Where I considered it possible, I interpreted or extrapolated their results at the IUCN timeframes (Table 1). In a couple of cases this was not possible and, consequently, I could not include them in my evaluation.

For the same taxa, I also checked which Red List categories were met when the assessment was based on criteria A–D (Gärdenfors 2000). Further, I compared the distribution of incidences of met criteria (A–E) of those vertebrate species having been the target of a PVA, with the corresponding distribution within 1) all threatened (CR, EN or VU) vertebrate species, and 2) all threatened species of all evaluated groups, identified by any of the five criteria (Table 2; Gärdenfors 2000).

Risk factors

Aiming at a characterisation of the threatened species, I also examined the distribution of incidences of the factors threatening the persistence, according to a classification made earlier at the Swedish Threatened Species Unit (Table 3). Because the current Red List (Gärdenfors 2000) was published very recently no analysis of risk or threat factors has yet been done for newly added species. Consequently, the analysis of threat factor was based on those 3105 species that were included both in the former Red Lists (Ehnström et al. 1993, Aronsson et al. 1995, Ahlén and Tjernberg 1996) and in the current: from this group, I compared factors threatening the persistence of 1) the red-listed species of all organism groups (n = 3105), 2) the subset of red-listed vertebrates except fishes (n = 99), and 3) those taxa with a PVA estimate of risk (n = 15). The classification was based on a list of 93 identified potential threat factors, of which 36 important factors are listed in Table 3. Each species could be assigned a maximum of 7 factors. In the original classification, a distinction was made between 1) important and active, and 2) marginal or potential future threats. That distinction was made in Table 1 (shown as a numeral suffix, e.g., Fx1, Fj2), but in the statistics presented in Table 3 the two categories of threat factors were treated together in order to reduce the number of numerals.

Results

Statistics

Eleven (73%) of the 15 conducted PVAs assessed vertebrates, two (13%) insects and two (13%) vascular plants. This may be compared with the Swedish Red List (Gärdenfors 2000) which lists 4120 species (of which 1953 in the categories Critically Endangered, Endangered and Vulnerable). The taxonomic break-down of the listed species is: 156 (4%) vertebrates, 2037 (49%) insects, 505 (12%) vascular plants, and 1422 (35%) other animal, plant and fungi groups.

Comparison of outcome of criteria A–D and criterion E

With two, or possibly three, exceptions the conducted PVAs, evaluated against criterion E, resulted in a Red List category with a lower risk level than the assessments using criteria A–D (Table 1). The wolverine *Gulo gulo* was estimated a 20% probability of extinction within 30 yr (generation length 6 yr, corresponding to EN) assuming the mortality rate observed in zoos. If mortality was assumed to be as high as observed in North American populations being subjected to hunting, the extinction risk well exceeds

Critically Endangered, i.e., a higher category than suggested by the criteria A–D (EN). A dunlin subspecies (*Calidris alpina* ssp. *schinzii*) has a probability of extinction of >10% within 100 yr if one assumes high synchrony between subpopulations, which corresponds to the same threat category (VU) as do the criteria A–D. The PVA of the white-backed woodpecker *Dendrocopos leucotos* was made five years ago, and since then the population in Sweden has decreased further. A new PVA would possibly meet criterion E for CR.

In at least a couple of the analyses, such as *Euphydryas maturna* (Fig. 1), the risk of extinction was relatively low within a short time frame (10–20 yr), but very high within 50–100 yr. Because the thresholds of criterion E operate with fixed timeframes that are 10 yr (or 3 gen.) and 20 yr (or 5 gen.) for CR and EN, respectively, *Euphydryas maturna* will – under criterion E – (only) be classified as Vulnerable.

The majority of the PVAs used demographic data (VORTEX – Lacy et al. 1995, Lacy 2000; RAMAS – Akçakaya 1997, 2000) or occupancy data (Metapop II and III – Ray et. al 1994, Sjögren-Gulve and Hanski 2000; The Incidence Function Model – Hanski et al. 1996, Hanski 1999, Sjögren-Gulve and Hanski 2000). Several simulations used several different assumptions for parameter values, either because of uncertainty about the correct level or to conduct a sensitivity analysis. With the exception of one of the two alternative analyses of the *Euphydryas maturna*, all of the analyses assumed independent variables (such as birth rate, survival, carrying capacity or number of available patches) were constant through time.

Of the 10 vertebrate species with a PVA at the national level (Table 1), the actual listing was based on criterion D (very small population) in all 10 cases (100%), and in one case also on criterion C (small population and decline) (Gärdenfors 2000). In no case was criterion A (declining population) or criterion B (small distribution and decline) the basis of listing. This contrasts with the distribution among all vertebrate species meeting any of the criteria A–E, where 31% of listings were based on criterion A and 65% on criterion D, and particularly with all threatened species from all evaluated groups were 31% were based on criterion A and only 35% on criterion D (Table 2).

Risk factors

The incidence of factors that actually or potentially threaten species differed quite substantially between the examined groups (Table 3). Of the entire group of 3105 examined red-listed species in Sweden, the most common threat factor was forest clear-cutting (affecting 44% of the species), followed by several other factors pertaining to a changing structure of old forests, cessation of grazing or afforestation of fields and stochastic processes. Among all red-listed vertebrate species the most common threat factor

Table 1. Populations of species in Sweden that recently have been the subject of PVAs, examining extinction risk, and the Red List category and criteria of these taxa in Sweden (Gärdenfors 2000). The table also shows the main risk or threat factors to these taxa according to a classification by the Swedish Threatened Species Unit (unpubl.); cf. Table 3 for explanation of abbreviations. The risk factors were classified into two classes: 1) important and active, and 2) marginal or potential future threats.

Species	Category according to criteria A–D[1]	Category according to criterion E (PVA)	Software/method[2]	Main risk factors	Reference
Canis lupus (wolf)	CR D	LR:lc/LR:nt[3] LR:lc (EN)[4]	VORTEX; Demographic, stage structured model	Fx1, Ft1, Fd1, Fj2, Z2	Ebenhard in Persson and Sand (1998), Ebenhard (2000) Angerbjörn et al. (1999)
Alopex lagopus (arctic fox)	EN D	VU	VORTEX; RAMAS	Fx1, Ft1, Oa2, Ok2	Guillou (1997)
Lynx lynx (lynx)	VU D1	LR:lc; LR:lc (VU)[5]	VORTEX; VORTEX	Fx1, Ft1, Fj2, Oa2	Andrén and Liberg (1999), Kontio (1998)
Ursus arctos (brown bear)	VU D1	LR:lc	Diffusion process; VORTEX	Fx1, Fj2, As2	Sæther et al. (1998), Lundblad (1995)
Gulo gulo (wolverine)	EN D	EN (CR)[6]	VORTEX	Fx1, Ft1, Ff2, Z2	Kindvall (1998)
Anser erythropus (lesser white-fronted goose)	CR D	EN	VORTEX	U1, Ff1, Op1, Fx2, Fi2, Vö2	Guillou unpubl.
Falco peregrinus (peregrine falcon)	VU[7] D	LR:lc[8]	VORTEX	Kp1, Kk1, Km1, Fx1, Fi1, Ff1, Fk2	Ebenhard (2000)
Dendrocopos leucotos (white-backed woodpecker)	CR C1+C2ab, D	EN[9]	VORTEX	Ad1, So1, Sb1, As1, Al1, Ah1, Vd1	Carlson and Stenberg (1995)
Calidris alpina ssp. schinzii (dunlin)	VU D1	VU/LR:nt[10]	RAMAS/GIS	Öb1, Vm1, Ff1, Op1, Ös2, Fb2, Vd2	Kindvall (1998)
Rana lessonae (pool frog)	VU D2	LR:lc/CR[11]	Metapop III	Vd1, Mt1, Op2	Sjögren-Gulve and Ray (1996), Sjögren-Gulve and Hanski (2000)
Rana esculenta (east Skåne, local pop.) (edible frog)	(CR C1)[12]	EN	Metapop III	—[13]	Kindvall (1998), Sjögren-Gulve and Hanski (2000)
Euphydryas maturna (scarce fritillary)	CR A1c, B1+2abcd	VU	Incidence function model; RAMAS/GIS RAMAS/GIS	Sp1, As1, Bk2, So2, Ar2, Öb2, Z2	Kindvall (unpubl.). Data from Eliasson (1994 and unpubl.).
Metrioptera bicolor (two-coloured bush-cricket)	VU D2	LR:lc LR:lc LR:lc	Metapop III Incidence function model	Sp1, Bl1	Kindvall (1998, 2000)

Table 1 .Cont.

Species	Category according to criteria A–D[1]	Category according to criterion E (PVA)	Software/method[2]	Main risk factors	Reference
Euphrasia rostkoviana ssp. *rostkoviana* (large eyebright)	EN B1+2cde+3d	VU	RAMAS/GIS	Sp1, Öb1, Ös2, Kg1	Kindvall (1998)
Gentiana campestris (Södermanland, local pop.) (field gentian)	(CR B1+3d)[14]	VU	Transition matrix model	Öb1, Kg1, Kp2	Lennartsson (2000 and unpubl.)

[1] The letters and figures following the threat category denote the criteria and subcriteria met according to the IUCN Red List system (IUCN 1994).

[2] Models are further described in Akçakaya (1997, 2000; RAMAS), Lacy (2000; VORTEX), Sjögren-Gulve and Hanski (2000; Logistic regression model of Metapop III, and the Incidence Function Model) and Hanski (1999; Incidence Function Model).

[3] Depending on assumed effect of inbreeding.

[4] One analysis, assuming a yearly kill by 40% of reproducing individuals, resulted in a high probability of extinction within 35 yr (5 gen.) corresponding to EN.

[5] An increased mortality by 3–6% would result in the population being Vulnerable.

[6] Depending on assumed mortality. If level as observed in Swedish zoos: EN (generation length 6 yr); if level as observed in wild populartions in North America: CR.

[7] Number of mature individuals in Sweden amount to <250 individuals, corresponding to EN D, but because the Swedish population is not isolated the threat category is downgraded to VU (cf. Gärdenfors et al. 1999).

[8] Analysis made in 1991 on different subpopulations in northern Europe, assuming a certain specified migration. Because a low risk of extinction was found for some of the subpopulations and the populations has continued to grow during the 1990s the analysis implies a Red List category corresponding to LR:lc.

[9] Since the analysis was made the population has declined further, thus, it cannot be excluded that a new analysis would result in CR.

[10] Depending on degree of synchrony assumed between subpopulations.

[11] If environment remains unaltered there is no measurable extinction risk, but if the entire area is affected by modern forestry (which currently is improbable) from yr 0 of the analysis the extinction risk will correspond to CR.

[12] This analysis was made on an isolated population in SE Sweden with the data available in 1979 (tested against the A–E criteria). The population went extinct ca 1990. Because it did not comprise the entire Swedish population there is no official Red List category.

[13] The species is not red-listed on a national level, consequently the Swedish Threatened Species Unit has not evaluated what factors might have threatened the persistence of the species.

[14] This analysis was made on an isolated population in Södermanland, E Sweden. Because it did not comprise the entire Swedish population there is no official Red List category. The threat factors refer to the entire Swedish population.

Table 2. Incidence (%, and within parenthesis numbers) of met Red List criteria A–E among 1) all taxa in Sweden meeting any criterion (1969, including 16 that were downgraded to NT, of 20 000 assessed), 2) all vertebrate (except fish) taxa meeting any criterion (83 of 330 assessed), and 3) the subset of those vertebrates having been the subject of a quantitative analysis on the national level (n = 10, see Table 1). Because the IUCN system requires that the highest category met, according to any of the A–E criteria, should be adopted, there are only two taxa (*Gulo gulo* and *Calidris alpina* ssp. *schinzii*, cf. Table 1) where criterion E is met.

	1. All threatened species in Sweden (n=1969)	2. All threatened vertebrates (excl. fishes) (n=83)	3. Vertebrates with PVA on national level (n=10)
A) decreasing population	31 (620)	31 (26)	0 (0)
B) small distribution and decrease	57 (1115)	10 (8)	0 (0)
C) small population and decrease	16 (320)	24 (20)	10 (1)
D) very small population	35 (690)	65 (54)	100 (10)
E) quantitative analysis	0.1 (2)	2 (2)	20 (2)

was ditching and draining of habitats, followed by various human disturbances, decreased grazing or afforestation of fields, pesticides and other toxins, and forest clear-cutting.

Among those 15 taxa (Table 1) for which a quantitative analysis of extinction risk has been conducted, two were local populations in Sweden, i.e., the extinction risk was not assessed for the entire part of the taxon occurring in Sweden. However, because the field gentian *Gentiana campestris* is red-listed as a whole in the country, the factors perceivably threatening that species can be found in the classification made by the Swedish Threatened Species Unit (unpubl.). Among these 14 taxa, the dominant risk factor is illegal hunting, affecting 50% of the species, compared with 16% and 0.5% of all red-listed vertebrates and all red-listed species, respectively. Around 30% of these 14 species were also considered to be affected by decreased grazing, various human disturbances and extreme weather conditions. About 20% were classified as affected by other stochastic processes, afforesting fields, ditching and draining, predation, and legal hunting.

Discussion

The objectives and presentation of results of population viability analyses vary greatly. PVAs can be used to evaluate the potential outcome of different management scenarios or to identify which stages during the life-cycle are most vulnerable to perturbations (sensitivity analyses) (Akçakaya 2000, Lennartsson 2000, Akçakaya and Sjögren-Gulve 2000). Frequently PVAs are used to predict extinction probabilities of populations by projection of current conditions. However, to date few have focused on assessing threat category according to the IUCN Red List system (IUCN 1994). Criterion E of the IUCN system requires not only that the PVA quantifies the absolute risk of extinction, but also that this is done at specific times (viz. 10 yr/3 generations, 20 yr/5 generations and 100 yr). The latter constraint means that it may be difficult to

evaluate published results of PVAs conducted with other objectives unless the full distribution is published.

Besides criterion E, the IUCN Red List system includes four other sets of criteria (A–D) that may be used to assess putative threatened species. All five sets of criteria are valued equally. Hitherto, the vast majority of the world-wide Red List assessments have applied criteria A–D, while very few seem to have utilised the criterion E option (cf. Baillie and Groombridge 1996, Oldfield et al. 1998). Our experience in Sweden is similar. We evaluated 20 000 species using the new IUCN criteria, of which 1953 met at least one criterion for being classified as threatened (Gärdenfors 2000). Only 13 (Table 1) were the subject of a PVA on the national level. In only two cases was criterion E referred to in the listing rationale, because in the remaining 11 the other criteria (A–D) resulted in a higher threat category.

Why is criterion E used so sparsely?

Two obvious reasons for the bias in applying the criteria are: a general unfamiliarity of the assessors with existing quantitative methods, and a deficiency of adequate data for conducting PVAs. However, there might be additional reasons, such as: 1) restricted flexibility or applicability of available PVA-packages, 2) a "tradition" among the PVA modellers to analyse certain categories (species or life histories) of taxa and consequently indirectly neglect others, 3) a tendency to ignore some of the significant processes affecting extinction risks, e.g., deterministic changes in the environment, catastrophes, regional stochasticity, genetic problems, or 4) a general lack of correspondence between the extinction risk expressed through the thresholds of criteria A–D and criterion E, respectively. Numbers 1 and 2 could result in the extinction risk of certain categories of species rarely being evaluated with PVAs, thus not being assessed against criterion E. If the third (3) explanation resulted in a frequent underestimation of extinction risk by conducted PVAs (Brook et al. 1997) and/or (4) the criteria A–D would tend to meet categories of greater risk than would criterion E, the latter criterion would more seldom be accounted for in the Red Lists.

1. PVA packages. An examination of capability of PVA packages for use across many taxa in different situations is beyond the scope of this paper and will not be explored further here. Availability of different packages might, however, also play a role. VORTEX (designed for analysing small populations) can be obtained for free, while, e.g., RAMAS packages (which are designed for a wider range of applications) must be bought.

2. A biased tradition. It is striking that the majority of the species that have been evaluated by PVAs in Sweden are vertebrates (Table 2). Also in 100% (10 out of 10 species) of the analysed vertebrate species, criterion D (very small population) was met, while criterion A (declining population) was never met. This distribution of listing categories contrasts with the group of all threatened vertebrates (excluding fishes), and even more if a comparison is made over all threatened organisms. This indicates a potential tendency to conduct PVAs on species with small populations, while species with larger declining populations are avoided or neglected.

Table 3. Main risk factors actually or potentially threatening red-listed species in Sweden (%). The table is based on classification from a list of 93 identified threat factors (Swedish Threatened Species Unit, unpubl.). The 36 factors affecting ≥ 15% of all red-listed species or any species that has been the subject of a PVA in Sweden, according to Table 1, are presented.

| | | | % affected of | |
Abb.	Threat factor	all red-listed species (n=3105)	red-listed vertebrates excl. fish (n=99)	species with a PVA (n=14)
As	Forest clear-cutting	44	22	21
Öb	Cessation or decreasing of grazing	23	32	29
Z	Stochastic events/processes	22	21	21
Ad	Removing dead or hollow trees, snags and logs	21	9	7
Sp	Field afforestation	19	27	21
Ag	Felling and removing old and big trees	17	10	0
Vd	Ditching and draining	14	38	21
Eb	Constructing buildings, roads, etc.	16	14	0
Al	Removal of deciduous trees from the forest	12	10	7
Ar	Forest thinning	12	10	7
Kg	Manuring and eutrophication	10	7	14
So	Exchanging deciduous forest with conifer forest	8	17	14
Ös	Ceased or decreased mowing	5	5	14
Bl	Insufficient bush clearing	5	4	7
Bk	Excessive bush clearing	4	3	7
Kp	Pesticides	4	16	14
Vm	Regulating water level, decreased fluctuation	3	8	7
Vö	Regulating water level, increased fluctuation	2	3	7
Ff	Disturbance through outdoor life and angling	2	33	29
Fi	Collection and similar	2	15	14
Fb	Disturbance from swimming/sunbathing people	2	8	14
Sb	Forest fire-fighting	2	4	7
Kk	Organic toxins	2	13	7
Ah	Removal of branches and debris in clearing areas	2	3	7
Ok	Interspecific competition	1	2	7
Km	Heavy metal pollution	1	10	7
Fx	Illegal hunting, persecution or controlling	0.5	16	50
U	Events during migration or wintering area	0.5	17	7
Fk	Mountaineering	0.5	1	7
Ft	Disturbance from cross-country driving or boat traffic	0.4	11	29
Ce	Extreme weather conditions	0.4	4	29
Mt	Soil/ground destruction caused by vehicles	0.4	1	7
Fj	Legal hunting	0.4	11	21
Op	Predation by other species	0.4	9	21
Oa	Diseases	0.3	4	14
Fd	Killing by cars or trains	0.2	7	7

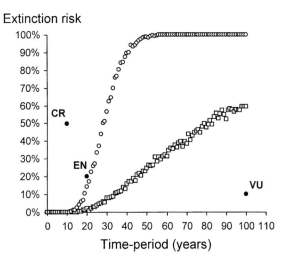

Fig. 1. Extinction probability trajectories for the scarce fritillary *Euphydryas maturna*. Results from simulations of an incidence function model (Kindvall unpubl.), based on data from Eliasson (1994, unpubl.). A scenario where all (30) habitat patches are assumed to become gradually less profitable (circles) is compared with a scenario with no environmental changes (squares). Each circle and square represent the fraction of 1000 simulation replicates that resulted in extinction during the specific time-period. Thresholds for Red List categories CR, EN and VU shown as black dots.

3. Ignoring deterministic changes and other future risks. Red-listed species can also be characterized by risk or threat factors. According to Table 3, there is a discrepancy as regards dominant risk factors between red-listed species or vertebrates in general and those species having been the subject of a PVA. The factors affecting most red-listed species pertain to changes in habitat structure, such as clear-cutting forests, removal of dead wood, cessation of grazing or field afforestation. These factors all tend to change in a deterministically negative way and will lead to a successive change in the birth and death rate, carrying capacity or available habitat. In contrast, the risk factors assigned to the PVA-species (hunting, various kinds of disturbance by human activities, and extreme weather conditions), tend to affect the species in such a way that their birth and death rates can be described with a constant mean over time.

Indeed, most conducted PVAs do use the recent or current demographic situation or amount of available habitat patches and observed turn-over in occupancy as non-varying parameters, add observed, estimated or probabilistic levels of demographic, genetic and environmental stochasticity, and project the future population through a large number of repeated simulations. Thus, scenarios where, e.g., birth rate or habitat availability are subjected to a deterministic change over time often seem to have been avoided. Including or excluding a projected deterministic change will, indeed, make a difference (Fig. 1).

Neglecting deterministic changes also contradicts the experiences by most people working with threatened species. Ninety of the above-mentioned 93 risk factors may affect the population in a deterministic way, while merely three (viz. extreme weather conditions, unknown threat factor and other stochasticity) must be accounted as stochastic variables (cf. Table 3). A majority of the most important of those 90 factors, such as forest clear-cutting, cessation of grazing, removal of dead wood and field afforestation, are subject to change over time and cannot be adequately described by a fixed figure with an added stochasticity in a PVA simulating a 100 yr time period.

The tendency to conduct PVAs on species having small populations (Table 2) conforms well with the observation by Caughley (1994) that conservation biologists have devoted much more attention to small populations affected by stochastic processes than to larger, deterministically declining ones. Why? One reason may be that it is not scientifically interesting to make a PVA assuming a rate of increase that is smaller than one or a deterministic decrease in carrying capacity or number of available habitat patches, because the result will always be that the population (sooner or later) will go extinct. Another reason may be that the long-prevailing small population paradigm has obscured the fact that also larger populations may be in risk of extinction.

Beside obvious changes in habitat availability and quality, risk factors such as catastrophes, regional stochasticity and genetic problems are often omitted in PVAs. In analyses conducted for the purposes of choosing between management options this may be appropriate, but it is certainly not appropriate when a PVA is done to estimate the absolute level of risk (Taylor 1995, Ralls and Taylor 1996).

4. Lack of congruence between the criteria? According to Table 1, PVAs evaluated against the IUCN criterion E often seem to suggest a category of less severe risk than do criteria A–D. This might be due to either 1) PVAs tend to underestimate the extinction risk of a population, or 2) to a lack of correspondence between the risk levels expressed by the numeral threshold of criteria A–D and E, i.e., A–D would tend to result in categories of higher risk.

1) A retrospective test of demographic PVAs based on 21 long-term ecological studies found that the PVA predictions were highly accurate (Brook et al. 2000). Still, one would expect that PVAs applied to systems where the resources are decreasing but assuming parameters with constant mean over time would overestimate mean persistence time.

2) It is of course conceptually impossible to make simple criteria like A–D always reflect the same risk as is expressed in criterion E. The question is, however, whether there is a consistent trend, e.g., that criteria A–D express a higher risk than criterion E. A general evaluation of that issue is beyond the scope of this paper. However, during the review process of the 1994 Red List Criteria it has been suggested that the threshold for VU under criterion A may be too low. As a consequence, that threshold will be raised to 30% for VU (IUCN 2000). If there were to be a general

tendency that criteria A–D estimate categories of greater risk, that could, however, be defended: they would incorporate the precautionary principle of making more conservative decisions with poorer data.

One additional complication regarding correspondence between criteria A–D and E should be mentioned. The extinction risk threshold for CR, EN and VU, respectively, of criterion E is fixed at a single, but different, time. If the PVA trajectory expressing extinction risk over time has a sigmoid shape, this may result in the Red List category possibly being non-precautionary. For instance, if a species according to a PVA has an almost 100% risk of becoming extinct within 100 yr, it may still only meet criterion E at VU level (≥ 10% risk at 100 yr) because the extinction risk at 20 yr is below 20% (the threshold for EN), and far below 50% within 10 yr (CR). Such sigmoid trajectories could be expected, e.g., when a population is fragmented into a number of subpopulations having asynchronic environmental stochasticity, or when there is a continuous reduction of heterozygosity in the population, or when a population under environmental pressure is initially large but decreases deterministically (Kindvall and Gärdenfors unpubl.). Particularly in the latter case, it may also intuitively be easy to underestimate the long term extinction risk, having the consequence that no PVA at all will be undertaken. Such a sigmoid pattern was obvious, e.g., in the case of the scarce fritillary (Fig. 1). The PVAs suggested a 60–100% (depending on degree of habitat change) risk of extinction within 100 yr, but the species would be classified as merely VU according to criterion E, because at 20 yr the risk was < 20% (Kindvall unpubl.).

Concluding remarks

In the debate about the IUCN criteria it has been suggested that criterion E should have predominance over criteria A–D. For instance, if criterion A suggests EN, while a PVA evaluated against criterion E suggests VU, the latter should be followed. From the discussion above, it is obvious that such a rule would be unfortunate and would probably often lead to an underestimation of the extinction risk.

In defence of hitherto conducted PVAs, it should be emphasized that most PVAs were not done with the aim of assessing Red List category, i.e., other aims guided the choice of species and analysis set-up. I do believe that PVAs have a potential to be a useful tool not only for management decisions but also in the classification of threatened species, and particularly so if more analyses are conducted that consider deterministic or stepwise changes in, e.g., birth rate or structural and qualitative habitat parameters (presently feasible with e.g. RAMAS and VORTEX). More experience must also be gained from analyses of larger, but declining populations. And more PVAs must be done with the explicit aim to explore the absolute risk of

extinction, including not only deterministic changes in resources, but also genetic risks, regional stochasticity and future catastrophes.

Finally, it will be essential to communicate to people not familiar with simulations both the strengths and weaknesses of PVAs and how to evaluate the output (and input) data. It is easy to take a figure, like 20% risk of extinction within 20 yr as "the truth" just because a computer has produced a graph saying so. Unfortunately, such an exact figure may be extremely uncertain, particularly if the input data were of low quality or if the model applied did not adequately reflect the real situation of the species.

Acknowledgements – Claes Eliasson, Dan Guillou, Tommy Lennartsson, Torbjörn Ebenhard and Oskar Kindvall provided unpublished PVA data. Barbara Taylor, David Keith, Oskar Kindvall, Per Sjögren-Gulve and Torbjörn Ebenhard provided many useful suggestions that substantially improved the paper. Nigel Rollison corrected the language. Thank you all!

References

Ahlén, I. and Tjernberg, M. (eds) 1996. Rödlistade ryggradsdjur i Sverige – artfakta. [Swedish Red Data Book of vertebrates 1996.] – Threatened Species Unit, Swedish Univ. of Agricult. Sci., Uppsala, in Swedish.

Akçakaya, H. R. 1997. RAMAS GIS: linking landscape data with population viability analysis (ver. 2.0). – Appl. Biomath., Setauket, New York.

Akçakaya, H. R. 2000. Population viability analyses with demographically and spatially structured models. – Ecol. Bull. 48: 23–38.

Andrén, H. and Liberg, O. 1999. Demografi och minsta livskraftiga population hos lodjur. [Demography and minimum viable population in lynx.] – CBM:s Skriftserie 1: 119–123, in Swedish.

Angerbjörn, A., Ehrlén, J. and Isakson, E. 1999. Vargstammen i Sverige 1977–1997 – en simulering av populationens utveckling. [The wolf population in Sweden 1977–1997 – a simulation of the population development.] – CBM:s Skriftserie 1: 75–84, in Swedish.

Aronsson, M., Hallingbäck, T. and Mattson, J.-E. (eds) 1995. Rödlistade växter i Sverige 1995. [Swedish Red Data Book of plants 1995]. – Threatened Species Unit, Swedish Univ. of Agricult. Sci., Uppsala, in Swedish with English summary.

Baillie, J. and Groombridge, B. 1996. 1996 IUCN Red List of threatened animals. – IUCN, Gland, Switzerland.

Brook, B. W. et al. 1997. Does population viability analysis software predict the behaviour of real populations? A retrospective study on the Lord Howe Island woodhen *Tricholimnas sylvestris* (Sclater). – Biol. Conserv. 82: 119–128.

Brook, B. W. et al. 2000. Predicitive accuracy of population viability analysis in conservation biology. – Nature 404: 385–387.

Carlson, A. and Stenberg, I. 1995. Vitryggig hackspett (*Dendrocopos leucotos*). Biotopval och sårbarhetsanalys. [White-backed woodpecker (*Dendrocopos leucotos*). Habitat choice and population viability analysis.] – Swedish Univ. of Agricult. Sci., Dept of Wildlife Ecology, Rep. 27, in Swedish.

Caughley, G. 1994. Directions in conservation biology. – J. Anim. Ecol. 63: 215–244.

Ebenhard, T. 2000. Population viability analyses in endangered species management: the wolf, otter and peregrine falcon in Sweden. – Ecol. Bull. 48: 143–163.

Ehnström, B., Gärdenfors, U. and Lindelöw, Å. 1993. Rödlistade evertebrater i Sverige 1993. [Swedish Red List of invertebrates 1993.] – Threatened Species Unit, Swedish Univ. of Agricult. Sci., Uppsala, in Swedish with English summary.

Eliasson, C. 1994. Projekt nätfjärilar 1992–1994. [Project fritillaries 1992–1994.] – WWF-Sweden, Solna, in Swedish.

Gärdenfors, U. 1995. The regional perspective. – In: Baillie, J., Callahan, D. and Gärdenfors, U. A closer look at the IUCN Red List categories. – Species 25: 30–36.

Gärdenfors, U. 1996. Application of IUCN Red List categories on a regional scale. – Guest essay in 1996 IUCN Red List of threatened animals. IUCN, pp. 63–66.

Gärdenfors, U. (ed.) 2000. Rödlistade arter i Sverige 2000 – The 2000 Red List of Swedish species. – Threatened Species Unit, Swedish Univ. of Agricult. Sci., Uppsala.

Gärdenfors, U. and Kindvall, O. 1999. Developing national Red Lists based on the new IUCN criteria. – Proc. of the XXIV Nordic Congr. of Entomol., Tartu, pp. 67–70.

Gärdenfors, U. et al. 1999. Draft guidelines for the application of IUCN Red List criteria at national and regional levels. – Species 31/32: 58–70.

Guillou, D. 1997. Simulated Arctic fox extinctions. PVA on *Alopex lagopus* in Fennoscandia. – Unpubl. report, Stockholm Univ.

Hanski, I. 1999. Metapopulation ecology. – Oxford Univ. Press.

Hanski, I. et al. 1996. The quantitative incidence function model. – Conserv. Biol. 10: 578–590.

IUCN 1994. IUCN Red List categories. – IUCN, Gland, Switzerland.

IUCN 2000. IUCN Red List categories. Prepared by the IUCN Species Survival Commission. – IUCN, Gland, Switzerland, in press.

Kindvall, O. 1998. Introduktion till sårbarhetsanalyser. [Introduction to population viability analyses.] – ArtDatabanken Rapporterar 2. Threatened Species Unit, Swedish Univ. of Agricult. Sci., Uppsala, in Swedish.

Kindvall, O. 2000. Comparative precision of three spatially realistic simulation models of metapopulation dynamics. – Ecol. Bull. 48: 101–110.

Kontio, T. 1998. Är den svenska lodjurspopulationen (*Lynx lynx*) livskraftig? En sårbarhetsanalys. [Is the Swedish lynx (*Lynx lynx*) population viable? A population viability analysis.] – Div. of Population Genetics, Stockholm Univ., in Swedish.

Lacy, R. C. 2000. Structure of the VORTEX simulation model for population viability analysis. – Ecol. Bull. 48: 191–203.

Lacy, R. C., Hughes, K. A. and Miller, P. S. 1995. VORTEX: a stochastic simulation of the extiction process. Ver. 7 users's manual. – IUCN/SSC/CBSG, Apple Valley, MN.

Lennartsson, T. 2000. Management and population viability of the pasture plant *Gentianella campestris*: the role of interactions between habitat factors. – Ecol. Bull. 48: 111–121.

Lundblad, C. 1995. Den svenska björnstammens (*Ursus arctos*) framtid. En populationsgenetisk studie av björnstammens sannolikhet för utdöende samt förlust av genetisk variation, baserad på datorsimuleringar. [The future of the Swedish brown bear (*Ursus arctos*) population. A population genetic study of extinction probability and loss of genetic diversity, based on computer simulations.] – Div. of Population Genetics, Stockholm Univ., in Swedish.

Mace, G. M. and Lande, R. 1991. Assessing extinction threats: toward a revaluation of IUCN threatened species categories. – Conserv. Biol. 5: 148–157.

Mace, G. M. and Collar, N. J. 1994. Extinction risk assessment for birds through quantitative criteria. – Ibis 137: 240–246.

Oldfield, S., Lusty, C. and MacKinven, A. 1998. The worlds list of threatened trees. – World Conservation Press, Cambridge, U.K.

Persson, J. and Sand, H. 1998. Vargen. Viltet, ekologin och människan. [The wolf. The wildlife, ecology and man.] – Swedish Hunters Assoc., in Swedish.

Ralls, K. and Taylor, B. L. 1996. How viable is population viability analysis? – In: Pickett, S. T. A. et al. (eds), Enhancing the ecological basis of conservation: heterogeneity, ecosystem function, and biodiversity. Chapman and Hall, pp. 228–235.

Ray, C., Sjögren-Gulve, P. and Gilpin, M. E. 1994. METAPOP (ver. 3.2). – Unpubl. computer program. Univ. of California, Davis.

Sæther, B.-E. et al. 1998. Viability of Scandinavian brown bear *Ursus arctos* populations: the effects of uncertain parameter estimates. – Oikos 83: 403–416.

Sjögren-Gulve, P. and Ray, C. 1996. Using logistic regression to model metapopulation dynamics: large-scale forestry extirpates the pool frog. – In: McCullough, D. R. (ed.), Metapopulations and wildlife conservation. Island Press, pp. 111–137.

Sjögren-Gulve, P. and Hanski, I. 2000. Metapopulation viability analysis using occupancy models. – Ecol. Bull. 48: 53–71.

Taylor, B. L. 1995. The reliability of using population viability analysis for classification of species. – Conserv. Biol. 9: 551–558.

Ecological Bulletins 48: 191–203. Copenhagen 2000

Structure of the VORTEX simulation model for population viability analysis

Robert C. Lacy

Lacy, R. C. 2000. Structure of the VORTEX simulation model for population viability analysis. – Ecol. Bull. 48: 191–203.

The structure of the VORTEX computer simulation model for population viability analysis is outlined. The program flow is described here in order to provide a detailed specification of the structure of a widely used population viability analysis model. VORTEX is an individual-based simulation program that models the effects of mean demographic rates, demographic stochasticity, environmental variation in demographic rates, catastrophes, inbreeding depression, harvest and supplementation, and metapopulation structure on the viability of wildlife populations. The model facilitates analysis of density-dependent reproduction and changing habitat availability, and most demographic rates can optionally be specified as flexible functions of density, time, population gene diversity, inbreeding, age, and sex. VORTEX projects changes in population size, age and sex structure, and genetic variation, as well as estimating probabilities and times to extinction and recolonization.

R. C. Lacy (rlacy@ix.netcom.com), Dept of Conservation Biology, Daniel F. & Ada L. Rice Center, Chicago Zoological Society, Brookfield, IL 60513, USA.

"VORTEX and like programs do exactly what they are told to do, as constrained by the static single-species models that provide their structure. They can be useful for various purposes so long as the user understands what the programs are doing …" (Caughley and Gunn 1996, p. 208).

The complexity and multiplicity of processes influencing the dynamics of natural populations of animals and plants means that population viability analysis (PVA) models are also frequently complex. Different models incorporate different population processes. Individual population processes can be modeled in various ways, requiring different sets of driving variables, using different equations to define the processes, and providing different output to describe the population dynamics. Users of PVA models should understand the basic structure of the models they use, and it is important that models used for scientific studies and conservation efforts can be examined and replicated. Yet often the details of PVA computer programs are not available to the users, because the code is proprietary information or otherwise not provided to users, or simply because the task of reading and understanding the source code for large and complex programs is formidable.

One possible remedy to the problem of PVA users needing to understand the models being used is for practitioners to develop their own computer programs. This would result in the user having a full understanding of a model that would be specifically designed for the analysis. Development of user-specific and case-specific models is usually not practical, however, as many population biologists are not skilled computer programmers, and the time required to develop a complex model is often prohibitive. Moreover, a complex computer program developed by and used by one person will sometimes contain serious programming errors. The testing of programs that are widely used may be a necessary prerequisite for reliable population viability analyses to be employed effectively in biodiversity conservation. Finally, the flexibility and expansive capabilities of generic PVA software to model a large diversity of population processes will often lead PVA practitioners to consider threats to population viability that would otherwise have been neglected.

Widely available PVA software can serve the same role as do statistical analysis packages. The ease of use, flexible application to diverse needs, and extensive prior testing facilitate many applications that would not otherwise be attempted. Ideally, perhaps, all users of statistical methods would write their own programs or otherwise study the code of the software entrusted for the analyses. More practically, confidence is gained in the reliability of generic software tools as more people use the programs and compare the generated results to expectations from statistical theory and to results for simple and well known cases. Also, users of statistical software are expected to be sufficiently familiar with the methods of statistical analysis to be able to choose appropriate models to apply to their problem, to be able to provide the proper input, and to be able to interpret the results.

Unlike the situation for statistical methods, however, there are not yet widely accepted and published accounts of standard methods for population viability analysis. The methods of population-based models (e.g., Starfield and Bleloch 1986, Burgman et al. 1993) are extensions of the methods of population ecology and demography (e.g., Pielou 1977, Caswell 1989), but many details of model construction require decisions about algorithms and methods that are not fully delineated in general treatments. The methods of individual-based PVA models have been only cursorily described in the scientific literature. Below is an outline of one widely used PVA software package, VORTEX, ver. 8.20. The basic approach taken in the VORTEX model is described in Lacy (1993), in Lindenmayer et al. (2000) and other papers describing applications of VORTEX, and in the software manual (Miller and Lacy 1999). Detailed documentation of the program flow and algorithms is provided here, so that users of VORTEX can confirm that the model is performing the analyses that are intended, and so that PVA practitioners in general can see an example of the structure of an individual-based PVA model. The VORTEX program is available at http://www2.netcom.com/~rlacy/vortex.html.

The pseudo-code presented below is an English-language outline of the program flow and primary algorithms used by VORTEX (which is written in the C programming language). This pseudo-code omits coding for: 1) input routines for reading parameters from files and/or keyboard; 2) output routines for writing results to files; 3) specification of default parameter values; 4) checks for illegal values, error handling; 5) memory management and initialization of memory; 6) details of C coding to achieve algorithms; 7) routines for on-line help; 8) routines for graphical display of functions specifying demographic rates, population sizes during simulations, simulation re-

sults; 9) routines for evaluating equations that specify demographic rates (e.g., breeding, mortality) as functions of population and individual variables (e.g., population size, gene diversity, year, age, sex, inbreeding) (see note 3 below); 10) tallies of mean within-population statistics and metapopulation summaries; 11) algorithms for calculating basic statistics, such as means, standard deviations, standard errors, and medians across years and across iterations.

Variables for storing input, intermediate calculations, and output are indicated in the pseudo-code by italicized labels. Many of the variables are arrays (e.g., a value stored for each population, or for each age class, or for each individual), as suggested by the loops within which they are calculated and used. The indices of such arrays are indicated within brackets (e.g., *MortalityRate[p][s][x]*, for each population, p, sex, s, and age, x). VORTEX uses many more variables (not shown in the pseudo-code) for facilitating calculations and accumulating sums, sums of squares, and other components needed for the basic statistics reported in the output.

In the pseudo-code, loops are indicated with FOR: and END LOOP statements, or by WHILE: and END WHILE statements. Conditional actions are indicated by IF: and END IF statements, or by IF:, ELSE:, and END IF/ELSE statements. BREAK indicates that program flow exits from the bottom of a loop. CONTINUE indicates that program flow jumps back to the next value at the top of the loop. Multiplication is indicated by the asterisk (*) symbol; ^ indicates exponentiation; SQRT indicates the positive square root.

Function modules defined outside of the main body of the pseudo-code program are labeled in the form **FUNCTION()**, and are specified below the main **VORTEX()** program. The actual C code is subdivided into many smaller functions; the pseudo-code shows only the flow of the overall program and its largest modules. The functions **RAND()** and **NRAND()** indicate, respectively, that a random number is generated from the uniform 0-1 distribution or from a unit normal distribution.

Explanatory comments, following pseudo-code sections, are preceded by //. More extensive explanations are given in notes following the code.

As an individual-based PVA simulation model, VORTEX represents each individual in memory, simulates life events (such as sex determination, breeding, mortality, and dispersal) which could occur to each individual, and monitors the status of each individual and the population as a whole. The characteristics tracked for each animal are: sex, alive/dead status, population membership, age, inbreeding coefficient, and two alleles at each of six loci. In addition, VORTEX maintains a matrix of kinship coefficients between all pairs of living animals, as this provides inbreeding coefficients for any offspring.

VORTEX models changes to a population as a series of discrete events that occur once per year (or other time interval). The annual sequence of demographic events is:

breeding; mortality; age 1 yr; migrate (disperse) among populations; harvest (managed removals); supplementation (managed additions); carrying capacity truncation; census (Fig. 1). Occurrences of events are probabilistic; demographic stochasticity emerges from chance variation in which individuals breed, die, and are of each sex. Environmental variation in demographic rates is imposed by sampling rates from specified distributions during each simulated year. Catastrophes, which occur with specified probabilities, cause one-year reductions in reproduction and survival. Genetic effects are modeled as reduced survivorship of inbred individuals.

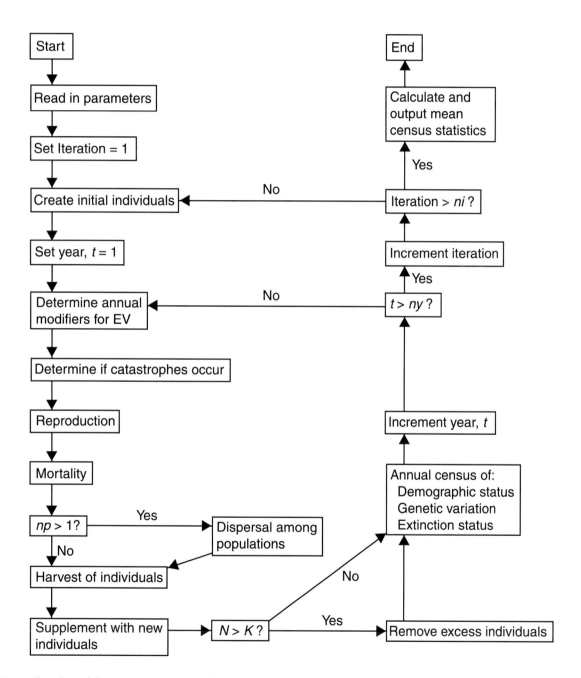

Fig. 1. Flow chart of the primary components of the Vortex simulation. Each step from "Create initial individuals" through "Annual census" is applied to each population in a modeled metapopulation. t = year; ny = number of years simulated; np = number of populations; ni = number of iterations; N = population size; K = carrying capacity; EV = environmental variation.

VORTEX program pseudo-code

BEGIN PROGRAM **VORTEX**():
Initialize random number generator; // See Note 1.
    FOR (each scenario):
        READ_SPECIES_PARAMETERS();
        IF (*NumberOfPopulations* > 1):
            READ_MIGRATION_PARAMETERS();
// VORTEX describes dispersal between populations as "migration."
        END IF
        FOR (each population, *p*):
            READ_POPULATION_PARAMETERS(*p*);
        END (population) LOOP
        FOR (each population, *pSource*):
// Calculate cumulative migration rates for each pairwise transition between populations.
            Set *CumMigrationProb[pSource][1]* = *MigrationProb[pSource][1]*;
            FOR (each destination population, *pDestination*, greater than 1):
                Set *CumMigrationProb[pSource][pDestination]* = *CumMigrationProb[pSource][pDestination -1]*+ *MigrationProb[pSource][pDestination]*;
            END (*pDestination*) LOOP
        END (*pSource*) LOOP
        Set *NumberLethals* = *InbreedingGeneticLoad* * *ProportionLoadDueToLethals*;
        Set *LethalEquivalents* = *InbreedingGeneticLoad* * (1 - *ProportionLoadDueToLethals*);
// See Note 2.
        FOR (each population, *p*):
            Set *GlobalBreedEV[p]* = *BreedEV[p]* * *EVConcordanceAmongPopulations*;
            Set *LocalBreedEV[p]* = SQRT(*BreedEV[p]* ^2 − *GlobalBreedEV[p]* ^2);
// Partition Environmental Variation in breeding (*BreedEV*) into the component that is common to all populations (*GlobalBreedEV*) and the component that is specific to each population. (*LocalBreedEV*); *TotalEV* ^ 2 = *GlobalEV* ^ 2 + *LocalEV* ^2
// Note: EVs are given as standard deviations.
            FOR (each sex, *s*):
                FOR (each age, *x*, up to age of breeding):
                    Set *GlobalMortEV[p][s][x]* = *MortEV[p][s][x]* * *EVConcordanceAmongPopulations*;
                    Set *LocalMortEV[p][s][x]* = SQRT(*MortEV[p][s][x]* ^2 − *GlobalMortEV[p][s][x]* ^2);
// Partition EV in mortality (*MortEV*) into the component that is common to all populations (*GlobalMortEV*) and the component that is specific to each population (*LocalMortEV*).
            END (age) LOOP
            END (sex) LOOP

            Set *GlobalKEV[p]* = *KEV[p]* * *EVConcordanceAmongPopulations*;
            Set *LocalKEV[p]* = SQRT(*KEV[p]* ^2 − *GlobalKEV[p]* ^2);
// Partition EV in carrying capacity (*KEV*) into the component that is common to all populations (*GlobalKEV*) and the component that is specific to each population (*LocalKEV*).
        END (population) LOOP
        FOR (each iteration):
            FOR (each population):
                Create initial individuals, assigning population, sex, age, alive/dead status, inbreeding coefficient and kinships (initially 0), and alleles at six loci;
                FOR (each of five non-neutral loci, *l*):
                    FOR (each founder allele, *a*):
// The probability that a given founder allele is a lethal is *NumberLethals* / 10, because there are 10 alleles across the five diploid loci.
                      IF (**RAND**() < *NumberLethals* / 10):
                        Set *Lethal[l][a]* = TRUE;
// Allele, *a*, of locus, *l*, is a recessive lethal.
                    ELSE:
                      Set *Lethal[l][a]* = FALSE;
                    END IF/ELSE
                  END (founder allele) LOOP
                END (locus) LOOP
                Display initial population sizes on screen and write to output files;
            END (population) LOOP
            FOR (each year):
                IF (*NumberPopulations* > 1):
                    GLOBAL_EV_RANDS();
// Generate random numbers for specifying environmental variation, concordant across populations, for the year.
                END (*NumberPopulations*) IF
                FOR (each population, *p*):
                    LOCAL_EV_RANDS();
                    CATASTROPHES(*p*);
// Determine if catastrophes occur that year.
                  Determine carrying Capacity (K) for year;
// See Note 3.
                  Add *LocalKEVNRand* * *LocalKEV[p]* to *CarryingCapacity[p]*;
// Adjust K for local EV.
                  Add *GlobalKEVNRand* * *GlobalKEV[p]* to *CarryingCapacity[p]*;
// Adjust K for global EV.
                  BREED(*p*);
// Go through breeding cycle to produce offspring.
                  MORTALITY(*p*);
// Determine who dies that year.
                END (population) LOOP
                Add 1 to the age of each animal;
                IF (*NumberPopulations* > 1):
                    MIGRATE();

// Determine which animals migrate between populations.
        END IF
        FOR (each population, p):
                IF (year during which animals are to be
                harvested):
                        HARVEST(p);
                END (harvest year) IF
        END (population) LOOP
        FOR (each population, p):
                IF (year during which animals are to be
                supplemented):
                        SUPPLEMENT(p);
                END (supplement year) IF
        END (population) LOOP
        FOR (each population, p):
                Tally *PopulationSize[p]*;
                IF (population is not extinct AND
                population was not extinct prior year):

// "Extinction" can be defined by the user as the absence of
one sex, or as the population size falling below a specified
lower limit.

                        $r[p]$ = log(*PopulationSize[p]* /
                        PopulationSizePriorYear[p]);

// Calculate population growth rate (r).
                END IF
                IF (not extinct AND *PopulationSize[p]*, N
                > *CarryingCapacity[p]*, K):
                        FOR (each living animal):
                                IF (RAND() > K / N):

// Stochastically kill excess above K.
                                        Animal dies;
                                END IF
                        END (each animal) LOOP
                END ($N > K$) IF
                Tally *PopulationSize[p]*;
                IF (population is extinct):
                        Decrement *NumberExtantPopulations[p]*;
                        IF (population was not extinct in prior
                        year):
                                Set *YearExtinct[p]* = *CurrentYear*;
                                IF (population has not been
                                recolonized):

// First extinction

                                        Set *TimeToExtinction[p]* =
                                        CurrentYear;
                                        Increment
                                        NumberOfExtinctions[p];
                                ELSE:

// Re-extinction of population
                                        Set *TimeToReextinction[p]* =
                                        CurrentYear −
                                        YearOfRecolonization[p];
                                END (recolonized) IF/ELSE
                        END (was not extinct) IF
                ELSE:

// Not extinct

                        IF (population was extinct in prior year):

// Recolonization

                                Set *YearRecolonized[p]* = *CurrentYear*;
                                Set *TimeToRecolonization[p]* =
                                CurrentYear − *YearExtinct[p]*;
                                Increment
                                NumberOfRecolonizations[p];
                        END (was extinct) IF
                        Set *YearExtinct[p]* = 0;

// Flag for not extinct
                END (extinct) IF/ELSE
                Display *PopulationSize[p]* on screen graph;
        END (population) LOOP
        FOR (each population, p):
                CALC_GENETIC_METRICS(p);
        END (population) LOOP
END (year) LOOP
END (iteration) LOOP
// At this point, the simulation is complete and summary
statistics can be calculated.
        FOR (each population, p):
                Calculate and report means, SDs, and SEs across
                iterations for
                        Population growth rate: $r[p]$ = N[*CurrentYear*]
                        / N[*PreviousYear*]
                        TimeToExtinction[p]
                        TimeToRecolonization[p]
                        TimeToReextinction[p]
                FOR (each year):
                        Calculate and output means, SDs, and SEs
                        across iterations for:
                                Probability of extinction, *PE[p]*

// SE = SQRT[*PE* * (1 - *PE*) / *NumberIterations*]
                                PopulationSize[p]
                                GeneDiversity[p]

// Gene Diversity = Heterozygosity expected under Hardy-
Weinberg equilibrium
                                ObservedHeterozygosity[p]

// = 1 - mean inbreeding coefficient
                                NumberAlleles[p]
                                LethalFrequency[p]
                END (year) LOOP
        END (population) LOOP
        Calculate and report within-population means of
        above summary statistics;
        Call program for displaying graphical displays of
        trends in:
                PopulationSize
                GeneDiversity
                Mean inbreeding coefficient (1 -
                ObservedHeterozygosity)
                Probability of population persistence to year
                Probability of extinction in that time interval
        Read in *DoAnotherScenario?*
        IF (*DoAnotherScenario?* is FALSE):
                BREAK from scenario LOOP

END IF
END (scenario) LOOP
END PROGRAM **VORTEX**()

BEGIN FUNCTION
READ_SPECIES_PARAMETERS():
// Get input parameters from keyboard or input file, describing simulation parameters, inbreeding effects, and basic species life history.
Read in Input/Output file names;
Read in *NumberOfIterations*;
Read in *NumberOfYears*;
Read in *ExtinctionDefinition*;
// Extinction can be defined as no animals of one sex, or as the population size falling below a specified minimum.
Read in *NumberOfPopulations*;
Read in *InbreedingGeneticLoad*;
Read in *ProportionLoadDueToLethals*;
Read in *EVCorrelationBetweenReproductionAndSurvival?*;
IF (*NumberOfPopulations* > 1):
Read in *EVConcordanceAmongPoulations*;
END IF
Read in *NumberTypesOfCatastrophes*;
Read in *Monogamous/Polygynous/Hermaphroditic?*;
Read in *FemaleBreedingAge*;
Read in *MaleBreedingAge*;
Read in *MaximumAge*;
Read in *SexRatio* at birth;
Read in *MaximumLitterSize*;
Read in *DensityDependentBreeding?*
// The pseudocode for modeling density dependent breeding is not given below.
END FUNCTION **READ_SPECIES_PARAMETERS**()

BEGIN FUNCTION
READ_MIGRATION_PARAMETERS():
// Get input population structure and migration patterns.
Read in *MigrationAges*;
Read in *MigrationSexes*;
Read in *MigrationSurvival*;
Read in *MigrationDensity*;
FOR (each population, *pSource*):
FOR(each other population, *pDestination*):
Read in *MigrationProb*[*pSource*][*pDestination*];
END LOOP
END LOOP
END FUNCTION
READ_MIGRATION_PARAMETERS()

BEGIN FUNCTION
READ_POPULATION_PARAMETERS(for population *p*):
Read in *ProportionFemalesBreeding*[*p*];
Read in *BreedEV*[*p*];
// Environmental variation is specified as a standard deviation.

Read in Litter size distribution (either as *MeanLitterSize[p]* and *SDLitterSize[p]* or as the fully specified distribution of *ProbLitterSize[p][n]*);
FOR (each age, *x*, up to *FemaleBreedingAge*):
Read in *FemaleMortality[p][x]*;
Read in *FemaleMortalityEV[p][x]*;
END (age) LOOP
FOR (each age, *x*, up to *MaleBreedingAge*):
Read in *MaleMortality[p][x]*;
Read in *MaleMortalityEV[p][x]*;
END (age) LOOP
FOR (each type of catastrophe, *c*):
IF (*NumberPopulations* > 1):
Read in *GlobalOrLocal[p][c]*;
END IF
Read in *CatastropheFrequency[p][c]*;
Read in *CatastropheBreedSeverity[p][c]*;
Read in *CatastropheSurvivalSeverity[p][c]*;
END (catastrophe) LOOP
CALC_DETERMINISTIC_GROWTH(*p*);
// Calculate deterministic population growth rate, generation time, and stable age distribution from mean birth and death rates. Effects of any catastrophes are averaged across years
Read in *ProportionMalesInBreedingPool[p]*;
// See Note 4.
IF (initial numbers of animals are to be distributed according to the stable age distribution):
Determine initial numbers of animals in each age-sex class;
// The stable age distribution would rarely assign whole numbers to each age-sex class. Integral numbers are assigned that most closely match the desired distribution.
ELSE (does not start at stable age distribution):
FOR (each sex):
FOR (each age up to *MaximumAge*):
Read in initial number of animals;
END LOOP
END LOOP
END (stable age distribution) IF/ELSE
Read in *CarryingCapacity[p] (K)*;
// K may be specified as a function of year or other parameters. See Note 3.
Read in *KEV[p]*;
Read in *Harvest[p]?*;
IF *(Harvest[p]? = Yes)*:
Read in *FirstYearHarvest[p], LastYearHarvest[p], HarvestInterval[p]*;
FOR (each age, *x*, up to *FemaleBreedingAge*):
// For harvest, all adults are treated in the same age category.
Read in *NumberFemalesToBeHarvested[p][x]*;
END LOOP
FOR (each age, *x*, up to *MaleBreedingAge*):
Read in *NumberMalesToBeHarvested[p][x]*;
END LOOP

END IF
Read in *Supplement[p]?*;
IF (*Supplement[p]?* = Yes):
   Read in *FirstYearSupplementation[p]*,
   LastYearSupplementation[p],
   SupplementationInterval[p];
   FOR (each age, *x*, up to *FemaleBreedingAge*):
      Read in *NumberFemalesToBeSupplemented[p][x]*;
   END LOOP
   FOR (each age, *x*, up to *MaleBreedingAge*):
      Read in *NumberMalesToBeSupplemented[p][x]*;
   END LOOP
END IF
END FUNCTION
READ_POPULATION_PARAMETERS()

BEGIN FUNCTION
CALC_DETERMINISTIC_GROWTH(for population *p*):
// Use standard life table analysis; solve the Euler equation to find the deterministic growth rate.
   Set fecundity, M = *MeanLitterSize[p]* * (1 - *SexRatio*);
// *SexRatio* is proportion males at birth.
   FOR (each type of catastrophe, *c*):
// Adjust M for catastrophes.
      Multiply M by *CatastropheFrequency[p][c]* *
      CatastropheBreedSeverity[p][c] + (1 −
      CatastopheFrequency[p][c]);
   END (catastrophe) LOOP
   FOR (each age, *x*):
      Set female survival, $P[x]$ = 1 − *FemaleMortality[p][x]*;
      FOR (each type of catastrophe, *c*):
// Adjust $P[x]$ for catastrophes.
         Multiply $P[x]$ by *CatastropheFrequency[p][c]* *
         CatastropheSurvivalSeverity[p][c] + (1 −
         CatastopheFrequency[p][c]);
      END (catastrophe) LOOP
      Multiply cumulative survivorship, $L[x]$, by $P[x]$;
      IF (*x* >= *FemaleBreedingAge*):
         Add $L[x]$ * M to *SumLxMx*;
         Add *x* * $L[x]$ * M to *SumAgeLxMx*;
      END IF
   END (age) LOOP
   Set $R0$ = *SumLxMx*;
   Set *GenerationTime[p]* = *SumAgeLxMx* / *SumLxMx*;
// Preliminary estimate
   Set *Lambda* = $R0 \wedge (1/GenerationTime)$;
// Preliminary estimate
   Solve Euler equation by iterative approximation, to yield precise *Lambda*;
   Set *r[p]* = log(*Lambda*);
   Set *GenerationTime[p]* = log(*R0*) / *r[p]*;
   FOR (each age, *x*):
// Determine stable age distribution.
      Set *StableAgeClassSize[p][x]* = (1 - *SexRatio*) * $L[x]$ /
      ($Lambda \wedge x$);

      Add *StableAgeClassSize[p][x]* to
      SumStableAgeClassSize[p];
   END LOOP
// Repeat age distribution calculations for males, but use female-based *Mx* and *Lambda*.
   FOR (each age, *x*):
      Set male survival, $P[x]$ = 1 − *MaleMortality[p][x]*;
      FOR (each type of catastrophe, *c*):
// Adjust $P[x]$ for catastrophes.
         Multiply $P[x]$ by *CatastropheFrequency[p][c]* *
         CatastropheSurvivalSeverity[p][c] + (1 −
         CatastopheFrequency[p][c]);
      END (catastrophe) LOOP
      Multiply cumulative survivorship, $L[x]$, by $P[x]$;
   END LOOP
   FOR (each age, *x*):
// Determine stable age distribution.
      Set *StableAgeClassSize[p][x]* = *SexRatio* * $L[x]$ /
      ($Lambda \wedge x$);
      Add *StableAgeClassSize[p][x]* to
      SumStableAgeClassSize[p];
   END LOOP
   FOR (each age, *x*, and sex, *s*):
      Divide *StableAgeClassSize[p][s][x]* by
      SumStableAgeClassSize[p];
   END LOOP
END FUNCTION
CALC_DETERMINISTIC_GROWTH()

BEGIN FUNCTION **GLOBAL_EV_RANDS()**:
   FOR (each type of catastrophe):
      Set *GlobalCatastropheRand* = **RAND()**;
// Select random number to determine if global catastrophe occur. See Note 5.
   END (catastrophe) LOOP
   Set *GlobalBreedEVRand* = **RAND()**;
// Select random number for specifying EV in breeding for that year.
   Set *GlobalBreedEVNRand* = **NRAND()**;
// Select random normal deviate for specifying EV in breeding for year. Whether *EVRand* or *EVNRand* will be used depends on the magnitude of EV. See Note 6.
   Set *GlobalBreedEVNRand* to same sign as
   GlobalBreedEVRand;
   IF (*EVCorrelationBetweenReproductionAndSurvival?* =
   No):
      Set *GlobalMortEVRand* = **RAND()**;
// Select random 0-1 number for specifying EV mortality for that year.
      Set *GlobalMortEVNRand* = **NRAND()**;
// Select random normal deviate for specifying EV in mortality for that year. Whether *EVRand* or *EVNRand* will be used depends on the magnitude of EV.
      Set *GlobalMortEVNRand* to same sign as
      GlobalMortEVRand;
      Set *GlobalKEVNRand* = **NRAND()**;

// Select random normal deviate for specifying EV in K for year.
    ELSE (EV in breeding is correlated with EV in mortality):
        Set *GlobalMortEVRand* = *GlobalBreedEVRand*;
        Set *GlobalMortEVNRand* = *GlobalBreedEVNRand*;
        Set *GlobalKEVNRand* = *GlobalBreedEVNRand*;
    END (EV correlation) IF/ELSE
END FUNCTION **GLOBAL_EV_RANDS()**

BEGIN FUNCTION **LOCAL_EV_RANDS()**:
    Set *LocalBreedEVRand* = **RAND()**;
// Select random number for specifying EV in breeding for year.
    Set *LocalBreedEVNRand* = **NRAND()**;
// Select a random normal deviate for specifying EV in breeding.
    Set *LocalBreedEVNRand* to same sign as *LocalBreedEVRand*;
    IF (*EVCorrelationBetweenReproductionAndSurvival?* = FALSE):
        Set *LocalMortEVRand* = **RAND()**;
// Select random number for specifying EV in mortality.
Set *LocalMortEVNRand* = **NRAND()**;
// Select random normal deviate for specifying EV in mortality.
        Set *LocalMortEVNRand* to same sign as *LocalMortEVRand*;
        Set *LocalKEVNRand* = **NRAND()**;
// Select random normal deviate for specifying EV in K.
    ELSE (EV in breeding is correlated with EV in mortality):
        Set *LocalMortEVRand* = *LocalBreedEVRand*;
        Set *LocalMortEVNRand* = *LocalBreedEVNRand*;
        Set *LocalKEVNRand* = *LocalBreedEVNRand*;
    END (EV correlation) IF/ELSE
END FUNCTION **LOCAL_EV_RANDS()**

BEGIN FUNCTION **CATASTROPHES(for population *p*)**:
    FOR (each type of catastrophe, *c*):
        IF (Catastrophe is local in effect):
            IF (**RAND()** < *CatastropheFrequency[p][c]*):
// See Note 5.
                Set *CatastropheFlag[c]* = TRUE;
// Catastrophe has occurred.
            ELSE:
                Set *CatastropheFlag[c]* = FALSE;
            END (catastrophe) IF/ELSE
        ELSE:
            IF (*GlobalCatastropheRand* < *CatastropheFrequency[p][c]*):
                Set *CatastropheFlag[c]* = TRUE;
            ELSE:
                Set *CatastropheFlag[c]* = FALSE;
            END (catastrophe) IF/ELSE
        END (Local/Global catastrophe) IF/ELSE
    END (catastrophe) LOOP

END FUNCTION **CATASTROPHES()**

BEGIN FUNCTION **BREED(for population *p*)**:
// Find breeders for the year …
    FOR (each living animal in the population):
        IF (sex = female AND age >= *FemaleBreedingAge*):
        Add female to breeding pool;
        END IF
        IF (not hermaphroditic):
            IF (sex = male AND age >= *MaleBreedingAge*):
                IF (**RAND()** < *ProportionMalesInBreedingPool[p]*):
                    Add male to breeding pool;
                END IF
            END IF
        END IF
    END (animal) LOOP
    IF (no males selected for breeding pool, but adult males do exist):
        Add one male at random to breeding pool;
    END IF
    IF (monogamous):
        FOR (each male in breeding pool, *m*):
            Set *MaleUsed[m]* = FALSE;
// Flag to indicate male is available for pairing
        END LOOP
    END IF
    IF (hermaphroditic):
        IF (only one breeding female AND *ProportionSelfing[p]* = 0):
            EXIT **BREED()**;
        END IF
    END IF
    FOR (each female, *Dam*, in breeding pool):
        Let *BreedRand* = **RAND()**;
        GETBREEDRATE();
// *BreedRate* is probability of breeding for the female, given by the user either as a constant, *ProportionFemalesBreeding*, or as a function of population size and other parameters. See Note 3.
        IF (*BreedRate* = 0):
            CONTINUE LOOP with next breeding female
        END IF
// Find a mate …
        IF (hermaphroditic):
            IF (**RAND()** < *ProportionSelfing[p]*):
                Let *Sire* = *Dam*
            ELSE
                Choose a *Sire* at random from breeding pool;
                WHILE (*Sire* is *Dam*):
                    Choose a new *Sire*;
                END WHILE
            END (selfing) IF/ELSE
        ELSE (not hermaphroditic):
            Choose a *Sire* at random from the male breeding pool;

IF (monogamous):
    WHILE (*MaleUsed[Sire]*):
      Choose a new *Sire*;
    END WHILE
    Set *MaleUsed[Sire]* = TRUE;
// Flag *Sire* as unavailable for future *Dam*s
    END IF
  END IF/ELSE
// Find the litter size for that pairing …
  IF (*MaximumLitterSize* > 0):
    Set *CumulativeProbLitterSize[0]* = 1 - *BreedRate*;
    FOR each possible litter size, *n*:
      Set *CumulativeProbLitterSize[n]* =
      CumulativeProbLitterSize[n - 1]
      + *ProbLitterSize[p][n]* * *BreedRate*;
    END LOOP
    FOR (each litter size, *n*, in decreasing order):
      IF (*BreedRand* > *CumulativeProbLitterSize*
      [n - 1]):
        Set *LitterSize* = *n*;
        BREAK from Litter Size LOOP
      END IF
    END LOOP
  ELSE:
// *MaximumLitterSize* = 0 is a code for using normal distribution of litter sizes.
    Set *LitterSize* = *MeanLitterSize[p]* +
    SDLitterSize[p] * **NRAND**();
    Set *LitterSize* = max(0, *LitterSize*);
    Set *LitterSize* = min(2 * *MeanLitterSize[p]*,
    LitterSize);
// Truncates symmetrically to avoid creating bias.
    Set *IntegerLitter* = Largest integer less than
    LitterSize;
    Set *Remainder* = *LitterSize* - *IntegerLitter*;
    IF (**RAND**() < *Remainder*):
// Round-off litter size probabilistically.
      Set *LitterSize* = *IntegerLitter* + 1;
    ELSE:
      Set *LitterSize* = *IntegerLitter*;
    END IF/ELSE
  END IF/ELSE
// Create the offspring …
  Set *Inbreeding* = Kinship between *Sire* and *Dam*;
  FOR (Offspring from 1 to *LitterSize*):
    Assign ID, age (= 0), population, alive (= TRUE);
    FOR (each of six loci):
// First locus is neutral, others can have lethals.
      Pick at random an allele from *Dam;*
      Pick at random an allele from *Sire*;
    END LOOP
    IF(not hermaphroditic AND **RAND**() <
    SexRatio):
      Assign sex as male;
    ELSE:
      Assign sex as female;

    END IF/ELSE
// Does offspring live? Offspring mortality is placed here in the code, rather than in the **MORTALITY**() function, for better speed and lower memory requirements.
    FOR (each non-neutral locus):
      IF (homozygous AND allele is a lethal):
        Offspring dies;
      END IF
    END LOOP
    IF (not yet dead):
      GETDEATHRATE();
      Set *SurvivalRate* = 1 - *DeathRate*;
      IF (*Inbreeding* > 0):
        Set *SurvivalRate* = exp(-0.50 *
        LethalEquivalents * *Inbreeding*);
      ENDIF
      IF (**RAND**() > *SurvivalRate*):
        Offspring dies;
      END IF
    END IF
    IF (not dead):
      Calculate kinship to every living animal;
// See Ballou (1983) for the method of calculating inbreeding and kinship coefficients.
    END IF
  END (offspring) LOOP
  END (breeding females) LOOP
END FUNCTION **BREED**()

BEGIN FUNCTION **GETBREEDRATE**():
  Obtain *BreedRate* by evaluating fecundity function for population and individual parameters;
// Most often, the fecundity function will simply return *ProportionFemalesBreeding* entered by the user. VORTEX provides the option, however, of making breeding a function of *PopulationSize*, *GeneDiversity*, *Inbreeding*, and other variables. See Note 3.
  ADJUSTRATE(*BreedRate*, *LocalBreedEV[p]*,
  LocalBreedEVRand, *LocalBreedEVNRand*);
// Adjust rate for local EV.
  ADJUSTRATE(*BreedRate*, *GlobalBreedEV[p]*,
  GlobalBreedEVRand, *GlobalBreedEVNRand*);
// Adjust rate for global EV.
  FOR (each type of catastrophe, *c*):
    IF (*CatastropheFlag[c]* = TRUE):
      Multiply *BreedRate* by
      CatastropheBreedSeverity[p][c];
    END IF
  END LOOP
END FUNCTION **GETBREEDRATE**()

BEGIN FUNCTION **MORTALITY**(**for population** *p*):
  FOR (each living animal in the population):
    IF (age > 0):
// Infant mortality occurs within the **BREED**() function, not here.

IF (at maximum age):
    Animal dies;
ELSE:
    GETDEATHRATE();
    IF (**RAND**() < *DeathRate*):
        Animal dies;
    END IF
END IF/ELSE
END (age > 0) IF
END (animal) LOOP
END FUNCTION **MORTALITY**()

BEGIN FUNCTION **GETDEATHRATE**():
    Obtain *DeathRate* by evaluating mortality function for
    population and individual parameters;
// Most often, the mortality function will simply return the
mortality rate entered by the user for the age and sex of the
current animal. VORTEX provides the option, however,
of making mortality a function of *PopulationSize*,
GeneDiversity, *Inbreeding*, and other variables.
    ADJUSTRATE(*DeathRate, LocalMortEV[p],
    LocalMortEVRand, LocalMortEVNRand*);
// Adjust rate for local EV.
    ADJUSTRATE(*DeathRate, GlobalMortEV[p],
    GlobalMortEVRand, GlobalMortEVNRand*);
// Adjust rate for global EV.
    FOR (each type of catastrophe, *c*):
        IF (*CatastropheFlag[c]* = TRUE):
            Let *DeathRate* = 1 -
            (*CatastropheSurvivalSeverity[p][c]* *
            (1 - *DeathRate*));
        END IF
    END LOOP
END FUNCTION **GETDEATHRATE**()

BEGIN FUNCTION **ADJUSTRATE**(*Rate, EV, EVRand,
EVNRand*);
    Determine binomial parameter *n* for modeling EV;
// See Note 6.
    IF (*n* < 26):
// Find the adjusted Rate from binomial EV.
        FOR (each *BinomialOutcome* 0 through *n*):
            Add *BinomialOutcome* / *n* to *BinomialProportion*;
            Calculate *BinomialProbability* for
            BinomialProportion;
            Add *BinomialProbability* to *CumulativeBinomial*;
            IF (*EVRand* < *CumulativeBinomial*):
                Set *Rate* = *Binomial Proportion*;
                BREAK from LOOP
            END IF
        END LOOP
    ELSE:
// Use Normal distribution for EV, and truncate symmetrically to avoid bias.
        IF (*Rate* > 0.5):
            Set *UpperLimit* = 1;

            Set *LowerLimit* = *Rate* - (1 - *Rate*);
        ELSE:
            Set *UpperLimit* = 2 * *Rate*;
            Set *LowerLimit* = 0;
        END IF/ELSE
        Add *EV* * *EVNRand* to *Rate*;
        Let *Rate* = max(*LowerLimit, Rate*);
        Let *Rate* = min(*UpperLimit, Rate*);
    END IF/ELSE
END FUNCTION **ADJUSTRATE**();

BEGIN FUNCTION **MIGRATE**():
    FOR (each living animal):
        IF (not in age range that migrates):
            CONTINUE LOOP with next animal;
        END IF
        IF (not a sex that migrates):
            CONTINUE LOOP with next animal;
        END IF
        Set *MigrationRand* = **RAND**();
        Set *pSource* to population of current animal;
        IF (*MigrationRand* > *CumulativeMigrationProb
        [pSource][NumberPopulations]*):
            CONTINUE LOOP with next animal;
// Does not migrate
        END IF
        Obtain *MigrationDensity* by evaluating function, or
        using specified constant parameter;
// See Note 3.
        IF (*PopulationSize[pSource]* / *CarryingCapacity
        [pSource]* < *MigrationDensity*):
            CONTINUE LOOP with next animal;
        END IF
        Obtain *MigrationSurvival* by evaluating function, or
        using specified constant parameter;
// See Note 3.
// Find to which population the animal migrates ...
        FOR (up to 10 attempts to enter another
        population):
// The limit of 10 attempts is imposed to prevent an infi-
nite loop from occurring when all populations are at carry-
ing capacity.
            IF (**RAND**() > *MigrationSurvival*):
                Animal dies;
                BREAK from LOOP, CONTINUE with
                next animal;
            END IF
            FOR (each population, *pDestination*):
                IF (*MigrationRand* < *CumulativeMigrationProb
                [pSource][pDestination]*):
                    BREAK from LOOP;
// Animal will try to migrate to *pDestination*
                END IF
            END LOOP
            IF (Population *pDestination* at carrying capacity):
                IF (tried 9 times before to find an open population):

Animal dies;
// Never found an open population into which to migrate.
BREAK from LOOP, CONTINUE with next animal;
END IF
IF (*CumulativeMigrationProb[pDestination]* *[NumberPopulations]* = 0):
Animal dies;
// Cannot migrate away from *pDestination*.
BREAK from LOOP, CONTINUE with next animal;
END IF
Set *MigrationRand* = **RAND**();
Set *pSource* = *pDestination*;
// Moves on from population *pDestination*, old *pDestination* becomes current *pSource*.
WHILE (*MigrationRand* > *CumulativeMigrationProb[pSource]* *[NumberPopulations]*):
Set *MigrationRand* = **RAND**();
// Must migrate somewhere, so draw a new random number.
END WHILE
END IF
END LOOP
Change animal's population to *pDestination*;
Adjust tallies of population sizes;
// Increment size of *pDestination*, decrement size of *pSource*.
END animal LOOP
END FUNCTION **MIGRATE**()

BEGIN FUNCTION **HARVEST**(for population *p*):
FOR (each age, *x*):
// HARVEST() lumps all animal above breeding age as a single class.
IF (*NumberMales[p][x]* <= *NumberMalesToBeHarvested[p][x]*):
All males age *x* die;
ELSE
WHILE (number harvested < *NumberMalesToBeHarvested[p][x]*):
Choose at random a living male in age class *x*;
Male dies;
END WHILE
END IF/ELSE
END LOOP
FOR (each age, *x*):
IF (*NumberFemales[p][x]* <= *NumberFemalesToBeHarvested[p][x]*):
All females age *x* die;
ELSE
WHILE (number harvested < *NumberFemalesToBeHarvested[p][x]* from age class):
Choose at random a living female in age class *x*;
Female dies;
END WHILE

END IF/ELSE
END LOOP
Adjust tallies of population size;
END FUNCTION **HARVEST**()

BEGIN FUNCTION **SUPPLEMENT**(for population *p*):
FOR (each age, *x*, up to *MaleBreedingAge*):
WHILE (number males created < *NumberMalesToBeSupplemented[p][x]*):
Create a male, assigning ID, age, sex, alleles, population;
Set kinships to all other animals = 0;
Set *Inbreeding* = 0;
END WHILE
END LOOP
FOR (each age, *x*, up to *FemaleBreedingAge*):
WHILE (number females created < *NumberFemalesToBeSupplemented[p][x]*):
Create a female, assigning ID, age, sex, alleles, population;
Set kinships to all other animals = 0;
Set *Inbreeding* = 0;
END WHILE
END LOOP
END FUNCTION **SUPPLEMENT**()

BEGIN FUNCTION **CALC_GENETIC_METRICS**(for population *p*):
FOR (each living animal in the population):
Increment *NumberAlleleCopies[a]* for each of the two alleles, *a*, at a neutral locus;
IF (allele 1 is same as allele 2):
Increment *NumberHomozygotes*;
END IF
END LOOP
FOR (each allele, *a*, of the neutral locus):
IF (*NumberAlleleCopies[a]* > 0):
Increment *NumberExtantAlleles*;
Add 0.25 * (*NumberAlleleCopies[a]* / *PopulationSize[p]*) * (*NumberAlleleCopies[a]* / *PopulationSize[p]*) to *ExpectedHomozygosity[p]*;
END IF
END LOOP
Set *GeneDiversity[p]* = 1 – *ExpectedHomozygosity[p]*;
Set *ObservedHeterozygosity[p]* = 1 - (*NumberHomozygotes* / *PopulationSize[p]*);
FOR (each living animal in the population):
FOR (each non-neutral locus):
IF (allele 1 at the locus is a lethal):
Increment *NumberLethals*;
END IF
IF (allele 2 at the locus is a lethal):
Increment *NumberLethals*;
END IF
END locus LOOP
END animal LOOP

Set *LethalFrequency[p]* = *NumberLethals* /
PopulationSize[p];
END FUNCTION **CALC_GENETIC_METRICS**()

Note 1: Random integers from 0 to 64K are generated by the algorithm given by Kirkpatrick and Stoll (1981). The C code was modified from Maier (1991). Random real numbers between 0 and 1 are produced by first generating a random integer between 0 and 64K, and then dividing that integer by 64K. Random numbers from a normal distribution, with mean = 0 and SD = 1 are generated by the polar algorithm supplied by Latour (1986). Binomially distributed numbers are generated by first calculating the cumulative probability distribution for the discrete outcomes of the desired distribution, then generating a random real number, and then assessing which binomial outcome covers the portion of the distribution encompassing the random real number.

Note 2: VORTEX asks for the effects of inbreeding to be entered as a number of "lethal equivalents" per diploid animal, with further specification of what proportion of this genetic load is due to recessive lethal alleles vs other genetic effects (such as overdominance). Recessive lethal alleles are modeled such that the death of animals homozygous for lethal alleles will reduce the frequency of the lethals and thereby reduce the average effects of future inbreeding. The proportion of inbreeding depression not due to lethal alleles is modeled as an impact on fitness that follows a negative exponential equation (Morton et al. 1956), and is not reduced during generations of inbreeding.

Note 3: For rates which can be specified as functions of age, sex, inbreeding, population size, gene diversity, year, and population, the rate to be used is determined by evaluating the function specified by the user. If the user enters a fixed constant for the rate, as is usually the case, then the function simply returns that constant. However, the user can specify a mathematical formula that defines a demographic rate as being density-dependent, or a function of other population parameters. For example, fecundity could be specified to decline in older age classes, adult mortality could be specified to increase with inbreeding, or habitat (carrying capacity) could be specified to decrease over time. The algorithms for parsing and evaluating user-defined rate functions (e.g., the first step of functions GETBREEDRATE() and GETDEATHRATE()) are not given in the pseudo-code.

Note 4: The proportion of males in the breeding pool can be entered directly, or indirectly in the form of the proportion of males that breed or as the average number of litters per breeding male. If the proportion in the breeding pool is given indirectly, VORTEX will assume that the distribution of male reproductive success follows a Poisson distribution. The proportion of males in the breeding pool is

then calculated by solving the following equations for the unknowns:

LittersPerMale = *ProportionFemalesProducingLitters* * (*NumberAdultFemales* /

NumberAdultMales) [Note: Adult sex ratio is determined from stable age distribution.]

LittersPerMale = *ProportionMalesInBreedingPool* * *LittersPerMaleInBreedingPool*

ProportionMalesBreeding = *ProportionMalesInBreedingPool* * (1 - exp(-*LittersPerMaleInBreedingPool*))

This last equation adjusts for the fact that in any given year some males in the breeding pool will not happen to be successful breeders (the zero class of the Poisson distribution).

Note 5: The occurrence of probabilistic events is determined by a random number generator. The event is deemed to occur if a random number between 0 and 1 is less than the probability of occurrence for that event.

Note 6: Environmental variation (EV) in breeding and in each mortality rate is modeled as a binomial distribution or as a normal distribution, depending on whether the magnitude of EV is large. The user specifies a mean and standard deviation for each rate. The binomial distribution that has a standard deviation closest to the desired EV is determined by solving the equation for the binomial variance, $V = p * (1 - p) / n$, for the parameter n when given the mean, p, and variance, $V = SD^2$. The parameter n is then rounded to the nearest whole number. If $n < 26$, the binomial distribution with parameters p and n is used for EV. Because of the rounding step necessary to produce the discrete binomial distribution, this distribution will often have a slightly different variance than that entered by the user. If $n > 25$, the normal distribution with mean p and variance V will be used to model EV. In such cases, the normal distribution very closely approximates the binomial distribution.

The binomial distribution is restricted to the interval 0 to 1, and it fits well the distribution of demographic rates across years observed in some natural populations (e.g., Lacy 1993). The PVA program INMAT (Mills and Smouse 1994) uses the related beta distribution for this purpose, and it too is restricted to the biologically meaningful 0 to 1 interval. In contrast, the normal distribution extends infinitely in both directions, although the tails beyond 0 and 1 are typically very small in those cases for which VORTEX uses a normal distribution to model EV. For example, if $p = 0.5$ and $SD = 0.1$ (so that the binomial parameter $n = 25$, the limiting case for VORTEX to use the normal approximation), then the area of the normal distribution outside of the 0-1 range is < 0.000001. When modelling EV as a normal distribution, the distribution must be truncated at 0 and 1. To avoid creating any bias in the mean demographic rate as a result of this truncation, VORTEX always truncates the distribution symmetrically.

For example, if the mean is $p = 0.3$, VORTEX truncates the distribution at 0.0 and 0.6. This truncation will cause the *SD* of the distribution to be very slightly less than that entered by the user.

Some PVA models use continuous distributions such as the normal or log normal to represent EV even when EV is large. In such cases, the necessary truncations can cause EV to be substantially less than intended by the user. Moreover, if the truncation is not symmetric, then the mean demographic rate generated by the model can be strongly biased away from the input parameter.

References

Ballou, J. D. 1983. Calculating inbreeding coefficients from pedigrees. – In: Schonewald-Cox, C. M. et al. (eds), Genetics and conservation: a reference for managing wild animal and plant populations. Benjamin/Cummings, Menlo Park, California, pp. 509–520.

Burgman, M. A., Ferson, S. and Akçakaya, H. R. 1993. Risk assessment in conservation biology. – Chapman and Hall.

Caswell, H. 1989. Matrix population models: construction, analysis and interpretation. – Sinauer.

Caughley, G. and Gunn, A. 1996. Conservation biology in theory and practice. – Blackwell.

Kirkpatrick, S. and Stoll, E. 1981. A very fast shift-register sequence random number generator. – J. Comp. Phys. 40: 517.

Lacy, R. C. 1993. VORTEX: A computer simulation model for population viability analysis. – Wildl. Res. 20: 45–65.

Latour, A. 1986. Polar normal distribution. – Byte, August 1986: 131–132.

Lindenmayer, D. B., Lacy, R. C. and Pope, M. L. 2000. Testing a simulation model for population viability analysis. – Ecol. Appl. 10: 580–597.

Maier, W. L. 1991. A fast pseudo random number generator. – Dr. Dobb's Journal, May 1991: 152–157.

Miller, P. S. and Lacy, R. C. 1999. VORTEX Ver. 8 users manual. A stochastic simulation of the simulation process. – IUCN/SSC Conservation Breeding Specialist Group, Apple Valley, Minnesota.

Mills, L. S. and Smouse, P. E. 1994. Demographic consequences of inbreeding in remnant populations. – Am. Nat. 144: 412–431.

Morton, N. E., Crow, J. F. and Muller, H. J. 1956. An estimate of the mutational damage in man from data on consanguineous marriages. – Proc. Nat. Acad. Sci. USA 42: 855–863.

Pielou, E. C. 1977. Mathematical ecology. - Wiley.

Starfield, A. M. and Bleloch, A. L. 1986. Building models for conservation and wildlife management. – MacMillan.

ECOLOGICAL BULLETINS

ECOLOGICAL BULLETINS are published in cooperation with the ecological journals Ecography and Oikos. Ecological Bulletins consists of monographs, reports and symposia proceedings on topics of international interest, often with an applied aspect, published on a non-profit making basis. Orders for volumes should be placed with the publisher. Discounts are available for standing orders.

ECOLOGICAL BULLETINS still available.

Prices excl. VAT and postage.

41. *The cultural landscape during 6000 years in southern Sweden - the Ystad project* (1991). Editor B. E. Berglund. Price DKK 250.00.

42. *Trace gas exchange in a global perspective* (1992). Editors D. S. Ojima and B. H. Svensson. Price DKK 50.00.

43. *Environmental constrains of the structure and productivity of pine forst ecosystems: a comparative analysis* (1994). Editors H. L. Goltz, S. Linder and R. E. McMurtie. Price DKK 50.00.

44. *Effects of acid deposition and tropospheric ozone on forest ecosystems in Sweden* (1995). Editors H. Staaf and G. Tyler. Price DKK 50.00.

45. *Plant ecology in the subarctic Swedish Lapland* (1996). Editors P. S. Karlsson and T. V. Callaghan. Price DKK 50.00.

46. *Boreal ecosystems and landscapes: structures, processes and conservation of biodiversity* (1997). Editor L. Hansson. Price DKK 250.00.

48. *The use of population viability analyses in conservation planning* (2000). Editors P. Sjögren-Gulve and T. Ebenhard. Price DKK 300.00.